Thermofluids

Thermofluids

From Nature to Engineering

David S-K. Ting
Mechanical, Automotive & Materials Engineering
Turbulence & Energy Laboratory
University of Windsor
Windsor, Ontario, Canada

ELSEVIER

ACADEMIC PRESS
An imprint of Elsevier
elsevier.com/books-and-journals

Academic Press is an imprint of Elsevier

125 London Wall, London EC2Y 5AS, United Kingdom
525 B Street, Suite 1650, San Diego, CA 92101, United States
50 Hampshire Street, 5th Floor, Cambridge, MA 02139, United States
The Boulevard, Langford Lane, Kidlington, Oxford OX5 1GB, United Kingdom

Notices

Knowledge and best practice in this field are constantly changing. As new research and experience broaden our understanding, changes in research methods, professional practices, or medical treatment may become necessary.

Practitioners and researchers must always rely on their own experience and knowledge in evaluating and using any information, methods, compounds, or experiments described herein. In using such information or methods they should be mindful of their own safety and the safety of others, including parties for whom they have a professional responsibility.

To the fullest extent of the law, neither the Publisher nor the authors, contributors, or editors, assume any liability for any injury and/or damage to persons or property as a matter of products liability, negligence or otherwise, or from any use or operation of any methods, products, instructions, or ideas contained in the material herein.

ISBN: 978-0-323-90626-5

For Information on all Academic Press publications visit our website at
https://www.elsevier.com/books-and-journals

Publisher: Charlotte Cockle
Acquisitions Editor: Rachel Pomery
Editorial Project Manager: Sara Valentino
Production Project Manager: Swapna Srinivasan
Cover Designer: Miles Hitchen

Working together
to grow libraries in
developing countries

www.elsevier.com • www.bookaid.org

Typeset by Aptara, New Delhi, India

Dedication

To all those who contemplate nature, savor its beauty,
and learn and live its simplicity.

"Nature is pleased with simplicity. And nature is no dummy."
–Isaac Newton.

Contents

Part 2
An Ecological View on Engineering Thermodynamics

Part 3
Environmental and Engineering Fluid Mechanics

Part 4
Ecophysiology-flavored Engineering Heat Transfer

Textbook Cover Photo

Christmas tree worms

Christmas tree worms have an array of photoreptors on eponymous feeding appendages. They absorb light maximally at 464 nm in wavelength, probably functioning as a silhouette-detecting intruder alarm [Bok et al., 2017]. This alert system apparently can be breached in turbulent waters, where Caribbean sharpnose puffers prey Christmas tree worms with their long snout and large fused front teeth [Hoeksema & ten Hove, 2017]. As such, this natural phenomenon illustrates the spectacular workings of engineering thermofluids; radiation, fluid turbulence, and others.

M. J. Bok, M. L. Porter, H. A. ten Hove, R. Smith, D-E. Nilsson, "Radiolar eyes of Serpuid Worms (Annelida, Serpulidae): structures, function and phototransduction," Biology Bulletin, 233: 39-57, 2017.

B. W. Hoeksema, H. A. ten Hove, "Attack on a Christmas tree worm by a Caribbean sharpnose pufferfish at St. Eustatius, Dutch Caribbean," Bulletin of Marine Science, 93(4): 1023-1024, 2017.

Caribbean sharpnose puffers have a long snout and large fused front teeth, well-suited to prey. Christmas tree worms possibly failed to dictate and retract in turbulent waters.

.

List of figures

List of tables

Preface

Some of the most insightful observations include those made by Henri Poincaré, "The scientist does not study nature because it is useful; he studies it because he delights in it, and he delights in it because it is beautiful." Nature is beautiful and it is a pleasure to appreciate it. Doing so, we will understand everything better. "Look deep into nature, and then you will understand everything better."—Albert Einstein. The author is convinced that approaching engineering thermofluids through the eye of nature can better inspire the freshman and sophomore into inquisitive minds.

This book is divided into four main parts. Part 1 is a general introduction of *Thermofluids in Nature and Engineering*. It reveals what the topic of thermofluids encompasses. This is followed by Part 2, *An Ecological View of Engineering Thermodynamics*. The laws of ecology are exploited to explain the first and second laws of thermodynamics, and vice versa. The basics of fluid mechanics are then conveyed as Part 3, *Environmental and Engineering Fluid Mechanics*. The essential transportation of thermal energy is elucidated in Part 4, *Ecophysiology-flavored Engineering Heat Transfer*.

Part 1 consists of four chapters. Chapter 1 introduces the subject of thermofluids, which encompasses three fundamental topics, thermodynamics, fluid mechanics, and heat transfer. Familiar occurrences of thermofluids in nature and everyday living are exploited to facilitate the appreciation of the subject. Thermodynamics is the study of energy, as disclosed concisely in Chapter 2. Conservation of energy asserts that the total quantity of energy in the universe remains the same. The quality of energy is most commonly illustrated by the natural transformation of energy to a progressively lower-quality form. All-important moving fluids are introduced in Chapter 3. Newtonian fluids, for which the viscosity is not a function of the flow shear, is the topic of interest for this introductory text. Chapter 4 discusses thermal energy, which is commonly known as heat. It is noted that heat transfer is thermal energy in transit, and conduction, convection, and radiation are the three fundamental heat transfer mechanisms.

Part 2 starts with The Four Laws of Ecology as Chapter 5. The importance of ecology in engineering is relayed. After highlighting thermoregulation in creatures, the parallel between the four laws of ecology and the first and second laws of thermodynamics is drawn. Chapter 6 delves into the first law of thermodynamics. It differentiates open, closed, and isolated thermodynamic systems.

Also presented are enthalpy and internal energy. Entropy is explained in terms of disorder in Chapter 7, the second law of thermodynamics. After defining a heat source and heat sink, a heat engine is framed, and the ideal Carnot heat engine is cast as the holy grail. With that, refrigerators, air conditioners, and heat pumps are elucidated as reverse heat engines.

Fluid Mechanics is the subject of Part 3. This part commences with non-moving fluid as Chapter 8, Fluid Statics. The importance of hydrostatic pressure is demonstrated in the working principles of manometers and barometers, and as forces acting on submerged surfaces. Archimedes' principle, buoyancy, and stability of floating objects are also put forward. Steady, inviscid, incompressible flow along a streamline is communicated in Chapter 9, Bernoulli Flow. The derivation, application, and limitations of the Bernoulli equation are detailed. Also disclosed are static, stagnation, dynamic, and total pressures, along with energy and hydraulic grade lines. Dimensional Analysis is delineated as Chapter 10. Dimensions and dimensional homogeneity cannot be undermined. The workings of the Buckingham Pi theorem are exhibited, which is logically followed by pragmatic dimensionless groups in fluid mechanics. Chapter 11 is about Internal Flow. Laminar, transition, and turbulent pipe flows are characterized in terms of Reynolds number. The pressure drop is related to the wall shear from the entrance to the fully developed pipe flow. Minor losses are differentiated from major losses, where the Moody chart is employed to determine the value of the friction factor. Part 3 concludes with External Flow as Chapter 12. Everyday external flows around common bodies are described, including the mechanisms behind lift and drag. Laminar, transition, and turbulent boundary layer development along a flat plate are elucidated. The familiar bluff body vortex shedding and streamlining are also described.

Part 4 is devoted to Heat Transfer, and it is composed of six chapters. Steady Conduction of Thermal Energy is the title of Chapter 13. It commences with Fourier's law of heat conduction, depicting the inverse relationship between thermal conductivity and temperature gradient. The analogy between electric current and heat is explained. After providing the equations for one-dimensional heat conduction in planar, cylindrical, and spherical coordinates, the chapter wraps up by differentiating the parallel-path method from the isothermal-plane method. Then, we move into transient heat conduction in Chapter 14, Transient Conduction of Thermal Energy. The concept of a lumped system is introduced, and the Biot number follows naturally. One-dimensional transient problems for a large plate, a long cylinder, and a sphere are described. The solution approach for a semi-infinite wall finishes off the chapter. Natural convection and thermals start off Chapter 15, Natural Convection. The underlying mechanisms of natural convection are revealed. The key factors are grouped into celebrated nondimensional parameters. The chapter also includes Rayleigh–Bernard convection, natural convection along with a vertical plate, continuous thermal plumes, and buoyant jets. More dominant is heat convection under the providence of a forceful flow, Forced Convection, the theme of Chapter 16. The indispensable convection heat transfer coefficient is construed and forced convection over a flat plate explicated. The relationship between Nusselt number and Reynolds and Prandtl numbers is formulated. The difference between constant-temperature and uniform-heat-flux boundary conditions is clarified. The correlation between forced convection and wall shear is illustrated. Thermal Radiation is conveyed in Chapter 17; real gray bodies are compared to ideal black bodies, in terms of emissivity. With that, radiation heat transfer calculations are performed. Absorptivity, transmissivity, and reflectivity are explained, and their relationship is disclosed. The book ends with Heat Exchangers as Chapter 18. Attention is drawn to the many intelligently designed heat exchangers in the animal kingdom. Heat exchangers can be categorized according to whether there is mixing between the hot and cold stream and/or the flow configuration, that is, counter-flow, parallel flow, and cross-flow. Both log mean temperature difference and number of transfer units methods are described in detail.

Acknowledgments

"Press forward. Do not stop, do not linger in your journey, but strive for the mark set before you."

—George Whitefield

This book would have remained but a wishful lingering thought without the unceasing encouragements and helping hands from numerous George Whitefields including the following.

Tachelle Z.-T. Ting, who exploited her dad as a dummy student when preparing for tests and examinations in biological and ecological sciences. The dummy student experienced numerous eureka moments witnessing intelligently designed creatures engineering and harnessing thermofluids to their fullness.

A good figure conveys more than a thousand words. It is preposterous to publish *Thermofluids: From Nature to Engineering* without the hundred-plus expository illustrations. A heartfelt thank you goes out to the Turbulence and Energy Laboratory artists. Their names are inscribed in the captions of their artwork.

The Elsevier publishing team, Peter Llewellyn, Maria Convey, Praveen Anand, Sara Valentino, and many others. It is a joy working with courteous collaborators.

Dr. Jacqueline A. Stagner, who scrutinized all 130,000-plus words, some of them twice. She also appears every here and there throughout the book as Dr. JAS, clarifying the point the author is struggling to put across.

The Turbulence and Energy Laboratory and its predecessor, Allinterest Research Institute, continue to provide the moral platform and fuel for arduous striving such as this.

Mom, dad, sisters, and brother, whose abundant thermofluids activities in the rainforest of Borneo are still reverberating in the author's heart.

Naomi, Yoniana, Tachelle, and Zarek Ting, who were there from the beginning, when the author first tested the water with Naomi Ting's Books.

This book is made possible by the downpouring grace from above.

Part 1

Introduction

Chapter 1

Thermofluids

"In all things of nature there is something of the marvelous."

Aristotle

Chapter Objectives
- Understand the subject of thermofluids.
- Comprehend the fundamentals of thermodynamics.
- Fathom the basics of fluid mechanics.
- Appreciate the fundamentals of heat transfer.
- Recognize everyday occurrences of thermofluids in nature.
- Introduce the book.

Nomenclature

A	area
a	acceleration
C	light dimension
E	energy; dE is the differentiable change of energy
F	force
g	gravity
h	height
h_{conv}	convection heat transfer coefficient
I	current
KE	kinetic energy
L	length dimension
M	mass dimension
m	mass
N	Matter dimension; amount of matter, or, number of moles
P	pressure
PE	potential energy
Q	heat; Q' is heat transfer rate
T	temperature; ΔT is the temperature difference
t	time; dt is the differentiable increase in time
V	velocity
W	work; W' is the work transfer rate or power
x	distance

Greek and other symbols

Δ	difference
ρ	density
$\forall$	volume

1.1   What is Thermofluids?

Thermofluids is the study of

1) Thermodynamics,
2) Fluid Mechanics, and
3) Heat Transfer.

The specific sequence, Thermodynamics, Fluid Mechanics, followed by Heat Transfer, is the most appropriate order for comprehending thermofluids. We will reveal the reasons behind this logical order after we define and say a few words about these three subjects that make up thermofluids.

1.1.1   Thermodynamics

Put succinctly, thermodynamics is the study of energy. If there were no energy, there would be no life, and you and I would not be here contemplating thermodynamics or thermofluids. Every living thing consumes energy to stay alive and, equally so, to thrive. Fig. 1.1 is a simplified depiction of the cycle of life. Energy from the sun is utilized by plants to grow, by building up and storing the converted energy in the leaves, fruits, seeds, stems, and roots. Grasshoppers represent creatures living off plants and acquiring energy from plants. Frogs feed on insects such as grasshoppers, and they are subsequently offered higher up the food chain for snakes to savor. Above the snakes, up in the sky, are the eagles; they are able to soar because of the protein and energy embodied within the slimy body of the snakes they savor. When these mighty birds' time is up, they decompose, providing whatever is left for mushrooms to bloom. While there is little energy in the mushrooms and the compost leftovers, they formulate essential nutrients for the plants to effectively harness solar energy and blossom, enabling all kinds of lives to flourish. It is important to note that we do not return any favor, energy, back to the sun. We do well when we make good use of the energy, minimizing its wastage, and savoring life to its full responsibly.

Like other plants, spinach exploits solar energy to convert water and carbon dioxide into energy-rich organic compounds. Spinach, in turn, grants cartoon superhero Popeye great power to overcome his rival, Bluto. It is worth highlighting that a more green-based diet can result in a more sustainable tomorrow (Ting and Stagner, 2020). To put it another way, solar energy is converted into hearty spinach, which energizes Popeye with extraordinary power. Recall that power is energy per unit time. A 0.1-kg can of spinach contains roughly 100 kJ of energy (Healthline, 2020). To lift the 200-kg Bluto up and down above Popeye's head,

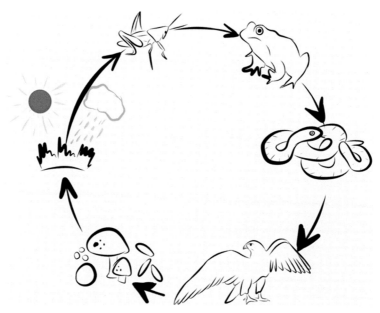

FIGURE 1.1   A simplified illustration of the cycle of life sustained via the orderly transfer of energy originating from the sun (created by S. Akhand).

2 m from the ground, the potential energy,

$$PE = mgh = 200 \text{ kg} \times 10 \text{ m/s}^2 \times 2 \text{ m} = 4000 \text{ J}, \qquad (1.1)$$

where m is mass, g is gravity, and h is height. Assuming 50% of the energy is used to overcome Bluto's struggling and losses, then, for Popeye to lift Bluto up and down thirty times, we need

$$2 \times 30 \times 4000 \text{ J} = 240 \text{ kJ} \qquad (1.2)$$

of energy. This is equivalent to two and one-half cans of spinach. Only a super-high-energy-density can of spinach is able to power Popeye to handle Bluto, as portrayed in the motion picture.

Example 1.1 Popeye paddling against Bluto
Given: Bluto grabs Olive Oyl (Popeye's love) and takes off in his boat. Hearing Olive calling for help, Popeye immediately jumps into a paddle boat and starts paddling at 3 m/s. Assume that the required force to maintain this speed, while overcoming drag, friction, and other losses, is 200 N.
Find: How many 0.1-kg cans of spinach does Popeye need to consume, if it takes him 10 min to catch up with Bluto?

Solution: The required energy can be deduced from the product of force and distance traveled. The distance traveled is,

$$x = V t = 3 \text{ m/s}(10 \times 60 \text{ s/minute}) = 1800 \text{ m},$$

where V is the velocity and t is the time. Accordingly, the required energy,

$$E = F x = 200 \text{ N} \times 1800 \text{ m} = 360 \text{ kJ},$$

where F is the force. With each 0.1-kg can of spinach containing only 100 kJ of energy, it takes 4 cans of spinach to fuel that 10-minute stretch!

According to Sweetpotato (2020), sweet potatoes are some of the most nutritious vegetables. Imagine Dr. JAS, of the Turbulence and Energy Lab and Allinterest Research Institute, invented a robot called T-E Keledek that feeds on sweet potatoes for energy to function. Every kilogram of sweet potato contains about 3600 kJ of energy. That being the case, T-E Keledek needs to consume less than one 0.1-kg can of sweet potato to teach Bluto the same lesson as Popeye did, that is, lifting Bluto up and down over one's head thirty times, rather than four 0.1-kg cans of spinach to perform the same maneuver as Popeye. In short, thermodynamically, one is able to get more work done from a machine, such as T-E Keledek, by feeding it sweet potatoes instead of the same mass of spinach. We should make time for a hearty sweet potato breakfast every morning before coming to school, reserving spinach for afternoon snack time.

1.1.2   Fluid mechanics

Fluid mechanics is the study of energy and forces in a fluid, that is, a liquid or a gas. For a giant squid, as portrayed in Fig. 1.2, to survive at more than 1000 m deep in the sea (Seasky, 2020), it has to be able to muddle through 1×10^7 Pa of hydrostatic (fluid at rest) pressure. The hydrostatic pressure is exerted by the column of water above the giant squid. Namely, the hydrostatic pressure is equal to the fluid density multiplied by gravity multiplied by depth or height of the fluid column crushing downward, that is,

$$P = \rho g h = 1000 \text{ kg/m}^3 \times 10 \text{ m/s}^2 \times 1000 \text{ m} = 1 \times 10^7 \text{Pa}. \qquad (1.3)$$

Comparatively, a dolphin can swim to roughly 300 m below sea level (Dolphin, 2020) and a typical scuba diver can dive up to about a 100-m depth (Humandive, 2020). This illustrates the weightiness of hydrostatic pressure when the involved fluid is a liquid. The significance of hydrostatic pressure in a gaseous fluid is seriously less noticeable.

Talking about pressure, the big heart of a giraffe has to push blood all the way up its long neck to its head; see Fig. 1.3. To strive upward against gravity, up a 7-foot-long neck, a roughly 2-foot-long heart, that can weigh over 25 pounds, is necessary (Giraffes, 2020). Thankfully, we only have to overcome somewhat

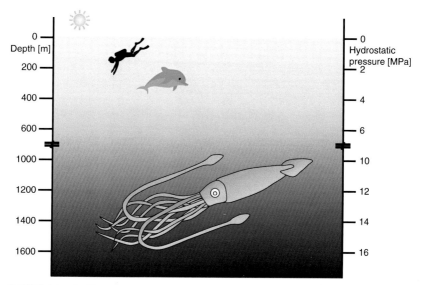

FIGURE 1.2 An illustration of the relative sizes and survival water depths of a giant squid, a dolphin and a human (created by Y. Yang).

over 1 foot of elevation from our heart to our brain, for which a relatively small, fist-sized heart is sufficient.

Example 1.2 Harnessing hydrostatic pressure to slide carts across a hallway

Given: A high-rise hotel has a water loop that includes 50-m vertical columns. Dr. JAS is hired to design a system that can make use of the potential energy to transport carts back and forth across a hallway on the main floor.

Find: What is the maximum velocity a 180 kg cart can move per 0.3 m³ of water released through the base of the column?

Solution: Without losses, all the potential energy

$$\text{PE} = \rho gh\forall = 1000 \text{ kg/m}^3 \left(10 \text{ m/s}^2\right)(50 \text{ m})\left(0.3 \text{ m}^3\right) = 150,000 \text{ J}$$

is transformed into kinetic energy, that is,

$$150,000 \text{ J} = \text{KE} = \frac{1}{2}\text{mV}^2 = \frac{1}{2}(180 \text{ kg})\text{V}^2.$$

Solving, we get V = 40.82 m/s = 147 km/h. This speed is dangerously high for moving a 180-kg cart down a hallway. Even with frictional and other losses, the speed is still too high for a moving cart in a hotel lobby. One way to resolve this problem is to use less water to push the cart. If 0.003 m³ of water is used instead, the resulting cart speed is reduced by an order of magnitude to 4.1 m/s. After accounting for involved losses, we end up in the ballpark of 1 m/s, and this is safe to operate.

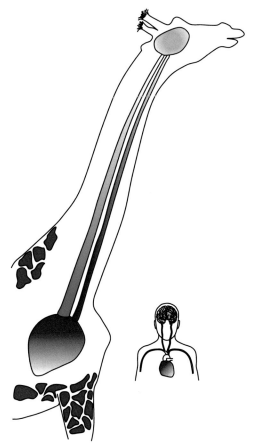

FIGURE 1.3   A giraffe's big heart versus a fist-sized human heart (created by Y. Yang). William Hazlitt stated that, "The seat of knowledge is in the head, of wisdom, in the heart."

Note that we have only employed a 50-m water column here. You can work out the numbers in terms of potential energy and cart-pushing power for the 1000-m water column that giant squids chill under.

1.1.3   Heat transfer

Bring two traditional hand fans to keep you cool when you take a stroll in the Sahara Desert. At least this is what appears to be the case for a fennec fox, as shown in Fig. 1.4. Even though its two huge ears are not for fanning, they do contain many blood vessels for circulating blood. The two eye-catching over-sized ears furnish large surfaces for dissipating heat from the circulating blood, in addition to hearing prey underground (Fennecfox, 2020). It is worth bringing up the surface-area-to-volume ratio. In biology, Bergmann's rule suggests that

FIGURE 1.4   A fennec fox equipped with fan-like ears abounding in heat-transferring blood vessels to stay cool in the Sahara Desert (created by S. Akhand). There are many blood vessels bringing warm blood to the big ears and returning cooler blood after dissipating heat through the extended surface area into the ambient air.

the average body size of a species population increases farther from the hot equator; see, for example, Vallejo (2018). One underlying assumption is that the metabolism rate increases with body size. As such, there is many times more heat to dissipate for an elephant than for a mouse. We may approximate a 3-m tall adult elephant as a 3-m tall, 1-m wide, and 4-m long rectangular box, for illustration purposes. This gives a volume of 12 m^3 and a total surface area of 38 m^2, that is, a volume-to-surface-area ratio of 0.32 m. An average mouse is roughly 0.04-m tall, 0.06-m wide, and 0.13-m long. The volume-to-surface-area ratio of a mouse is thus 0.01 m. That being the case, to mitigate heatstroke, elephants are equipped with large ears. Unlike fennec foxes, elephants do fan their ears to boost heat transfer (Wright, 1984).

We will revisit some of these points in more detail as we more explicitly invoke various principles of thermofluids in later chapters. Now, let us explain why it is most logical to take up Thermodynamics first, followed by Fluid Mechanics, and lastly Heat Transfer.

1.2   Thermodynamics, fluid mechanics, and heat transfer

Thermodynamics deals with energy. Energy is the ability to perform work or to heat. Some of the most common forms of energy that we deal with in engineering thermofluids are thermal energy, potential energy, and kinetic energy. Potential energy, in terms of hydrostatic pressure, has been explicitly invoked when explaining fluid mechanics earlier. As such, it is clear that we need to have a reasonable understanding of thermodynamics so that we can better appreciate fluid mechanics. Fluid mechanics is not simply concerned with stagnant fluids but, more frequently, with moving fluids. That being the case, there is a lot of kinetic energy involved. It is said that diamonds are forged under pressure. While the magnitude is substantially less, pressure is also needed to move a fluid; this is because friction is omnipresent on earth[1]. Case in point, external energy is applied via a pump to move a fluid along a pipe. To that end, most of the electrical energy for running the pump is converted into kinetic energy associated with the moving fluid. The other fallout of thermodynamics is that losses are inevitable, that is, not all energy input is transformed into the kinetic energy of the flowing fluid. Pressure is force per unit area, which in SI units are N/m^2, and this can be expressed as $J \cdot m/m^2$ or J/m^3. Namely,

$$N/m^2 = J \cdot m/m^2 = J/m^3. \tag{1.4}$$

We see that pressure is energy per unit volume and, for flows in pipes or channels, it is the energy per unit volume of the moving fluid.

The big-ear heat transfer of fennec foxes and elephants we discussed earlier primarily deals with heat convection. Convection heat transfer is the transfer of heat from the surface of the hot ears to the surroundings via the movement of air. On that account, it is clear that heat transfer deals with the transfer of thermal energy; that is thermodynamics in transient mode. For convection, it heavily involves fluid mechanics; that is, heat is transferred by flow motion. For the latter, we appreciate it when fanning on a warm day, increasing the air movement and, thus, convective heat transfer from our body to the ambient air. The rate of convection heat transfer is proportional to the surface area available for transferring heat and the temperature difference between the surface and the surroundings. Namely,

$$Q' \propto A\Delta T, \tag{1.5}$$

where A is the area and ΔT is the temperature difference or gradient. We can recast Eq. (1.5) into an equality expression by introducing a proportionality constant, that is,

$$Q' = h_{conv}A\,\Delta T. \tag{1.6}$$

The proportionality constant, h_{conv}, is primarily dictated by fluid mechanics. In general, the faster the fluid moves, the larger this convection heat transfer coefficient. We will expound on this later when we deal with heat transfer.

1.3   Dimensions and units

The simple illustration of the mouse versus elephant volume-to-surface-area ratio hints at the importance of size. To appreciate size, we first need to fathom dimensions and units. Dimensions deal with measurable physical quantities, that is, a dimension is a measure of a physical variable. Table 1.1 lists the seven primary dimensions along with their standard notations and symbols. Primary dimensions are independent dimensions from which secondary dimensions are

---

1. In contrast, there is very little friction or drag in space. For this reason, space shuttles require very little energy to cruise, except in the presence of other forces such as gravity.

TABLE 1.1 The seven primary dimensions and their standard notations and units.

Dimension	Symbol	SI units	English units
Current	I	A	–
Length	L	m	ft, in, miles
Light	C	cd	–
Mass	M	kg	lbm
Matter	N	mol	-
Temperature	T	K	R
Time	t	s	-

FIGURE 1.5 Height versus length (created by Y. Yang). Dr. JAS measures a lamp post to the desired length and asks T.-E. Baru to cut off the excess length. T.-E. Baru sees "height" but not "length."

formed. For instance, length is a primary dimension from which both area and volume are formed. Thereby, area and volume are secondary dimensions.

Fig. 1.5 illustrates a conversation between Dr. JAS and novice T.-E. Baru. T.-E. Baru is one prospect who will make the Turbulence and Energy Lab proud one day; that is, there is hope for anyone who can learn from mistakes. Richelle E. Goodrich put it more elegantly, "Many times what we perceive as an error or failure is actually a gift. And eventually, we find that lessons learned from that discouraging experience prove to be of great worth." The key here is to have the humility to learn. T.-E. Baru simply thought that height and length, or distance, are two different dimensions.

Although we may consider the above example silly, we are reminded about incidents such as Gimli Glider: Air Canada Flight 413 (CBC83, 2020). Specifically, on July 23, 1983, a Boeing 767 heading from Montreal to Edmonton ran out of fuel at an altitude of 12,000 m. It is fortunate that the skillful pilots maneuvered the plane, gliding it to a safe emergency landing at Gimli Industrial Park Airport, a former airbase in Manitoba. Why did it run out of fuel? It was a metric mix-up; the fueling was done in units of liters, while the request was in gallons (1 liter = 0.254 US gallon). This is not an isolated incident, every so often there is a similar mistake. Thankfully, not all mistakes are life-threatening.

Before we set "making the case concerning the importance of dimensions and units" to rest, let us be enlightened by an historic mess-up in North American history. According to sources such as Columbus (2020), Christopher Columbus miscalculated the circumference of the earth. Columbus used Roman miles instead of nautical miles, causing him to end up in the Bahamas on October 12, 1492, instead of Asia, the intended destination.

Concerning primary dimensions, Table 1.1 shows that electric current, I, is quantified in Amperes, A, and the intensity of light, C, is expressed in units of candelas, cd. Accordingly, amperes and candelas are primary units. It is evident that units are for specifying the amount of a dimension. For the primary dimension of matter, such as water, every 6.022×10^{23} molecules of it is quantified as 1 mol.

The prevailing dimensions that are involved in thermofluids are length, mass, temperature, and time. That being the case, the system of dimensions that we adopt is mass {m}, length {L}, temperature {T} and time {t}, where curly brackets, { }, embrace the corresponding notation of the primary dimension. The SI units for the secondary dimension of energy is joules. One joule of energy is spent when you apply 1 N of force to push a lazy sloth[2] over a distance of 1 m. According to Newton's second law of motion, force is equal to mass times acceleration, that is,

$$F = ma = \left[kg \cdot m/s^2 \right] = [N] = \left\{ ML/t^2 \right\}. \qquad (1.7)$$

In context, square brackets, [], are used here to enclose units. Applying a force, F, over a distance, x, we have

$$F\,x = (ma)(x) = \left[kg \cdot m/s^2 \right][m] = [N \cdot m] = [J] = \left\{ ML^2/t^2 \right\}. \qquad (1.8)$$

Power, or rate of work performed, is simply energy per unit time, that is,

$$W' = dE/dt = \left[kg \cdot m^2/s^2 \right]/[s] = [N \cdot m/s] = [J/s] = [W] = \left\{ ML^2/t^3 \right\}. \qquad (1.9)$$

---

2. According to One Kind Planet (2020) and Scienceabc (2020), the slow-moving sloth sleeps up to 20 hours a day. Even so, this first runner-up lost the crown to koalas, who are only awake 2 to 6 hours a day. Koalas feed on eucalyptus leaves that contain toxins and are very high in fibre. The double whammy takes most of the energy out of koalas.

The above illustrates the dimensions of three common secondary dimensions encountered in thermofluids. The saying, "Do not compare apples to oranges" is true in life and also in engineering. Thermodynamically, the energy you acquire by eating five identical apples and two twin oranges of the same size is not equal to seven analogous apples, that is,

$$5 \text{ apples} + 2 \text{ oranges} \neq 7 \text{ apples}. \tag{1.10}$$

According to Healthline (2020), one medium-sized, 0.1-kg apple contains 216 J of energy, while a 0.1-kg orange has 197 J of energy. Therefore, we can say that a 0.1-kg orange gives Dr. JAS 91% of the energy of that of a 0.1-kg apple. If Dr. JAS munches five 0.1-kg apples and two 0.1-kg oranges for lunch, that is,

$$5(216 \text{ J}) + 2(197 \text{ J}) = 1474 \text{ J}. \tag{1.11}$$

In other words, Dr. JAS would have 1474 J of energy to coach T.-E. Baru. The point here is to make sure that the equation we use is dimensionally homogeneous. SI units are standardized for a good reason; they are easier to use than conventional units. Because of that, the consistent use of SI units for all terms in an equation will save a lot of headaches. For Eq. (1.11), every term has the dimension of energy in joules; under those circumstances, the equation is sound.

Example 1.3 A round trip to and from Proxima Centauri
Given: Dr. JAS is ready to take her crew to the nearest star, other than the sun, in her T&E Spaceship at warp speed.
Find: The time it takes to make the round trip.

Solution: Proxima Centauri is 4.2 light years away from earth.

$$1 \text{ light year} = 9.461 \times 10^{12} \text{ km}.$$

In Star Trek, Captain Picard can cruise the USS Enterprise at 299,792 km/s, the warp speed.

The time it takes Dr. JAS to make the round trip between Earth and Proxima Centauri is

$$\text{distance/speed} = 2(4.2)\left(9.461 \times 10^{12} \text{ km}\right)/299,792 \text{ km/s} = 2.65 \times 10^8 \text{ s}.$$

In other words, it will take Dr. JAS

$$2.65 \times 10^8 \text{ s}/60 \text{ seconds/minute}/60 \text{ minutes/hour}/24 \text{ hours/day} = 3068 \text{ days}$$

or

$$3068 \text{ days}/365 \text{ days/year} = 8.4 \text{ years}.$$

According to relativity, the faster Dr. JAS moves, the more mass she has. When she travels at the speed of light, her mass will become infinite. While warp speed is the speed of light, "warp" describes the situation that Dr. JAS is still in "normal" and "non-relativistic" space. Another way to travel fast without undergoing relativistic effects is for Dr. JAS to channel through a wormhole.

While the push to use SI units for everything has been around and adopted by many nations around the world for many years, the reality is that the English (or the slightly different American) Engineering System that uses pound, Fahrenheit, etc., is not going to go away for a while. Patriotism is a major reason behind the opposition to SI-unit adoption. Apparently, it is also more convenient, in some cases, to use English units. For example, most of us find it easier to have a sense of how tall a person is if the height is given in feet and inches, instead of in meters and centimeters. For weight watchers, there is more than double the encouragement when we measure weight loss in pounds as compared to kilograms. Nonetheless, if you wish to avoid making unnecessary calculation mistakes, stay faithfully with SI units.

1.4 Organization of the book

The first four chapters aim to introduce Thermofluids: From Nature to Engineering. In this chapter, we have been briefed on thermofluids. In Chapter 2, we will familiarize ourselves with the study of energy. Life-sustaining moving fluids will be presented in Chapter 3, while the transfer of thermal energy, heat transfer, will be delivered in Chapter 4.

Part 2 is a comprehensive coverage of engineering thermodynamics with an ecological flavor. The importance of ecology in engineering is stressed. The four laws of ecology, along with some ingenious animal thermoregulation, will be introduced in Chapter 5. This is followed by the first law of thermodynamics in Chapter 6, where common forms of energy and the transfer from one form to another are explained. Chapter 7 will conclude Part 2, Engineering Thermodynamics, with the second law of thermodynamics. The quality of energy and the one-way street for energy flow are expounded with the help of entropy, a thermodynamic property that signifies disorder. Also elucidated is the heat engine, the essential device for converting thermal energy into useful work.

Environmental and Engineering Fluid Mechanics is the subject of Part 3, which provides a thorough disclosure of basic fluid mechanics. This part starts with motionless fluid, that is, fluid statics, as Chapter 8. The importance of fluid pressure or hydrostatic force on flat surfaces is accentuated. Frictionless moving fluid is covered in Chapter 9 under Bernoulli flow. The Bernoulli equation is illustrated in terms of static, dynamic, and hydrostatic pressures, and also as energy and hydraulic grade lines. Dimensional analysis is communicated in Chapter 10. The classical Buckingham Pi theorem is put to use and the key non-dimensional fluid mechanics parameters are presented. Part 3 ends with internal and external flows, Chapters 11 and 12, respectively. Only Newtonian fluids are considered in this introductory textbook. The transformation of flow from laminar to transition to turbulent in a conduit with increasing Reynolds number is introduced. Also put forward is the flow development at a pipe entrance. The crux of Chapter 11 is the energy equation, including major and minor head losses in pipe flow. For external flow, in Chapter 12, viscosity, boundary layer, and boundary layer development are conveyed. Bluff body aerodynamics is one of the highlights of the chapter.

Heat transfer is systematically disclosed in Part 4, Ecophysiology-flavored Engineering Heat Transfer. Chapter 13 delineates the steady conduction of thermal energy. The analogy between electric current flow and heat flow is drawn. The scope is confined to one-dimensional heat conduction and this is expressed in planar, cylindrical, and spherical coordinates. The concept of a lumped system is presented in Chapter 14, simplifying the analysis of transient heat conduction. After establishing the Biot number, one-dimensional transient heat conduction in a large plate, a long cylinder, a sphere, and a semi-infinite wall is disclosed. Natural heat convection is the topic of Chapter 15. Both natural thermals in a boundless environment and confined Rayleigh–Bernard convection are described. Also covered are natural convection along a vertical plate, continuous thermal plumes, and buoyant jets. Chapter 16 delineates heat convection powered by an external force, that is, forced convection. The many common parameters underlying forced convection are collapsed into the non-dimensional Nusselt, Reynolds, and Prandtl numbers. External flow forced convection along a constant temperature and uniform heat flux flat surface and around a cylinder are discussed. Also addressed is forced convection inside a conduit. Thermal radiation is put forward in Chapter 17. Following the definition of the ideal black body radiation, the Stefan–Boltzmann law is modified, with the introduction of emissivity, for gray surfaces encountered in real life. Energy conservation in terms of absorptivity, transmissivity, and reflectivity is portrayed. The importance of view or shape factor is also furnished. The book wraps up with heat transfer in applications, heat exchangers, as Chapter 18. In addition to detailing the standard types of heat exchanger analyses and application approaches, readers are encouraged to appreciate and mimic the intelligently-designed heat exchangers exploited by animals to thrive.

Problems

1.1 Spinach energy.
   Given that a 0.1-kg can of hearty spinach gives Popeye 100 kJ of energy, how many calories is that? Suppose only 30% of that energy is used to do work and the rest is dissipated as metabolic heat. How much metabolic heat is generated by Popeye over a 10-min period?

1.2 Sheep versus cow.
   The metabolic rate of a typical 45-kg adult sheep is approximately 50 W, while that of a 400-kg cow is around 270 W. Assume that the energy density of hay and corn are 8 MJ/kg and 6 MJ/kg, respectively. How much hay is need to feed a 45-kg sheep every day? How much corn is needed for a 400-kg cow on a daily basis?

1.3 Corn energy for plowing.
A plow requires 1000 lbf to till a field. How much corn does an ox need to consume to supply the energy for two hours of plowing?

1.4 Ascending energy.
A 5-kg American bald eagle ascends 3,000 m in the air (Animal Fact Guide, 2021). How much energy is needed to do so? Providentially, birds like eagles take advantage of convective thermal plumes and use little energy of their own when soaring.

1.5 Sharkskin for saving pumping energy.
Sharkskin is intelligently designed for drag (friction) minimization; see Dean and Bhushan (2010), Wu et al. (2018), and Tian et al. (2021). How much can you save, in terms of pumping power, if a shark-skin design is implemented on the inner surface of the water distribution pipes of a school? Assume that the piping network consists of 150 m of 1.27-cm iron pipe with an average water velocity of 0.5 m/s.

1.6 Heat convection from an elephant ear.
Suppose a 0.03-m^2 ear of an elephant at 36°C is losing 30 W of heat to the ambient air, which is at 21°C. What is the corresponding convection heat transfer coefficient?

1.7 Titanium for deep-sea diving.
Dr. JAS is designing a deep-sea diving suit that will allow her to study giant squids in their natural habitat at 1000 m below sea level. Approximate the suit as a 40-cm diameter spherical helmet, a 50-cm cylinder for the torso, two 20-cm diameter cylinders for the legs, and two 10-cm diameter cylinders for the arms. What should the thicknesses be if they are made of Grade-6 titanium?

References

Animal Fact Guide, https://animalfactguide.com/animal-facts/bald-eagle/, (accessed October 18, 2021) 2021.

CBC83, Gimli glider, posted July, 2018, https://www.cbc.ca/archives/when-a-metric-mix-up-led-to-the-gimli-glider-emergency-1.4754039, (accessed July 23, 2020) 2020.

Columbus, http://www.christopher-columbus.eu/navigation-longitude.htm, (accessed on July 22, 2020) 2020.

Dean, B., Bhushan, B., 2010. Shark-skin surface for fluid-drag reduction in turbulent flow: a review. Philosophical Trans. Royal Soc. A: Mathematical, Phys. Eng. Sci. 368, 4775–4806.

Dolphin, https://www.dolphincommunicationproject.org/, (accessed July 16, 2020) 2020.

Fennecfox, https://www.nationalgeographic.com/animals/mammals/f/fennec-fox/, (accessed July 20, 2020) 2020.

Giraffes, https://bpsfuelforthought.wordpress.com/2012/08/14/why-giraffes-don't-have-brain-damage/, (accessed July 20, 2020) 2020.

Healthline, https://www.healthline.com/nutrition/foods, (accessed July 21, 2020) 2020.

Humandive, https://thecoastalside.com/how-deep-can-a-human-dive/, (accessed July 20, 2020) 2020.

One Kind Planet, https://onekindplanet.org/top-10/top-10-laziest-animals/, (accessed July 21, 2020), 2020.

Scienceabc, https://www.scienceabc.com/nature/why-sloths-slow-makes-lazy-damn-slow-move.html, (accessed July 21, 2020) 2020.

Seasky, http://www.seasky.org/deep-sea/giant-squid.html, (accessed July 20, 2020) 2020.

Sweetpotato, http://www.foodreference.com/html/sweet-pot-nutrition.html, (accessed July 21, 2020) 2020.

Tian, G., Fan, D., Feng, X., Zhou, H., 2021. Thriving artificial underwater drag-reduction materials inspired from aquatic animals: progresses and challenges. Royal Soc. Chem. Adv. 11, 3399–3428.

Ting, T.Z-T., Stagner, J.A., 2020. Sustainable food for thought. In: Stagner, J.A., Ting, D.S-K. (Eds.), Sustaining Resources for Tomorrow: Green Energy and Technology. Springer, Cham, pp. 99–108.

Vallejo, D. "Are species larger in cold environments? The principles of Bergmann's rule," Zoo Portraits, www.zooportraits.com, (accessed September 18 2018), 2018.

Wu, L., Jiao, Z., Song, Y., Liu, C., Wang, H., Yan, Y., 2018. Experimental investigations on drag-reduction characteristics of bionic surface with water-trapping microstructures of fish scales. Sci. Rep. 8, 12186.

Wright, P.G., 1984. Why do elephants flap their ears? South Afr. J. Zoology 19 (4), 266–269.

Chapter 2

Energy and thermodynamics

"Thermodynamics is a funny subject. The first time you go through it, you don't understand it at all. The second time you go through it, you think you understand it, except for one or two small points. The third time you go through it, you know you don't understand it, but by that time you are so used to it, it doesn't bother you any more."

Arnold Sommerfeld

Chapter Objectives

- Understand that thermodynamics is the study of energy.
- Comprehend the conservation of energy.
- Appreciate the quality of energy.
- Recognize the three fundamental types of thermodynamic systems.
- Fathom thermodynamic state, equilibrium, and properties.

Nomenclature

c_P specific heat capacity at constant pressure
E energy; E_{in} is the energy entering a system, E_{out} is the energy leaving a system, E_{stored} is the energy stored in a system, E_{work} is the energy for performing work
g gravity
h height
KE kinetic energy
m mass
n amount of substance in moles
PE potential energy
Q thermal energy or heat; Q_{out} is the heat leaving a system
T temperature; ΔT is the temperature difference
V velocity

2.1 The study of energy

The study of energy and energy conversion from one form to another constitutes thermodynamics; see Table 2.1. Energy can be viewed as the capacity for heating and for doing work. Thermal, potential, kinetic, mechanical, chemical, electrical, and magnetic are common forms of energy. Thermal energy is the energy

Thermodynamics: From Nature to Engineering https://doi.org/10.1016/B978-0-323-90626-5.00018-5

TABLE 2.1 Thermodynamics, energy, and forms of energy.

Thermodynamics	The study of energy and energy conversion
Energy	The capacity to heat or do work
Common forms of energy	Thermal, potential, kinetic, mechanical, chemical, electrical, magnetic

associated with temperature, and it is a form of energy of paramount concern in thermodynamics. Within context, solar energy is the starting point. A critical part of this solar energy enables the growth of a myriad of plants via photosynthesis. To rephrase it, vegetation harnesses solar energy for fuel to sustain hosts of living creatures. The good sun does not only lavish us with energy in terms of food, it also supplies us with other essential elements such as vitamin D3 (Karsten et al., 2009). Artificial ultraviolet light falls short of duplicating the sun in providing adequate vitamin D for vitality. The familiar sun-basking creatures, iguanas, have been found to suffer vitamin D deficiency when kept indoors under artificial lighting (Laing et al., 2001).

Concerning food as energy for fueling and sustaining life, food provides the needed energy, via metabolism to keep a creature warm and to empower it to thrive. For instance, black-tailed jackrabbits in the Mojave Desert expend the most energy, at 172 kcal/kg/day, in the winter when the ambient temperatures are consistently below the zone of thermoneutrality (Shoemaker et al., 1976). This is sustained by improved metabolism and increased digestive tract throughput. The latter physiological adaptation is also exploited by creatures such as grasshoppers when the ambient temperature fluctuates more than desirable (Harrison and Fewell, 1995). The adaptation to survive and thrive also includes additional feeding at lower temperatures.

As explained by Çengel and Boles (2015), thermodynamics is a compound word from "therme" and "dynamis." In Greek, "therme" means heat, and "dynamis" denotes power. To that end, thermodynamics bespeaks earlier human efforts in harnessing heat, thermal energy, to do work via a heat engine. Our body is a biological heat engine; we harness the energy in the food we savor via chemical reactions to enable us to do work such as studying and exercising. The remaining energy is dissipated as metabolic heat, as depicted in Fig. 2.1. The fuel for this biological engine is the food that we wallow in. Energy is generated via metabolism, the biological process of converting what we eat and drink into energy. The most familiar and classic manmade heat engine is the internal combustion engine that powers transportation. About 40% of the chemical energy from the fuel is harnessed for moving the vehicle, powering the lights, and keeping the driver and passengers comfortable. In the neighborhood of 60% of the energy is wasted as heat. Similarly, the energy from our food intake is converted into work. This includes the operation of our cells and organs and

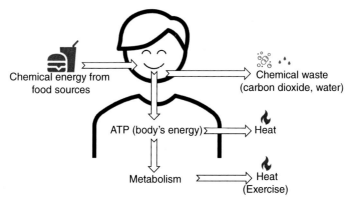

FIGURE 2.1   A living human body is a heat engine (created by O. Imafidon). Food provides the fuel to generate work and heat. The unused energy from the excess food is stored as fat for future usage.

mechanical work such as walking. A significant portion is dissipated as heat, and this amount can be significantly increased via vigorous physical work such as running, aerobics, swimming, and dancing. There is also storage of unused energy from the excess food we consumed in the form of fat. When the usage is above and beyond the energy intake, our body resorts to body fat to make up for the shortfall.

Example 2.1 Bike ride for a low-calorie ice cream

Given: Grandma Nenek aspires to inspire her 4-year-old grandson, Cucu, to be active and healthy. The ice cream shop a couple of kilometers away from grandma's house is one of Cucu's favorite hangouts. There are a few varying-distance paths between grandma Nenek's house and the ice cream shop.

Find: Help grandma Nenek figure out the distance Cucu must bike to get the ice-cream so that all the calories consumed are burned up on the journey to and back from the ice cream shop.

*Solution:*Assume grandma Nenek is able to trick Cucu into getting the small, low-calorie ice cream cone containing only half of a typical ice cream cone. Note that a typical ice cream contains about 400 calories of energy.

According to Caloriesburnedhq (2021), for a typical 4-year-old weighing 18 kg going at the beginner speed of 16 km/h, it takes about 2 h to burn the energy content in the ice cream cone. Being a good grandma is a challenge, and we have verified this thermodynamically.

Thermodynamics is an essential subject for all disciplines of engineering, and it is also a requisite for students in science. As a result of that, there are numerous monographs dedicated to disclosing thermodynamics in detail. Some of the most recognized and comprehensive undergraduate textbooks dedicated to engineering thermodynamics include Borgnakke and Sonntag (2019), Moran et al. (2018), Reynolds and Colonna (2018), Çengel and Boles (2015), and Balmer (2011). Radulescu (2019) and Eastop and McConkey (1993) are two of the lighter introductory books on engineering thermodynamics. Readers who are enthusiastic about thermodynamics after being briefed about thermofluids may also wish to check out Granet et al. (2021), Reisel (2021), Turns and Pauley (2020), Whitman (2020), Luscombe (2018), Ozilgen and Sorgüven (2017), Steane (2017), Kondepudi and Prigogine (2015), Jacobs (2013), Koretsky (2013), Klein and Nellis (2012), Wong (2012), Kreuzer and Tamblyn (2010), Lemons (2009), Kondepudi (2008), Shavit and Gutfinger (2008), O'Connell and Haile (2005), and Gyftopoulos and Beretta (2005). Biological thermodynamics is covered in Demirel and Gerbaud (2019) and Haynie (2008).

2.2 The conservation of energy

The energy from the food we feed on is equal to the sum of energy for doing work, dissipating as heat, storing as fat, and letting out not fully digested or utilized, as illustrated in Fig. 2.1. The conservation of energy can be expressed as

$$E_{in} = E_{out} + E_{stored}, \tag{2.1}$$

where E_{in} is the energy entering a system such as a human body, E_{out} is the energy leaving the system, and E_{stored} is the energy stored in the system. For a human body, the energy equation can be written as

$$E_{in} = E_{work} + Q_{out} + E_{stored} + E_{out}. \tag{2.2}$$

Here, E_{in} is the energy input via food intake, E_{work} is the energy used for doing all kinds of work for keeping a person alive and for enabling the individual to enjoy life to its fullness, Q_{out} is the energy dissipated as heat[1], E_{stored} is the energy stored as fat, and E_{out} is the unused energy in our feces and urine. Note that E_{stored} is negative when an individual controls the amount of food intake such that more energy is used than consumed.

The principle of energy conservation is a cornerstone of thermodynamics and, hence, it is set as the first law of thermodynamics; see Table 2.2. We see that the energy in food is converted into work, heat, fat, and/or passed out. To this end, the first law of thermodynamics states that energy cannot be created from nothing, nor can it be destroyed or made to disappear; energy can be transformed from one form into another.

---

1. Because of this metabolic heat, every living soul is warm and, thus, an effective way to gauge how long a person has been dead is to measure the body temperature. Knowledge from the third element of thermofluids, heat transfer, enables an inspector to determine the time of death from the variation of body-environment temperature difference as a function of time.

TABLE 2.2 The conservation of energy.

The first law of thermodynamics states that energy can neither be created nor destroyed, but it can be converted from one form into another form

The first law of thermodynamics is the law concerning the conservation of energy

Example 2.2 The conservation of energy of a falling durian

Given: In some parts of the world, durians are considered to be the king of fruits. The potent smell seems to heighten its flavor for durian lovers. For those who do not appreciate the king of fruits, durians smell like some combination of rotten onions, turpentine, a gym sock, and sewage (Stromberg, 2012). They grow on tall durian trees and tend to fall at night when they are ripe. Consider a 3-kg durian at 7 m above the ground that falls freely to the ground.

Find: How much potential energy does the 3-kg durian have when it is hanging at 7 m above the ground? What is its maximum velocity, if there are no losses such as drag and noise when falling? What happens to the energy (associated with the potential energy at a height of 7 m) when (or after) the durian hits the ground.

Solution: The potential energy of the 3-kg durian at a height of 7 m is

$$PE = mgh = 3\,\text{kg}\left(10\,\text{m/s}^2\right)(7\,\text{m}) = 210\,\text{kg}\cdot\text{m}^2/\text{s}^2,$$

or 210 J. Here, m is mass, g is gravity, and h is height.

The first law of thermodynamics says that energy is conserved. Therefore, if there were no losses, all the potential energy will be converted into kinetic energy just before the durian hits the ground, noting that the potential energy of the durian is zero at ground level. In other words,

$$PE = 210\,\text{J} = KE = \frac{1}{2}mV^2 = \frac{1}{2}(3\,\text{kg})V^2,$$

where V is the velocity. Solving for velocity, we get

$$V = 11.83\,\text{m/s}.$$

This is the terminal velocity if there is no air to impose drag on the falling durian. In reality, there are significant losses in the presence of air and, hence, the terminal velocity is considerably less than 11.83 m/s. Even at half this terminal velocity, it is important for durian lovers to stay clear of falling durians. It is one thing to be, like Newton, hit by a falling apple, it is quite another thing to be hit by a falling durian.

After the durian hits the ground, its potential energy and kinetic energy are both zero. Other than losing a considerable amount of energy to the air, most of

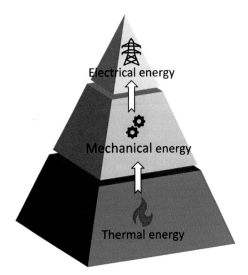

FIGURE 2.2 Comparing the quality of electrical, mechanical, and thermal energy (created by O. Imafidon). The ease in transforming the form of energy to other forms indicates the quality of the energy.

the energy is lost on impact, transforming into vibration and noise that disperse into the ground and the surroundings.

2.3 The quality of energy

To function effectively, we need a certain amount of quality sleep every day. It is better to have a few hours of good sleep than many hours of bad dreams. Such is also the case when it comes to energy. The ease of transforming one form of energy to other forms has been used to measure energy quality (Midžić et al., 2014). For example, when comparing thermal, electrical, and mechanical forms of energy, the order of decreasing quality, as depicted in Fig. 2.2, is from electrical to mechanical to thermal. In other words, electrical is the highest quality energy and thermal is the lowest quality. We see electrical energy running fans and compressors for refrigerators and air conditioners everywhere. To convert mechanical energy into electricity, a relatively sophisticated mechanical-to-electrical generator is required. Water and wind turbines are such engineering inventions. Also, a complex heat engine is required to convert thermal energy into mechanical work. A familiar example is the spark-ignition engine for mechanically driving a car. In this case, the thermal energy is generated from rapid chemical reactions of fuel and air. A mechanical-to-electricity generator is an addition required to further convert the mechanical energy into electricity. The transformation from thermal to mechanical to electrical energy is not easy, and this is expressed by the significant losses at each stage. On the other

TABLE 2.3 The quality of energy.

Energy has both quantity and quality

The quantity of energy is conserved

The quality of energy deteriorates

Energy naturally flows in the direction of decreasing quality

The second law of thermodynamics affirms the decreasing quality of energy in all real processes

hand, electricity is easily converted into thermal energy, as illustrated in electric resistance heaters, electric ranges, and electric kettles. Note that not all the heat is used in boiling the water or cooking the food, in the case of an electric kettle or electric stove, but all the electricity is converted into heat. To put it another way, the electricity-to-thermal energy conversion efficiency is 100%. The reverse is substantially less than 100% efficient.

The quality of energy is covered under the second law of thermodynamics which states that energy only flows in one direction naturally, that is, from high to low quality; see Table 2.3. A common example is the flow of thermal energy, heat, from high to low temperature. It follows that, even within the form of thermal energy, the quality of the heat decreases with decreasing temperature. This spontaneous flow of heat from high to low temperature is familiar to everyone who has enjoyed the thermal comfort provided by air conditioning in the hot summer. The heat from the ambient transfers naturally into the living or occupied space such as a house, an office, or a vehicle. The removal of some of the heat from the occupied space to the higher-temperature ambient requires energy to operate an air conditioner. Here is the catch of the second law of thermodynamics that we cannot beat; the energy used for driving the air conditioner, along with the thermal energy to be removed from the lower-temperature occupied space, equal to the amount of thermal energy dumped into the ambient according to the conservation of energy, that is, the first law of thermodynamics. The second law of thermodynamics that concerns with the quality of energy exhibits itself by ensuring that we always end up with lower quality energy. In the air conditioning case, the transport of some thermal energy from the lower-temperature space leads to more than this transported amount of thermal energy into the environment. Some notable amount of quality electricity for running the air conditioning ends up as low-quality thermal energy in the environment.

Let us look at the quality of energy from the physiological or trophic perspective. Consider one million joules of sunlight, as shown in Fig. 2.3. Only 1% of those million joules is converted into vegetation, the primary energy and/or food producers (Mammoth Memory, 2021). Primary consumers such as grasshoppers can tap into 10% of the available energy in the greens they

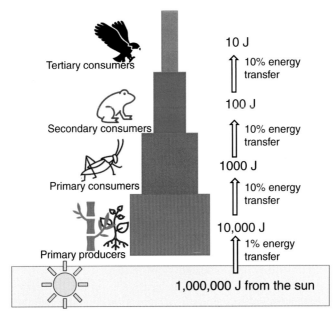

FIGURE 2.3 A hierarchy of food energy chain (created by O. Imafidon). Consuming the food of the lower level is more efficient and environmentally friendly because 90% losses are compounded up the pyramid.

consume. According to that, 10,000 J of energy in the grass is converted into a thousand joules of energy in grasshoppers. Secondary consumers such as frogs need to eat a thousand joules of grasshoppers to create 100 J of energy in yummy and muscular frog meat for feeding the next level up. Birds that prey on frogs can be the tertiary consumers. Birds must consume 100 J of energy-containing frogs per 10 J of energy requirement. Note that there can be more levels on the hierarchy. For example, snakes can make the tertiary level of consumers and snake-savoring birds could be the quaternary level. Furthermore, the actual food chain is more complex than what this illustrated here. Humans, for example, can fit in at different levels. Vegetarians are primary consumers, making the most use of the energy lavished from the sun. If you enjoy tasty frog meat, you will be on the highest consumer level, the same as birds such as the long-legged, long-necked herons.

So, how do we relate the food-energy hierarchy with the quality of energy? We can, for example, let thermal energy form the base of the pyramid. It takes a lot of thermal energy to boil water and form steam. A large quantity of steam energy is required to spin a steam turbine and produce mechanical energy. Many joules of mechanical energy are necessary to generate a few joules of electricity. In short, the quality of energy decreases as we move down from the top of the pyramid to its base. The lowest form of energy, thermal energy, forms the base of the energy-quality pyramid.

Example 2.3 Quality of sensible, latent, and chemical energy

Given: Dr. JAS takes her class to a cattle farm. Seeing the cattle passing a lot of gas and recalling hearing about people harnessing farts for energy, an idea pops into T-E Baru's mind. He wishes to harness the sensible energy of the warm methane gas. He recalls that sensible energy is the part of the internal energy that is associated with molecular motion including random movement, translation, and rotation.

Find: If T-E Baru is on the right track or is there a better way to tap into cattle's farts.

Solution: Internal energy is the energy associated with the motion of the molecules. It may exist in sensible, latent, chemical, and nuclear forms. The energy density, the amount of energy per unit quantity of the substance, increases from sensible to latent to chemical to nuclear form. The cattle's flatulence contains some amount of low-quality sensible heat. Harnessing this heat is generally not a worthwhile effort. Having said that, capitalizing the chemical energy of the methane gas produced by the cattle is not only a viable endeavor but also an environmentally friendly option. Furthermore, capturing the methane from cattle's farts, burps, and manure for good use is more humane than many human interventions of cattle's digestive systems; see, for example, Beil (2015).

The chemical energy associated with combusting methane with air is more than 800 kJ per mole of methane. The sensible heat of one mole of methane at $37°C$ with respect to a room temperature of $20°C$ is

$$Q = nc_P\Delta T = 1 \, mol(36 \, J/(mol \cdot K))(37°C - 20°C) = 612 \, J.$$

Here, n is the amount in moles, c_P is the specific heat capacity at constant pressure, and ΔT is the temperature difference. There is a three-orders-of-magnitude difference between the harnessable sensible heat and the heat of combustion. In short, T-E Baru should capture the bio-methane and capitalize the significant heat of combustion for good use.

2.4 Thermodynamic systems

A thermodynamic system is a specified region of interest. That region can encompass a quantity of matter or a space with fixed or moveable boundaries. The body of a creature, such as a human body, is a thermodynamic system. Applying the conservation of energy to such a system, we observe that the energy into the body minus the energy out of the body is equal to the energy stored in the body. For this reason, we can manage our weight, which indicates energy storage, by balancing energy intake with energy usage. Table 2.4 summarizes the three types of thermodynamic systems.

TABLE 2.4 The three fundamental thermodynamic systems.

Closed system (control mass)	A thermodynamic system consisting of a fixed amount of mass (matter). No mass (matter) enters or exits a closed system. Energy can cross the boundary of a closed system
Open system (control volume)	A thermodynamic system defined by a volume in space of interest. Both matter and energy can enter or exit an open system
Isolated system	A completely sealed thermodynamic system or a physical system so far removed from other systems that it does not interact with them. Neither matter nor heat can transfer to or from the system

An unopened can of soda is a closed system that consists of a fixed amount of mass. No mass can cross the boundary of a closed system. Energy, however, can enter or exit a closed system. A human body is an open system in the sense that both energy and mass can cross its boundary. The entire universe is an isolated system in which the total mass and total energy remain the same because neither mass nor energy can cross its boundary. We will expound on thermodynamic systems in a later chapter.

2.5   Thermodynamic state, equilibrium, and properties

A thermodynamic system can either be in an equilibrium state or a non-equilibrium state (Balmer, 2011). A state is the condition of a system or a subsystem[2] at a specific time. A thermodynamic state can be defined by a set of thermodynamic properties such as pressure and temperature. In this book, we only deal with thermodynamic systems that are in an equilibrium state where the thermodynamic characteristics are well defined. A thermodynamic system is in an equilibrium state when its characteristics do not undergo a net change with time. In *statics*, the study of *mechanical equilibrium*, the mechanical forces within the system of interest are balanced so that there is no acceleration. A system in which the temperature is uniform throughout is in *thermal equilibrium*. *Chemical equilibrium* is achieved when each forward reaction is balanced by the respective reverse reaction. That being the case, there is no net change in any chemical specie with respect to time. Thermodynamics deals with static,

---

2. A subsystem can be a specified region within a larger system. For example, an air distributing duct is a subsystem of the heating, ventilation, and air conditioning system. A steady flow of cooled air in an air duct continues to lose heat and, thus, the air temperature increases. Namely, the thermodynamic state of the air changes. For this steady case, however, the state at any location along the air duct is constant, defined by the values of the corresponding thermodynamic properties such as temperature.

thermal, chemical, and other phenomena, that is, thermodynamic equilibrium encompasses all these equilibriums.

The identifiable characteristics of a thermodynamic system in equilibrium are the thermodynamic properties. Thermodynamic properties are characteristics of thermodynamic systems whose values vary only with the thermodynamic equilibrium state of the systems. In other words, thermodynamic properties are independent of the process or path taken to reach the state. Consider a can of soda; we can identify its characteristics such as mass, volume, temperature, and pressure. A 355 ml can hold 42 g and/or 355 ml of soda. How the 42 g of soda is filled into the can does not affect the state property, mass of soda in the can.

Thermodynamic properties can be grouped into two classes: extensive and intensive. Mass, volume, and energy content are extensive properties, as their values are a function of the size of the system. A 500-ml can of ginger ale contains twice the amount of kcal than the same ginger ale in a 250-ml can. On the other hand, temperature and pressure are intensive properties because they are independent of the mass of the system. We can tell if a thermodynamic property is extensive or intensive by recognizing if size matters or not. When size matters, the property is an extensive property; when size does not matter, it is an intensive property. Some extensive properties can be converted into intensive ones. For example, volume is an extensive property while specific volume is an intensive property.

Problems

2.1 Work to potential energy.

When visiting his friend who lives on the 17th floor, which is 60 m above ground level, T-E Kaya brings his new electric bike in the elevator with him. How much work does the elevator have to do to bring the 25-kg bike up from the ground floor to the seventeenth floor?

2.2 Drop tower thermodynamics.

One of Dr. JAS' favorite rides at amusement parks is the drop tower. She loves it so much that she approached the owner of an amusement park to propose an engineering solution to harness the energy during the drop. What can you say about the amount of recoverable energy, according to the first law of thermodynamics? What does the second law of thermodynamics have to say about this recovery?

2.3 Half-diet for burning fat.

During a pandemic lock-down, a man gained 4 cm of waistline, and this is equivalent to 5 kg of body mass. He sets a resolution to burn off the fat by halving his diet from 4000 to 2000 calories per day. How long does it take to lose that 5 kg? Note that an average man needs about 2500 calories per day and there are 9 calories per gram of human fat. In other words, his body will resort to using stored energy, that is, fat, to make up for the 500 calories per day of shortfall. If he stays with 4000 calories-per-day diet, how long does he need to jog every day, for him to lose 5 kg of fat in a year?

2.4 Siting for a wind turbine.

A wind turbine is to be installed on a campus to offset part of the energy demand. The wind speed at one site is 11 m/s but the wind only blows 2200 h per year. The wind at another suitable site blows at 7 m/s for 3300 h a year. Which one is a better site for harnessing the wind energy?

2.5 Meatballs for sustaining a city.

The entire population of a city lives on a meatball diet. The population is 300,000 and an average person consumes 1 kg of meatballs per day. Assume the meatballs are made of 100% grass-fed cattle. The energy conversion efficiency from grass to beef is 10%. The photosynthesis efficiency of converting solar energy into hearty grass is 1%. How many acres of land in Edmonton, Alberta, Canada is needed to feed this city? If the beef comes from Vernon, Texas, USA instead, what is the corresponding required cattle-rearing land?

2.6 Energy for breathing.

Part of the energy from metabolism is used for breathing. A typical adult exhales 0.5 L of air per breath. How much energy is used when exhaling air into the open atmosphere which is at 101 kPa? Hint: assume a constant pressure process similar to blowing air to form a soap bubble, where work is equal to pressure times the change in volume.

2.7 Waste heat by aliens.

As the T&E Spaceship was accelerating to warp speed, Captain JAS registered an usual amount of thermal energy on a distant planet. She immediately commanded the T&E Spaceship to head toward the planet. What is Dr. JAS after? Hint: Since the heat is not associated with star formation, it is likely that an intelligent species has been engineering an avant-garde civilization. A large amount of waste heat is being generated to power and sustain that civilization.

References

Balmer, R.T., 2011. Modern Engineering Thermodynamics. Academic Press, Amsterdam.

Caloriesburnedhq, https://caloriesburnedhq.com/calories-burned-biking/ (accessed September 10, 2021), 2021.

Beil, L., 2015. Greener cows: research rounds up less burpy bovines. Sci. News 188 (11), 22–25.

Borgnakke, C., Sonntag, R.E., 2019. Fundamentals of Thermodynamics, Tenth Edition Wiley, Hoboken, NJ.

Çengel, Y.A., Boles, M.A., 2015. Thermodynamics: An Engineering Approach, 8th ed. McGraw-Hill, New York, NY.

Demirel, Y., Gerbaud, V., 2019. Nonequilibrium Thermodynamics: Transport and Rate Processes in Physical, Chemical and Biological Systems, 4th ed. Elsevier, Oxford.

Eastop, T.D., McConkey, A., 1993. Applied Thermodynamics for Engineering Technologists, 5th ed. Pearson /Princtice Hall, New York, NY.

Granet, I., Alvarado, J.L., Bluestein, M., 2021. Thermodynamics and Heat Power, 9th ed. CRC Press, Boca Raton, FL.

Gyftopoulos, E.P., Beretta, G.P., 2005. Thermodynamics: Foundations and Applications. Dover Publications, Inc., Mineola, NY.

Harrison, J.F., Fewell, J.H., 1995. Thermal effects on feeding behavior and net energy intake in a grasshopper experiencing large diurnal fluctuations in body temperature. Physiol. Zool. 68 (3), 453–473.

Haynie, D.T., 2008. Biological Thermodynamics, Second Illustrated Edition Cambridge University Press, Cambridge.

Jacobs, P., 2013. Thermodynamics. Imperial College Press, London.

Karsten, K.B., Ferguson, G.W., Chen, T.C., Holick, M.F., 2009. Panther chameleons, furcifer pardalis, behaviorally regulate optimal exposure to UV depending on dietary vitamin D3 status. Physiol. Biochem. Zool.: Ecol. Evolutionary Approaches 82 (3), 218–225.

Klein, S., Nellis, G., 2012. Thermodynamics. Cambridge University Press, New York, NY.

Kondepudi, D., 2008. Introduction to Modern Thermodynamics. Wiley, Chichester.

Kondepudi, D., Prigogine, I., 2015. Modern Thermodynamics: From Heat Engines to Dissipative Structures, 2nd ed. Wiley, Chichester.

Koretsky, M.D., 2013. Engineering and Chemical Thermodynamics, Second Edition Wiley, Hoboken, NJ.

Kreuzer, H.J., Tamblyn, I., 2010. Thermodynamics. World Scientific, Singapore.

Laing, C.J., Trube, A., Shea, G.M., Fraser, D.R., 2001. The requirement for natural sunlight to prevent vitamin D deficiency in Iguanian lizards. J. Zoo Wildl. Med. 32 (3), 342–348.

Lemons, D.S., 2009. Mere Thermodynamics. John Hopkins University Press, Baltimore.

Luscombe, J.H., 2018. Thermodynamics. CRC Press, Boca Raton, FL.

Mammoth Memory, https://mammothmemory.net/biology/organisms-and-their-environment/ecosystems-organisms-and-their-environment/energy-flow.html, (accessed September 8, 2021), 2021.

Midžić, I., Štorga, M., Marjanović, D., 2014. Energy quality hierarchy and "transformity" in evaluation of product's working principles. Procedia CIRP 15, 300–305.

Moran, M.J., Shapiro, H.N., Boettner, D.D., Bailey, M.B., 2018. Fundamentals of Engineering Thermodynamics, 9th ed. Wiley, Hoboken, NJ.

O'Connell, J.P., Haile, J.M., 2005. Thermodynamics: Fundamentals for Applications. Cambridge University Pressure, Cambridge.

Ozilgen, M., Sorgüven, E., 2017. Biothermodynamics: Principles and Applications,. CRC Press, Boca Raton, FL.

Radulescu, L.F., 2019. A Concise Manual of Engineering Thermodynamics. World Scientific, Singapore.

Reisel, J., 2021. Principles of Engineering Thermodynamics, 2nd ed. Cengage Learning, Boston, MA.

Reynolds, W.C., Colonna, P., 2018. Thermodynamics: Fundamentals and Engineering Applications. Cambridge University Press, Cambridge.

Shavit, A., Gutfinger, C., 2008. Thermodynamics: From Concepts to Applications, 2nd ed. CRC Press, Boca Raton, FL.

Shoemaker, V.H., Nagy, K.A., Costa, W.R., 1976. Energy utilization and temperature regulation by Jackrabbits (Lepus californicus) in the Mojave Desert. Physiol. Zool. 49 (3), 364–375.

Steane, A.M., 2017. Thermodynamics: A Complete Undergraduate Course. Oxford University Press, Oxford.

Stromberg, J., 2012. Why does the durian fruit smell so terrible? Ask Smithsonian 2017, A Smithsonian magazine special report. November 30, 2012.

Turns, S.R., Pauley, L.L., 2020. Thermodynamics: Concepts and Applications, 2nd ed. Cambridge University Press, New York, NY.

Whitman, A.M., 2020. Thermodynamics: Basic Principles and Engineering Applications. Springer, Cham.

Wong, K.V., 2012. Thermodynamics for Engineers, 2nd ed. CRC Press, Boca Raton, FL.

Chapter 3

Moving fluids

"Whatever is fluid, soft, and yielding will overcome whatever is rigid and hard. What is soft is strong."

Laozi

Chapter Objectives
- Differentiate a fluid from a solid.
- Understand what a continuum fluid is.
- Appreciate the beautiful and all-important fluid motions in nature.
- Comprehend viscosity.
- Fathom Newtonian fluids.
- Recognize the general categories of fluid motions.

Nomenclature

A	area
D	diameter
d_{gap}	gap distance
F	force
g	gravity
h	height
L	length
m	mass; $m_{molecule}$ is mass per molecule
P	pressure
Re	Reynolds number
R_i	radius of the inner cylinder
R_o	radius of the outer cylinder
S_{rpm}	rotational speed in revolutions per minute
T	temperature
t	time; δt is a very short time
T_q	torque
U	velocity or velocity in the x direction; U_{plate} is the velocity of the plate, $U|_{y=0}$ is the velocity at $y = 0$
V	velocity or velocity in the y direction; V_t is the tangential velocity
W	width
x	x direction; δx is a very small displacement in the x direction, x_{plate} is the velocity of the plate.

The Engineers from Nature in Engineering. DOI: https://doi.org/10.1016/B978-0-323-90626-5.00011-2

y   y direction or distance in the y direction
z   z direction

Greek and other symbols

β angle; $\delta\beta$ is angular displacement of shearing strain
γ change in angle or shearing strain; γ' is rate of change in angle
Δ difference
δ a small difference or change
μ viscosity
ρ density
τ shear
ω rotational speed in radians per second
$\mathfrak{R}$ gas constant
$\forall$ volume

3.1 What is a fluid?

Intuitively, a fluid is a liquid or a gas that flows easily. It has no shape of its own; a gas occupies the entire container while a liquid fills up the container up to the free surface. Atmospheric air is a moving fluid that is essential to life on Earth. Equally indispensable is the vast volume of moving water that occupies more than seventy percent of Earth's surface.

In general, we can differentiate a fluid from a solid in the following manner; see Fig. 3.1.

Fluids tend to flow under a force. Water flows in a river due to the force of gravity. It follows that a fluid is a substance that deforms continuously under shear. The rate of deformation is a function of the viscosity of the fluid. It is much

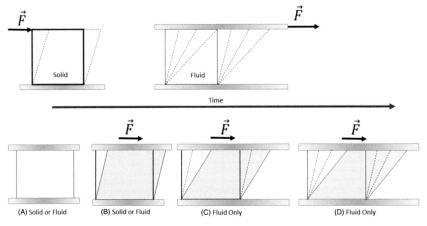

FIGURE 3.1 Fluids versus solids (created by F. Fashimi). Fluids cannot sustain a shear force, that is, a fluid continues to move under a shear force. A solid deforms under a shear force to a certain amount and stops.

easier to pour a cup of water into a sink than a cup of honey. On that account, we recognize that the viscosity of honey at room conditions is much higher than that of water.

Solids tend to deform, but not flow, under force. A cup of hair gel deforms explicitly when it is being poured into a sink. Once it is in the sink, it stays on the sink surface, unless you flush it down the drain with hot water. To put it another way, the gel deforms and stops, that is, it does not flow under pressure. The amount of deformation per unit force of a solid is a function of the moduli of rigidity and elasticity.

Imagine Dr. JAS takes you into the T&E Hovercraft that is glued to the surface of an elastic solid mat. Turning on the hovercraft engine and fan would deform the elastic mat similar to that illustrated in Fig. 3.1. The fixed finite deformation is proportional to the applied shear force induced by the hovercraft fan and the modulus of elasticity of the solid mat. Turning off the engine would restore the elastic mat to its initial shape. When you turn on the hovercraft on the surface of a layer of water instead of the elastic mat, the hovercraft would slide, shearing the liquid water all the way. The layer of water would not restore to its original state after the engine is turned off, that is, the deformation is permanent. To that end, we see that a fluid deforms continuously under shear.

Mechanics, or more explicitly, classical mechanics, is a branch of physics concerned with motions of bodies and the forces acting on them; note that motionless is included under "motions." The Merriam-Webster dictionary defines mechanics as "a branch of physical science that deals with energy and forces and their effect on bodies" (Merriam-Webster, 2020). To rephrase it, mechanics is the study of forces acting on bodies and the resulting motions they cause. In fluid mechanics, these bodies are bodies of fluids. That being the case, fluid mechanics is the study of fluids at rest and in motion. Newtonian (or classical) mechanics deals with the motion of bodies under the action of forces. For completeness, let us concisely differentiate statics, kinematics, and dynamics.

Statics is the branch of mechanics that deals with the loads on bodies at rest, where the forces are in static equilibrium. The static load induced by the pressure of water due to gravity on a giant squid at 1000 m underwater is an order of magnitude larger than that on a diver at a depth of 100 m. The giant squid is able to exert a force equal and opposite to 1000 m of seawater, plus the added dynamic force when it maneuvers around in the water.

Kinematics is the branch of mechanics that studies the motion of bodies without reference to the forces involved. An inspiring example is an Olympic athlete's disciplined and polished motion in terms of displacement, velocity, and acceleration of different body parts. Key points include ankles, knees, and hips.

Dynamics is the branch of mechanics that is concerned with the motion of bodies, and the equilibrium of bodies, under the action of forces. A dynamic kid playing on a trampoline is an everyday illustration of dynamics in action.

3.2   The continuum fluid

When we ascend in altitude, rising above where commercial airplanes cruise, the air becomes progressively thinner. This thin-air condition can also be achieved via vacuuming a container of gas to a low absolute pressure. With decreasing pressure, the molecules of the gas become far apart and the gas as a whole does not behave as a continuum. Accordingly, a microscopic approach, realized via statistics, is suited for describing the behavior of the molecules. Rarefied gas dynamics deals with the flows of thin gases.

Closer to Earth, the everyday fluids that concern us follow continuum behavior. The fluid can be treated as a continuous medium because the density of the molecules is high and the space is populated with many molecules. The overall properties at any region do not change moment to moment because of the movements of the molecules. Let us evaluate the minimum volume, δV, for which this is true. For air at atmospheric conditions, for a mathematically differentiable volume making up a cube of 0.1-mm side, $\delta V = 0.001 \text{ mm}^3$, there are 2.5×10^{13} molecules. Fluctuations of a few hundred molecules due to molecular motions into or out of the differentiable volume of interest amount to nothing. That being so, it is a continuous medium and its thermodynamic properties such as density, pressure, and temperature are continuous functions of time and space.

Example 3.1   Continuum fluid or rarefied gas.
Given: Dr. JAS says that a fluid should be treated as a rarefied gas if it consists
         of less than 10^{12} molecules per mm^3.
Find: The air pressure for this threshold at 20°C.

Solution: Assuming air behaves as an ideal gas, where the pressure, P, is related to the density, ρ, gas constant, $\Re$, and temperature, T, according to:

$$P = \rho \Re T.$$

The density:

$$\rho = m_{molecule}\left(10^{12} \text{ molecules/mm}^3\right),$$

where the mass per molecule:

$$m_{molecule} = \text{Molecular weight/Avogadro's number}$$

$$m_{molecule} = 28.97 \text{ kg/kmol}/6.023 \times 10^{23} \text{ molecules/mol} = 4.81 \times 10^{-26} \text{ kg}$$

Therefore, the density:

$$\rho = (4.81 \times 10^{-26} \text{ kg})\left(10^{12} \text{ molecules/mm}^3\right) = 4.81 \times 10^{-5} \text{ kg/m}^3$$

The corresponding air pressure:

$$P = \rho \Re T = (4.81 \times 10^{-5})(287 \text{ m}^2/(\text{s}^2 \cdot \text{K})(293 \text{ K}) = 4.0 \text{ Pa}$$

This pressure is extremely low compared to the standard atmospheric pressure of 101.325 kPa. In other words, atmospheric air is surely a continuum fluid.

3.3 Nature thrives in moving fluids

Dandelions are a nuisance to those who see a perfect lawn as a single-specie soft grass[1]. In spite of the fact that these folks try to exterminate them, dandelions have many potential health benefits (Dandelion, 2020). Thanks to natural moving fluids, the ever-blowing atmospheric air, the seeds of this salubrious perennial flowering plant are spread far and wide; see Fig 3.2. The details underlying parachuting dandelion seeds are simply breathtaking; see Cummins et al. (2018). Ledda et al. (2019) found that the intelligently designed porosity of the dandelion's "parachute" is just right for promoting a steady wake regime. The porosity threshold to have an axisymmetric and stable separated vortex ring is also confirmed by the computational fluid dynamics work conducted by Qiu et al. (2020).

What about blackfly larvae? They secure themselves to the bed of a stream and slander slightly toward downstream; see page 217 of Vogel (1994). As the water current passes around their cylindrical-shaped bodies, flow vortices are created as shown in Fig. 3.3, sweeping up food into their cephalic fans. While blackfly larvae secure their tail into the stream bed, some stream invertebrates bow to the incoming water current and, in so doing, generate the appropriate vortices for digging and/or feeding (Soluk and Craig, 1990, 1988).

Talking about taking advantage of fluid vortices, starfish larvae have this mastered. Gilpin et al (2017) found that up to nine vortices are detected around the larvae. These vortical flow structures, along with ciliary tangles, apparently assist starfish larvae in balancing feeding and swimming. While fluid vortices are intriguing, nonvortical moving fluids are equally essential for sustaining life on earth.

3.4 What is viscosity?

Viscosity is the property that makes fluid sticky. The stickier a fluid is, the less flowing it does. Honey is a sweet but sticky fluid under room conditions. While in everyday living we do not consider water and air sticky, both have a finite viscosity. It is an accepted fact that viscosity changes with temperature. Honey sitting in a container chilled in a fridge for some time is stickier than when it is at room temperature. That being the case, the viscosity of honey, like most fluids, decreases with increasing temperature.

---

1. This kind of perfect lawn is bad for the environment and the ecosystem; see, for example, Learn (2021).

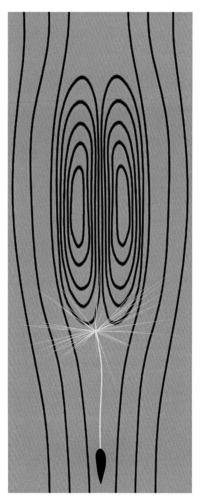

FIGURE 3.2   The moving fluid with a stable vortex ring above a flying dandelion seed (created by X. Wang). The appropriate porosity, shape, density, or weight and its distribution, and dimensions furnish the right conditions for a stable vortex ring above the seed, empowering the seed to travel far and wide riding on the omnipresent atmospheric wind.

One of the key reasons why viscosity is of great interest and concern to engineers is because power is required to move a fluid. The sticker the fluid is, the larger the required power to push the fluid at a desirable rate. To this end, viscosity is the resistance of a fluid to motion. Another familiar application of viscosity is in lubrication. If the viscosity of the lubricating fluid is too thick, it imposes resistance rather than reducing it. On the other hand, a lubricant that is too thin loses its lubricating ability because the fluid does not stick to the surfaces of the moving parts.

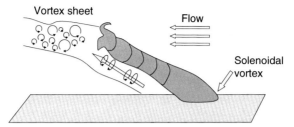

FIGURE 3.3 A blackfly larva capitalizing on flow vortices and flow-induced vibrations to scoop up food and convect it into its cephalic fans (created by Y. Yang). The cephalic fans dance in harmony with the organized flow structures that sweep up food, and synchronously conveying the food to the fans.

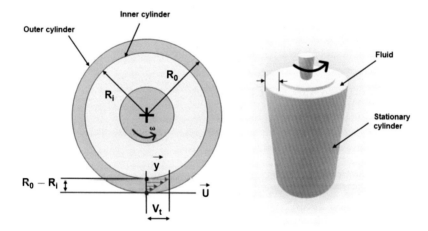

FIGURE 3.4 A rotational rheometer for measuring fluid viscosity (created by F. Fashami). With known gap, $d_{gap} = R_o - R_i$, inner radius, R_i, inner cylinder length, L, the torque, T_q, required to spin the inner cylinder at a prescribed speed, S_{rpm}, yields the fluid viscosity, μ.

Example 3.2 Measuring viscosity.

Given: Dr. JAS presents to you three different fluids and wishes to know their respective viscosity. She tasks T-E Baru to design an experimental apparatus for determining the viscosities of the three fluids.

Find: A simple way for T-E Baru to measure the viscosities of the fluids of interest.

Solution: The classical approach is to use a rotational rheometer. The fluid is placed between two concentric cylinders, as depicted in Fig. 3.4. The amount of torque required to spin the inner cylinder at a desirable speed can be used to accurately determine the viscosity of the fluid. For the inner rotating cylinder, the torque,

$$T_q = FR_i,$$

where F is the force and R_i is the radius of the inner cylinder. The tangential velocity of the surface of the inner cylinder,

$$V_t = \omega R_i,$$

where ω is the rotational speed in radians per second.

The relationship between shear, τ, and velocity gradient, dU/dy, for a Newtonian fluid, explained in the next section, can be expressed as

$$\tau = \mu dU/dy,$$

where μ is the fluid viscosity, which typically varies with temperature. Force is equal to shear times area, that is,

$$F = \tau A = \mu(2\pi R_i L)dU/dy.$$

We note that the involved area, $A = 2\pi R_i L$, where L is the length of the cylinder. The torque,

$$T_q = FR_i = \mu(2\pi R_i^2 L)(\omega R_i/d_{gap}) = \mu 2\pi R_i^3 L\omega/d_{gap},$$

where d_{gap} is the gap between the inner and outer cylinders. We can convert the rotational speed to revolutions per minute, S_{rpm}, by noting that $\omega = 2\pi\,S_{rpm}\,/\,(60\ s/minute) = \pi\,S_{rpm}/30$. With that, we have

$$T_q = \mu\pi^2 R_i^3 LS_{rpm}/(15d_{gap}).$$

As we are after the viscosity, we can rearrange the above equation into

$$\mu = 15d_{gap}T_q/(\pi^2 R_i^3 LS_{rpm}).$$

For a known gap, d_{gap}, a known radius of the inner cylinder, R_i, a known inner cylinder length, L, and a specified rotational speed in revolutions per minute, S_{rpm}, the measured torque, T_q, corresponds to the fluid viscosity, μ.

3.5 Newtonian fluids

Newtonian fluids are fluids whose viscosity does not change with respect to the amount of shear. Consider a long rectangular tank of width W with a movable vertical wall on one side as depicted in Fig. 3.5. Let us fill the tank with an elastic material, say natural rubber, which can be realized by filling the tank with fresh sap from rubber trees and letting it solidify. Then, we apply a finite but small force, F, to pull the moveable plate slightly to the right. The solidified rubber would deform according to the amount of force applied. Once displaced by the applicable amount, δx, the movable plate and the rubber stop. At equilibrium, the applied force,

$$F = \tau A, \tag{3.1}$$

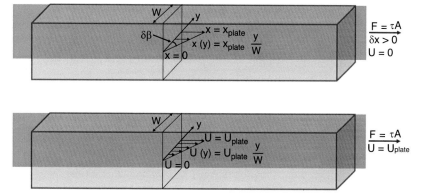

FIGURE 3.5   Shearing a rubber block versus a Newtonian fluid, of width W (created by Y. Yang and D. Ting). The rubber block, being a solid, deforms, while water, being a fluid, moves continuously.

where τ is the shearing stress and A is the surface area of the rubber in contact with the movable wall. The solidified rubber is an elastic solid and the angular displacement, or the shearing strain, is proportional to the shear, that is,

$$\delta\beta \propto \tau. \tag{3.2}$$

Once the applied force is stopped, the rubber block would return to its initial undeformed state.

Instead of rubber, let us fill the tank with water and apply a small force, F, to the moveable plate. For a Newtonian fluid, the plate will move continuously at velocity U_{plate}. Looking down, in plan view, the water moves with velocity U, which is a linear function of y, the distance of the moving plate from the parallel fixed wall. Specifically,

$$U = (y/W)U_{plate}, \tag{3.3}$$

where W is the narrow width of the tank.

The *no-slip condition* asserts that at any solid surface the relative velocity of the (Newtonian) fluid with respect to the surface is zero. That is to say that, due to finite viscosity, recall that all fluids are intrinsically sticky, the fluid particles[2] next to the solid surface stick to the surface. This is one cornerstone in fluid mechanics, at least as far as Newtonian fluids are concerned.

Invoking the no-slip condition at the moving and nonmoving sidewalls of the tank in Fig. 3.5, we have two boundary conditions for Eq. 3.3. Namely, at the nonmoving wall, the fluid velocity,

$$U|_{y=0} = 0. \tag{3.4}$$

---

2. Fluid particles are very small volumes or parcels of fluid that encompass a great number of molecules and at the same time are infinitely small.

At the moving wall, the fluid moves with the plate, that is,

$$U|_{y=W} = U_{plate}. \tag{3.5}$$

With these two boundary conditions, along with the fact that the velocity profile in the y direction is linear, we can solve the problem. In other words, the velocity of the fluid at any location can be determined.

By definition, a fluid deforms continuously under the shear-induced by the moving plate. In the absence of losses, the moving wall would continue to move at velocity U under the action of force F. Accordingly, the water will continue to deform. Therefore, let us look at an instant shortly after the force F is applied, as portrayed in Fig. 3.5. Let us mark a line AB across the width at time zero, when F is about to be applied. After a short time, δt, Line AB would become Line AB'. The corresponding small angle,

$$\delta\beta \approx tan\delta\beta = \delta x/W. \tag{3.6}$$

The rate of shearing strain,

$$\gamma' = \lim_{\delta t \to 0} \frac{\delta\beta}{\delta t} = \frac{U_{plate}}{W} = \frac{dU}{dy}. \tag{3.7}$$

The shearing stress, τ, increases with increasing applied force, F. That being the case, we have

$$\tau \propto \gamma' \propto dU/dy. \tag{3.8}$$

We can introduce a proportionality constant, μ, to convert Eq. (3.8) into a bona fide equation, that is,

$$\tau = \mu dU/dy. \tag{3.9}$$

The proportionality constant, μ, is known as the absolute viscosity, dynamic viscosity, or simply, viscosity. Dividing μ by the fluid density, ρ, gives the kinematic viscosity, v. For Newtonian fluids, the shearing stress, τ, is *linearly* related to the rate of shearing strain. In other words, a plot of τ versus dU/dy gives a constant slope; see Fig. 3.6. This indicates that the fluid is a Newtonian fluid and the slope corresponds to the viscosity of the fluid. Fig. 3.6 shows that the slope is typically rather sensitive to temperature, that is, viscosity decreases with increasing temperature. A distinctive feature of non-Newtonian fluids is a nonlinear τ-versus-dU/dy slope. Everyday examples of non-Newtonian fluids are honey, toothpaste, ketchup, yogurt, and paint. Non-Newtonian fluids are categorized based on how τ varies with dU/dy. Blood is a non-Newtonian fluid that behaves as a Newtonian fluid when the shear rate is relatively high, i.e., greater than 100 s^{-1} (Berger and Jou, 2000). We only deal with Newtonian fluids in this book. Common fluids such as air, water, glycerol, alcohol, mineral oil, and thin motor oil are Newtonian fluids.

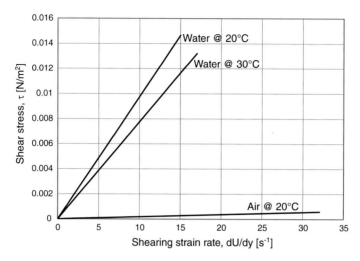

FIGURE 3.6 Shear stress versus shearing strain rate for typical Newtonian fluids (created by O. Imafidon). The slope corresponds to the viscosity. With increasing temperature, from 20°C to 30°C, the slope corresponding to water decreases. In other words, the viscosity of most Newtonian fluids decreases with increasing temperature.

3.6 A classification of fluid motions

To recap, the scope of this introductory text is on continuum Newtonian fluids such as air and water. Within this scope, we can categorize fluid motions according to:

1) fluid viscosity,
2) fluid compressibility,
3) flow space,
4) steady versus unsteady, and
5) laminar versus turbulent.

3.6.1 Fluid viscosity

While all fluids have finite viscosity, the viscosity of common fluids such as air and water in motion can often be neglected except within a very thin layer next to solid surfaces called the viscous or boundary layer. Inviscid flow problems in which the viscosity is zero are a lot easier to solve. For this reason, many fluid mechanics frontiers have derived elegant solutions to a handsome set of inviscid flow problems. These idealized inviscid flows are referred to as potential flow[3] problems.

---

3. Potential flow signifies an idealized, inviscid, incompressible, and irrotational flow.

3.6.2 Fluid compressibility

When the velocity of a moving gaseous fluid such as air exceeds about 30% of the speed of sound, the effect of fluid compressibility can no longer be neglected. At 20°C and standard atmospheric pressure, sound travels at roughly 343 m/s or one kilometer every 2.9 s. For most flows that we deal with in our everyday living, their speeds are significantly below the threshold where compressibility becomes appreciable. One obvious exception is in turbomachinery, where compressibility is prevailing and engineers must resort to gas dynamics, where fluid mechanics and thermodynamics are strongly coupled. In this book, we only deal with *incompressible flow*.

3.6.3 Flow space

Atmospheric air tends to move around freely. The fluid mechanics of a bird flying in the air is classified as *external flow* because, at a short distance from the flying bird, the air is unbounded. Flows around manmade structures such as stay-cables, buildings, and vehicles are also external flows. As we mentioned earlier, the effect of viscosity is largely limited to a thin layer around a solid surface. In external flow, this viscous layer around the body of the structure of interest plays a critical role in the drag it experiences and the wake formed downstream of the structure.

Blood flowing in a blood vessel is an *internal flow*, so are air flowing in an air duct and water flowing in a pipe. As internal flows are bounded on all sides, the effects of walls and viscosity propagate inward from all sides. The region where this happens is called the entrance region, beyond which the boundary layers merge and the flow becomes fully developed and viscosity is important everywhere.

Between unbounded external flow and bounded internal flow is another condition where the flowing fluid is partly bounded. Water flowing in natural and manmade channels is possibly the most prevalent example. This type of flow is called *open-channel flow*. The free surface can play a crucial role under some conditions.

3.6.4 Steady versus unsteady flow

The standard context for steady versus unsteady flow pertains to the flow behavior at a specific spatial location. Fig. 3.7 shows that the flow downstream of a square cylinder is steady at a Reynolds number, $Re = \rho UD/\mu$, of 45, where ρ is density, U is velocity, and D is a characteristic dimension or diameter of the cylinder. The two counter-rotating vortices in the near wake rotate steadily with no change in their features with respect to time. Increasing the Reynolds number to 50 results in unsteady flow, where the two counter-rotating vortices continuously change size and position with time. If we focus on a spot in the near wake, we will see the flow varies with respect to time, that is, the flow is unsteady.

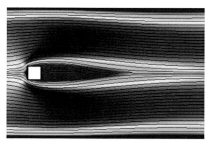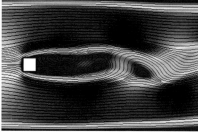

FIGURE 3.7 Laminar steady versus unsteady wake downstream of a square cylinder (created by X. Wang). The local flow is steady when its features do not change with respect to time, and it is unsteady if the flow characteristics vary with time. The two counter-rotating vortices remain stationary behind the cylinder at a Reynolds number of 45. These vortices shed alternatively at a slightly higher Reynolds number of 50.

3.6.5 Laminar versus turbulent flow

Both cases shown in Fig. 3.7 are laminar flows, where the flow appears smooth and is predictable. The flow behind the square cylinder at Re = 50 is unsteady but smooth and predictable. It is predictable in this case because the vortical flow is periodic and we know exactly what the general flow features are going to be with respect to time. If we continue to increase the Reynolds number, the wake will become transitional where the regular, large-scale, organized vortices are still largely discernable but the smaller-scale velocity fluctuations are random. The wake will eventually become predominantly turbulent with further increase in the Reynolds number. Turbulent flows are random with respect to both time and space. The average overall flow features are describable but not the specifics at a particular location or instance in time.

3.7 Fluid mechanics textbooks

Fluid mechanics is everywhere. That being the case, the subject is covered in many disciplines. These include different branches of engineering, meteorology, geology, and sciences, especially environmental and hydrological sciences. Being a well-established field, there are many comprehensive textbooks published over the years. Some of the author's favorite introductory fluid mechanics books are Cengel and Cimbala (2021), Fox et al. (2020), Gerhart et al. (2018), and White (2021). In addition to these, Alexandrou (2001), Bansal (2008), Douglas et al. (2005), Fang (2019), de Nevers (2021), Elger et al (2019), Finnemore and Franzini (2002), Graebel (2001), Granger (1995), Hibbeler (2018), Janna (2020), Massey and Ward-Smith (2012), Mory (2013), Morrison (2013), Nakayama (2018), Potter et al. (2015), Rajput (2019), Sawhney (2011), Schobeiri (2010), Song (2020), and Street et al. (1995) also cover the fundamentals with their unique perspective and emphasis.

Problems

3.1 Compressible versus incompressible flow.
Give a familiar example of compressible flow and contrast that with an incompressible flow.

3.2 Steady versus unsteady internal flow.
Provide an example of an everyday, steady, internal flow. How can you make that internal flow into an unsteady one?

3.3 Laminar, transition, and turbulent flow.
Water is to be delivered through a straight circular pipe at a specified volumetric flow rate. The Reynolds number corresponding to a 2.5-cm-diameter pipe is 1500 and the flow is laminar. What is the pipe size for the flow to be transitional at Re = 2000? What is the threshold diameter below which the flow is laminar, and beyond which the flow is turbulent, with Re larger than 2300?

3.4 Rising incense smoke.
Describe the flow of rising incense smoke in a largely quiescent room, as depicted in Fig. 3.8. What is driving the change from laminar to transition, to turbulent flow as the smoke rises?

3.5 Fluid mechanics frontiers.
Name and describe one important fluid mechanics contribution made by (1) Archimedes [287-212 BC], (2) Leonardo da Vinci [1452-1519], (3) Evangelista Torricelli [1608-1647], (4) Isaac Newton [1642-1726], (5) Daniel Bernoulli [1700-1782], (6) Claude-Louis Navier [1785-1836], (7) William Froude [1810-1879], (8) George Gabriel Stokes [1819-1903], (9) Osborne Reynolds [1842-1912], and (10) Ludwig Prandtl [1875-1953].

3.6 Conservation of mass.
A water supply pipe is 2.54 cm in diameter and the velocity is 0.03 m/s. What is the corresponding velocity after it is converged into a 1.27 cm diameter pipe?

3.7 Torricelli's Theorem.
Torricelli's Theorem states that the water leaving a sharp-edged hole of a water tank wall is at the same speed as an object free falling from a height equal to the water column above the hole. What is the speed of the water jet leaving a hole punctured through the wall at 7 m below the free surface of a water tank? Hint: the free-falling velocity, V, is related to gravity, g, and height, h, as $g = \frac{1}{2} V^2/h$.

3.8 Honey flows faster than water.
Dr. JAS is pondering how to convey the fact that under certain conditions honey can move faster than water. Based on Vuckovac et al. (2020), enlighten her on making droplets of honey flow faster than water.

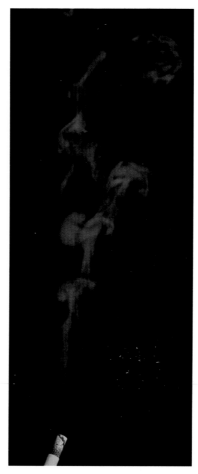

FIGURE 3.8 Rising incense smoke changing from laminar to transition, to turbulent flow as it rises (created by X. Wang).

References

Alexandrou, A.N., 2001. Principles of Fluid Mechanics. Prentice Hall, Hoboken, NJ.

Bansal, R.K., 2008. A Textbook of Fluid Mechanics. Laxmi Publications (P) Ltd., New Delhi.

Berger, S.A., Jou, L-D., 2000. Flows in stenotic vessels. Annu. Rev. Fluid Mech. 32, 347–382.

Cengel, Y., Cimbala, J., 2021. Fluid Mechanics: Fundamentals and Applications, Fourth Edition McGraw-Hill, New York, NY.

Cummins, C., Seale, M., Macente, A., Certini, D., Mastropaolo, E., Viola, I.M., Nakayama, N., 2018. A separated vortex ring underlies the flight of the dandelion. Nature 562, 414–418.

Dandelion, https://www.healthline.com/nutrition/dandelion-benefits, (accessed August 5, 2020), 2020.

Douglas, J.F., Gasiorek, J.M., Swaffield, J.A., Jack, L.B., 2005. Fluid Mechanics. Pearson /Prentice Hall, Hoboken, NJ.

Fang, C., 2019. An Introduction to Fluid Mechanics. Springer, Cham.

de Nevers, N., 2021. Fluid Mechanics for Chemical Engineers, Fourth Edition McGraw-Hill, New York, NY.

Elger, D.F., LeBret, B.A., Crowe, C.T., Roberson, J.A., 2019. Engineering Fluid Mechanics, Twelfth Edition Wiley, Hoboken, NJ.

Finnemore, E.J., Franzini, J.B., 2002. Fluid Mechanics with Engineering Applications. McGraw-Hill, New York, NY.

Fox, R.F., McDonald, A.T., Mitchell, J.W., 2020. Fox and McDonald's Introduction to Fluid Mechanics, Tenth Edition Wiley, Hoboken, NJ.

Gerhart, P.M., Gerhart, A.L., Hochstein, J.I., 2018. Munson, Young and Okiishi's Fundamentals of Fluid Mechanics, Eighth Edition Wiley, Hoboken, NJ.

Gilpin, W., Prakash, V.N., Prakash, M., 2017. Vortex arrays and ciliary tangles underlie the feeding–swimming trade-off in starfish larvae. Nat. Phys. 13, 380–386.

Graebel, W., 2001. Engineering Fluid Mechanics, International Student Edition Taylor & Francis, New York, NY.

Granger, R.A., 1995. Fluid Mechanics. Dover Publications, Inc., New York, NY.

Hibbeler, R.C., 2018. Fluid Mechanics, Second Edition Pearson, London.

Janna, W.S., 2020. Introduction to Fluid Mechanics, Sixth Edition CRC Press, Boca Raton, FL.

Learn, J.R., 2021. "Your perfect lawn is bad for the environment. Here's what to do instead," Discover, May 29, 2021. https://www.discovermagazine.com/environment/your-perfect-lawn-is-bad-for-the-environment-heres-what-to-do-instead. (accessed August 26, 2021).

Ledda, P.G., Siconolfi, L., Viola, F., Camarri, S., Gallaire, F., 2019. Flow dynamics of a dandelion pappus: a linear stability approach. Phys. Rev. Fluids 4 (7), 071901.

Massey, B.S., Ward-Smith, J., 2012. Mechanics of Fluids, Ninth Edition Spon Press /Taylor & Francis, London.

Merriam-Webster Dictionary, 2020. https://www.merriam-webster.com/dictionary, (accessed August 6, 2020).

Mory, M., 2013. Fluid Mechanics for Chemical Engineering. Wiley, Hoboken, NJ.

Morrison, F.A., 2013. An Introduction to Fluid Mechanics. Cambridge University Press, New York, NY.

Nakayama, Y., 2018. Introduction to Fluid Mechanics, Second Edition Butterworth-Heinemann, Oxford.

Potter, M.C., Wiggert, D.C., Ramadan, B.H., 2015. Mechanics of Fluids, SI Edition Cengage Learning, Boston, MA.

Qiu, F-S., He, T-B., Bao, W-Y., 2020. Effect of porosity on separated vortex rings of dandelion seeds. Phys. Fluids 32, 113104.

Rajput, R.K., 2019. A Textbook of Fluid Mechanics. S. Chand Publishing, New Delhi.

Sawhney, G.S., 2011. Fundamentals of Fluid Mechanics, Second Edition I.K. International Publishing House Pvt. Ltd., New Delhi.

Schobeiri, M.T., 2010. Fluid Mechanics for Engineers: A Graduate Textbook. Springer, Berlin.

Soluk, D.A., Craig, D.A., 1988. Vortex feeding from pits in the sand: a unique method of suspension feeding used by a stream invertebrate. Limnol. Oceanogr. 33 (4), 638–645.

Soluk, D.A., Craig, D.A., 1990. Digging with a vortex: flow manipulation facilities prey capture by a predatory stream mayfly. Limnol. Oceanogr. 35 (5), 1201–1206.

Song, H., 2020. Engineering Fluid Mechanics. Metallurgical Industry Press /Springer, Singapore.

Street, R.L., Watters, G.Z., Vennard, J.K., 1995. Elementary Fluid Mechanics, Seventh Edition Wiley, Hoboken, NJ.

Vogel, S., 1994. Life in Moving Fluids—The Physical Biology of Flow, 2nd ed. Princeton University Press, New Jersey.

Vuckovac, M., Backholm, M., Timonen, J.V.I., Ras, R.H.A., 2020. Viscosity-enhanced droplet motion in sealed superhydrophobic capillaries. Sci. Adv. 6 (42), eaba5197.

White, F.M., 2021. Fluid Mechanics, Ninth Edition McGraw-Hill, New York, NY.

Chapter 4

The transfer of thermal energy

"The light and heat of the universe comes from the sun, and its cold and darkness from the withdrawal of the sun."

Leonardo da Vinci

Chapter Objectives

- Understand what thermal energy or heat is.
- Fathom specific heat.
- Differentiate heat transfer from thermodynamics.
- Appreciate the three heat transfer mechanisms.
- Comprehend and distinguish conduction, convection, and radiation.

Nomenclature

A   area, the surface area involved in transferring heat

c   heat capacity; c_P is the specific heat capacity at constant pressure; c_v is the specific heat capacity at constant volume

h   heat transfer coefficient

h_{conv} convection heat transfer coefficient

k   thermal conductivity

m   mass

Nu   Nusselt number, the ratio of convection heat transfer over conduction heat transfer

Q   heat; Q' is the heat transfer rate, Q_{cond}' is the conduction heat transfer rate, Q_{conv}' is the convection heat transfer rate, Q_{emit}' is the radiation rate emitted by a surface above absolute zero, $Q_{emit,\ max}$' is the maximum radiation rate emitted by a blackbody

T   temperature; ΔT is the temperature difference, T_H is the higher temperature, T_L is the lower temperature, T_s is the surface temperature

t   time; t_{sand} is the time associated with sand, t_{water} is the time associated with water

x   distance or thickness

Greek and other symbols

Δ difference

σ the Stefan-Boltzmann constant

ε emissivity

4.1 What is thermal energy?

Energy is the capacity to cause changes. It is the quantitative property for doing work or for heating. Thermal energy is the portion of the energy of a system that

Thermodials. From Nature to Engineering. DOI: https://doi.org/10.1016/B978-0-323-90626-5.00005-7

FIGURE 4.1   Creating disorder by melting and evaporating ice cubes with heat in a frying pan (created by X. Wang). Ice is more rigid and thus orderly than water, and water is more orderly than steam. The progressively more energetic H_2O is driven by temperature, the higher the temperature, the more disorder are the H_2O molecules.

varies with temperature. Temperature is a measure of the average kinetic energy of the particles in a system. Heat is the form of energy that is transferred by virtue of a temperature difference, from the more energetic region to a less energetic one. That being the case, heat is thermal energy in transition. Snow monkeys, macaques, are known for savoring thermal energy in volcanic hot springs. For this case, thermal energy from the center of the earth is transmitted via molten magma. The hot molten magma subsequently transfers its thermal energy to heat the water, forming the hot springs for the snow monkeys to bask in.

Thermal energy is considered to be the energy associated with creating disorder in a system. This is because the higher the temperature, the higher the kinetic energy of the particles of a system. Consider the ice cubes in the frying pan, as shown in Fig. 4.1. Assume that the environment is at atmospheric pressure, that is, 101 kPa, and that the temperature of the ice cubes is initially at −4°C. The ice cubes stay at their respective spot in their cubical shape at this subzero temperature. One of the unique features of H_2O is that the molecules are more loosely packed in the solid phase[1], ice, than in the liquid phase, water; see Fig. 4.2. Be that as it may, for an ice cube, the H_2O molecules are very much organized, and they hold together rigidly as a cube. To put it another way, there is very little kinetic energy associated with these H_2O, or lumps of H_2O, particles. With thermal energy transferring from the room to the ice cubes, eventually the temperature of the ice cubes increases to that of the melting temperature of 0°C.

---

1. This intelligently ordained feature results in lower density ice to float and/or to stay on the top layer so that underwater plants and creatures are not frozen to dead but thrive through the cold winter. As a matter of fact, the warmer water under the layer of frozen ice or snow keeps a multitude of living things warm, shielding them from the bone-chilling wintery wind.

H2O Molecules

Disorder with increasing Temperature

FIGURE 4.2 H_2O molecules become progressively more disordered from ice to water to steam, with increasing temperature (created by M. Abbasi). A unique feature about H_2O is that the molecules are more loosely packed in the solid phase, ice, than in the liquid phase, water. At the same time, the molecules are more rigidly held together in ice than in water. The crystalline lattice in ice is dominated by a regular array of hydrogen bonds, spacing the molecules farther apart than they are in liquid water. Liquid water contains the intermolecular force, that is, the hydrogen bond occurs between the partially negative oxygen of a water molecule and the partially positive hydrogen on a neighboring water molecule.

At this temperature, the ice cubes melt and form liquid water, which spreads all over the frying pan. It is clear that the H_2O molecules are more energetic and disordered at room temperature, in liquid phase, compared to when they were frozen, in the solid phase as rigid cubes. The water in the pan will vaporize into steam after turning on the burner for a while. In its gaseous phase, the H_2O molecules from the original ice cubes spread everywhere in the house. This illustrates why thermal energy is regarded as the energy for creating disorder.

Associating thermal energy with disorder alone is like appreciating an elephant from the perspective of one of the blind men in the parable of the blind men and the elephant[2]. Thermal energy has many positive facets. Imagine what

---

2. The parable says that an elephant was brought to a group of blind men who had no prior knowledge of elephants. The person whose hands landed on the trunk described the elephant as a chunky snake. The one who groped an ear concluded that the elephant is like a fan. The chap who embraced a leg surmised that it is a pillar like a tree trunk. The individual who touched its side conjectured that it is like a wall. The man who felted a tusk asserted that the elephant is similar to a spear.

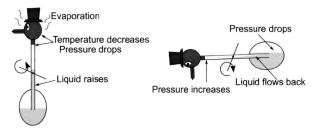

FIGURE 4.3 A drinking bird operates as a heat engine (created by A. Raj). The ambient air supplies the heat to operate this heat engine. A portion of this supplied heat is utilized to evaporate water from the head, causing the liquid to rise up via the inner stem to the head. The remaining heat is wasted, for example, by heating the head when the bottom end of the stem is opened up above the liquid surface, or by heating the other parts of the bird's body.

earth would be like without thermal energy continuously being lavished from the sun? In one word, frozen. Under the circumstances, there would also be no hot springs for the snow monkeys to bask in. Nor would there be any forms of clean and sustainable solar energy for us to tap into. We need thermal energy every day for keeping us warm, for cooking, toasting, making coffee, drying and ironing clothes, etc. High-temperature thermal energy is required for making ceramics, tiles, bricks, cement, and, more so, for manufacturing metallic products. Thermal energy is used for producing useful work via a heat engine. Take for example, a power plant is a heat engine for powering a city, an internal combustion engine is a heat engine for moving a bus, a steam engine is a heat engine to locomote a train, a turbojet engine is a heat engine for flying a plane, a marine engine is a heat engine for cruising a ship. The drinking or dipping bird as illustrated in Fig. 4.3 is an ingenious heat engine that is credited to Sullivan (1946). As water evaporates from the fabric making up the long nose and wrapping around the head, the temperature of the head decreases. The temperature drop causes some of the methylene chloride inside the glass head to condense, leading to roughly a 605 times reduction in volume. The resulting pressure drop in the head sucks up the liquid at the base along the inner stem. With the rising liquid creeping up to the head, the head becomes heavier, and the bird takes a bow. When the bird kowtows, the bottom end of the inner tube opens up above the liquid surface and thus some liquid slides down to the bottom bulb due to gravity. The increased weight of the bottom bulb restores the bird to its vertical position. The cycle repeats as the bottom bulb is heated by ambient air, causing liquid to rise up along the inner stem to the cooler head.

4.2 Specific heats

Specific heat is the amount of energy required to increase the temperature of a unit mass of a substance by one degree. It signifies the energy storage

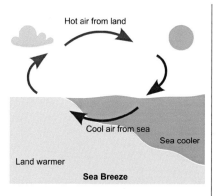
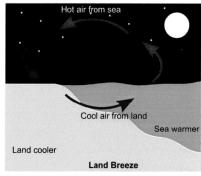

FIGURE 4.4   Modulating temperature via sea breeze and land breeze (created by A. Raj). The high heat capacity of water buffers and dampens temperature fluctuations in near-water neighborhood. During the day, the land, due to its lower heat capacity, is heated up faster than the waters, by the rising sun, as the hot air rises on land, the cooler air above the waters breezes and cools the land. After sunset, the lower heat capacity land cools off faster than the higher heat capacity water, and the wind circulation reverses into land breeze.

capacity. The larger the amount of thermal energy required to increase a unit substance by a unit temperature, the higher its capacity in storing thermal energy. Mathematically, the specific heat can be expressed as:

$$c = Q/(m\Delta T), \tag{4.1}$$

where Q is heat, m is mass, and ΔT is temperature change. Water, with a specific heat capacity of about 4.2 kJ/(kg·°C) around atmospheric conditions, has the highest heat capacity of all liquids. For this reason, oceans heat up and cool down a lot slower than the land, where the heat capacity of sand is approximately 20% that of water. Fish and other living organisms are protected because of water's resistance to temperature fluctuations. Sea breezes during the hot afternoon and land breezes during the cold night modulate the local temperature all thanks to the high heat capacity of water; see Fig. 4.4. When the seasonal temperature approaches the freezing point, farmers make use of water to protect their crops from frost by spraying water on them or by growing their crops in arrays of shallow water. Understandably, this technique only works when the minimum temperature is not too much below freezing and it does not stay below freezing for too long.

Example 4.1 Wall-of-water season extenders for gardening.

Given: A wall-of-water is a plastic wall that holds water for surrounding and protecting a plant from thermal stresses. It is particularly applicable for preventing frost damage during the transition seasons such as early spring and late autumn.

Find: How much longer does it take to lower the temperature of the wall-of-water from 5°C to -5°C, when it holds water versus sand.

Solution: Assume the same rate of thermal energy or heat transfer, then, the relative time it takes to drop 10°C is a function of the respective heat capacity. The amount of thermal energy transferred:

$$Q = mc_P \Delta T,$$

where m is the mass of the system of interest, that is, wall-of-water or wall-of-sand, c_P is the specific heat capacity at constant pressure, and ΔT is the temperature difference or drop.

For 1 kg of water, the amount of thermal energy transferred from the wall-of-water to the surroundings to bring about a 10°C temperature drop in the wall-of-water is:

$$Q = (1 \text{ kg})(4200 \text{ J}/(\text{kg} \cdot °C)(10° \text{ C}) = 42 \text{ kJ}.$$

The time it takes to cool 1 kg of water by 10°C is as follows:

$$t_{water} = Q/Q' = 42 \text{ kJ}/Q',$$

where Q' is the heat transfer rate. As a first approximation, we can assume Q' to be constant and the same for both water- and sand-filled wall-of-water.

For 1 kg of sand, the amount of thermal energy transferred from the wall-of-sand to the surroundings to bring about a 10°C temperature drop in the wall-of-sand is:

$$Q = (1 \text{ kg})(840 \text{ J}/(\text{kg} \cdot °C)(10°C) = 8.4 \text{ kJ}.$$

The time it takes to cool 1 kg of sand by 10°C is

$$t_{sand} = Q/Q' = 8.4 \text{ kJ}/Q'.$$

If we assume the same heat transfer rate, Q', then, it takes $t_{water}/t_{sand} = 42/8.4$ or 5 times longer to lower the 1 kg of water by 10°C. On that note, it is clear that water is the preferred choice.

We see 1 kg of water stores about five times the amount of thermal energy as 1 kg of sand. It follows that the specific heat of water is roughly five times that of sand. Fig. 4.5 illustrates that the specific heat at constant volume, c_V, is the amount of heat required to raise 1 kg of the matter of interest at a constant volume. For air at 20°C and atmospheric pressure, the specific heat at constant volume, c_V, is approximately 0.717 kJ/(kg·K). To put it concisely, it takes 718 J of thermal energy to raise 1 kg of air at constant volume by 1°C. If the thermal energy addition takes place under constant pressure, say one atmosphere, 1006 J of thermal energy is needed to raise 1 kg of air by 1°C. To put it another way, the specific heat at constant pressure, c_P, of air at atmospheric condition is equal to 1.006 kJ/(kg·K). We see that it takes 289 J more thermal energy to raise a

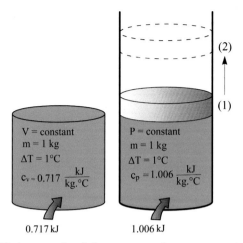

FIGURE 4.5 Specific heat capacity of air at constant volume versus constant pressure (created by F. Kermanshahi). For compressible fluid like air the specific heat capacity at constant pressure is larger than the specific heat capacity at constant volume by the expansion work of pushing the system boundary such as a movable piston.

kilogram of air under a constant pressure process than under a constant volume process. This 289 J is used to raise the piston, that is, expansion work, under the constant-pressure heat addition. For an incompressible or in-expandable substance such as water, there is no expansion work, and it follows that c_V is equal to c_P. Note that c_V and c_P change with temperature and less so with pressure.

4.3 Heat transfer versus thermodynamics

Thermodynamics deals with energy in equilibrium. Heat transfer describes thermal energy in transit. The rate of heat transfer can be expressed in the generic form:

$$Q' = hA\Delta T, \tag{4.2}$$

where h is the heat transfer coefficient, A is the heat transfer area, and ΔT is the temperature difference between the hot and cold region that leads to heat transfer. The natural or spontaneous flow of thermal energy is always from high to low temperature. Heat transfer is the study of the mechanism involved in the flow of thermal energy of interest and the rate at which it takes place. The key differences between heat transfer and thermodynamics are summarized in Table 4.1. To highlight, thermodynamics is concerned with the amount of heat a system possesses from one equilibrium state to another. Heat transfer complements thermodynamics by revealing the physical mechanisms and the

TABLE 4.1 Heat transfer versus thermodynamics.

Heat transfer	Thermodynamics
Nonequilibrium phenomenon. Thermal energy in transit	Equilibrium phenomenon. Energy, including thermal energy, in equilibrium
Rate of thermal energy transfer	Concerned with quantity but not rate
Mechanisms of thermal energy transfer	Does not deal with thermal transfer mechanisms

rate of the thermal energy transfer involved during the nonequilibrium, transient process.

Example 4.2 Heat transfer versus thermodynamics.

Given: Dr. JAS is an environmental steward who tries to reduce her energy usage in her everyday living. In the morning, before she boils water to make her tea using an electric kettle, she washes the dishes in the sink. This allows her to boil warmer water than otherwise. Her student, T-E Kaya, believes in speed and, thus, makes his coffee using the first cup of water from the tap and boils it in the microwave.

Find: The element associated with heat transfer and that corresponding more closely with thermodynamics.

Solution: Dr. JAS boils from warmer water because she is concerned with energy usage and not the rush in getting her tea made at the expense of using more energy. This is thermodynamics. It is noted that it takes a shorter time to boil from warmer water. On the flip side, the total time it takes from warming up the tap water by doing the dishes to the warm water boiling in the electric kettle is longer than that required for boiling the cold tap water.

T-E Kaya is interested in speeding up the water-heating process and, thus, utilizes what he sees as the faster heating approach. Incidentally, a microwave happens to be a very effective appliance for cooking. To save energy and time, Dr. JAS can boil her warm water using a microwave. There are empirical studies, see, for example, Treehugger (2021), that have compared electric kettles versus microwaves for boiling a cup of water. The results indicate that an electric kettle has an edge over a microwave because the microwave loses more energy, some of which is for heating the cup.

4.4   The three heat transfer mechanisms

The three fundamental modes of transferring thermal energy are conduction, convection, and radiation. These three heat transfer mechanisms are covered in all introductory heat transfer textbooks including Balaji et al. (2021), Bayaztoglu et al. (2012), Bergman et al. (2018), Cengel and Ghajar (2020), Holman (2010), Kreith et al. (2010), Lienhard IV and Lienhard V (2019), Nellis and Klein (2009), Rathakrishnan (2012), Sawhney (2008), Sukhatme (2005), and Sunden (2012). The author would recommend those who are new to the field to start with Bergman et al. (2018) and/or Cengel and Ghajar (2020). Let us go through the three fundamental heat transfer mechanisms one at a time.

4.4.1   Conduction

Conduction heat transfer is the transfer of thermal energy between two objects by direct contact. For instance, a metal cooking pot sitting in a air-conditioned house feels cold when you touch it. This is because your warmer hand loses heat to the colder pot via conduction. If the pot is being used for cooking on the stove, it would feel hot when you touch it. Thankfully, our reflexes save our hand from being cooked by conduction heat transfer from the hot pot. Pancakes are cooked almost exclusively via conduction heat transfer. For this reason, we have to remember to flip a pancake timely. Cooking a sunny-side-up egg, on the other hand, takes advantage of conduction on the bottom side. The lighter side of this is a cheerful lesson, that is, to remain positive even when your bottom is being seared.

Fundamentally speaking, conduction is achieved at the molecular level via the transfer of kinetic energy. Conduction can take place in solids, liquids, and gases. The thermal energy is transferred from more energetic particles to adjacent, less energetic, ones. Typically, conduction is most effective via solids, as the molecules are most closely packed. This can be visualized by lining up a row of students standing shoulder to shoulder and having Dr. JAS shake the first one side to side in line with the row. With the first student swaying side to side, the next student starts to sway, and this motion propagates down the line. On the grounds of this, the kinetic energy and, thus, also thermal energy is transferred.

For engineering applications, it is convenient to have a user-friendly mathematical expression for determining the conduction heat transfer rate. We note that the temperature gradient is driving the heat transfer, as shown in Fig. 4.6. A surface at the same temperature as our hand feels neither cold nor hot. This is because there is no temperature difference and, thus, no heat transfer. Many animals curl their bodies into a ball when it is cold. They do so to reduce their exposed body surface. To put it another way, the heat transfer rate is proportional to the surface area available for transferring heat. Putting these parameters into an expression, we have the rate of heat conduction (Fig. 4.6):

$$Q'_{cond} \propto A\,dT/dx, \tag{4.3}$$

where A is the area available for heat transfer and dT/dx is the temperature gradient in the x-direction, the direction of heat conduction. This is illustrated in

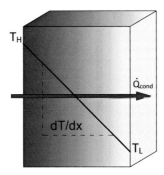

FIGURE 4.6   The rate of heat conduction is proportional to the surface area available for heat transfer and the (negative) temperature gradient (created by F. Kermanshahi). Note that heat transfer spontaneously from high to low temperature, that is, in the negative-temperature-gradient direction.

Fig. 4.6. We can replace the proportional sign with an equal sign by introducing an empirical constant, k, which is called thermal conductivity. Namely,

$$Q'_{cond} = -kAdT/dx. \tag{4.4}$$

The negative sign signifies that heat is transferred from high to low temperature, that is, in the negative-temperature-gradient direction. We can rewrite this expression as:

$$Q'_{cond} = -kA(T_H-T_L)/\Delta x, \tag{4.5}$$

where T_H is the higher temperature, T_L is the lower temperature, and Δx is the involved material thickness. Metals typically have substantially larger thermal conductivity than non-metals. For instance, the thermal conductivity values of aluminum and steel are 239 W/(m·K) and 50 W/(m·K), respectively, whereas for cotton fabric, it is in the neighborhood of 0.04 W/(m·K). This is the reason why metallic pieces of furniture feel a lot colder than fabric ones.

4.4.2   Convection

Convection heat transfer is the transfer of thermal energy via fluid motion. To put it another way, heat convection is realized via mobile fluid particles. The most obvious everyday example is boiling water. The hot fluid from the heated base of an electric kettle or cooking pot convects the thermal energy to the cooler fluid. The convection of hot water, predominantly upward because hot fluid tends to expand, becoming lighter, and rise under gravity, causes cooler fluid to sink and fill in the void it creates and, hence, come into contact with the heated base. The more rigorous the fluid motions are the faster the heat transfer rate. The transfer of heat from the solid base to the very thin layer of fluid that 'sticks' to the solid surface because of viscosity, however, is via conduction. In the scenario where there is no fluid motion, such as a very thin layer of fluid sandwiched between a hot and a cold surface, the heat is transferred via conduction.

In short, convection consists of molecular motion plus macroscopic motion. The molecular motion is the same as what has been described for conduction. The macroscopic motion represents the convecting fluid motion. Random fluid motion characterized by flow turbulence, is known to promote convection heat transfer. In engineering, the standard expression of the convection heat transfer rate is:

$$Q'_{conv} = h_{conv}A(T_H - T_L), \tag{4.6}$$

where h_{conv} is the convective heat transfer coefficient. As heat convection is applicable only when fluid is involved, the high temperature, T_H, is associated with the participating fluid, and the low temperature, T_L, is that of the concerned surface exposed to the fluid, or vice versa. The convection heat transfer coefficient can be conveniently expressed in its normalized form in terms of the Nusselt number. On account of the fact that, in the absence of flow motion, the heat transfer through a fluid is via conduction, it is, therefore, logical to normalize the convection heat transfer coefficient with the conduction heat transfer coefficient. In other words, the Nusselt number can be defined as

$$Nu = h_{conv}/(k/\Delta x). \tag{4.7}$$

4.4.3 Radiation

Radiation is the transfer of energy via photons or electromagnetic waves. Thermal radiation or radiation heat transfer is electromagnetic radiation with a wavelength between 0.1 and 100 μm. All surfaces or matter at a temperature above absolute zero, 0 K, emit thermal radiation. Radiation heat transfers best in a vacuum, in the absence of a interfering medium. A sunburn is an explicit, negative fallout of radiation heat transfer from the sun. Notwithstanding the fact that too much radiation is harmful, most living things survive and thrive on some optimum level of solar radiation. The intelligently designed butterflies know how to accurately maintain the temperature of their wings for optimum performance (Tsai et al., 2020). Advani et al. (2019) found that Glanville Fritillary, a heliophilic (one that is attracted or adapted to sunlight) butterfly, orient itself perpendicular to the sun's rays to gain heat for taking off. To compensate for convection heat loss during flight, especially in cooler ambient air, Finnish butterflies warm up their wings and take off with a higher wing temperature than Alpine butterflies that can benefit from more intense radiant heat after takeoff.

For a perfectly emitting body, a blackbody, the Stefan-Boltzmann law states that its radiation heat transfer rate:

$$Q'_{emit,max} = \sigma A T_s^4 \tag{4.8}$$

where the Stefan-Boltzmann constant, $\sigma = 5.67 \times 10^{-8}$ W/(m^2·K^4) and the surface temperature, T_s, is in absolute units that is, in Kelvin or Rankine. All

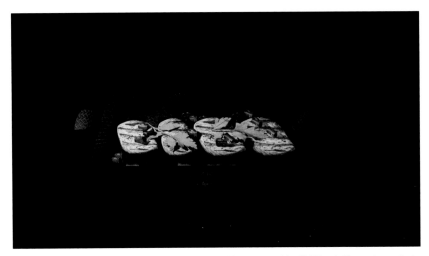

FIGURE 4.7   The heat transfer of barbequing fish fillets (created by X. Wang). To put it concisely, heat is conducted from the hot grill to the fish as witnessed by the beautiful grill lines. Heat is convected from the fire to the fish via mostly invisible energetic hot plumes of different sizes. At the same time, heat is radiated from the sooty fire and the hot inner surfaces to the respective fish surface that is in view.

real surfaces emit less than the ideal blackbody and, thus, we have:

$$Q'_{\text{emit}} = \varepsilon \sigma A T_s^4, \tag{4.9}$$

where the emissivity, ε, takes a value between zero and one. Most nonshiny or nonreflective surfaces have an emissivity of around 0.9. If the surface of interest only sees another surface, then we can write:

$$Q'_{\text{rad}} = \varepsilon \sigma A \left(T_H^4 - T_L^4 \right). \tag{4.10}$$

A serene example is a horizontal roof open to the cool night sky, where the higher-temperature roof radiates heat to the lower-temperature sky. Two parallel surfaces facing each other, such as the face brick and common brick enclosing an air gap in a building envelope, is another familiar example. Note that in the presence of air as the medium in an air gap, there is also convection, in addition to radiation.

Example 4.3 Barbeque fish fillet.

Given: It is customary for Dr. JAS to have her group of researchers over near the end of summer to enjoy a nice afternoon in her big backyard. Other than a variety of healthy vegetables, Dr. JAS always beefs up the dinner with appetizing barbeque fish fillets.

Find: The heat transfer mechanisms underlying the barbequing of the fish fillets, as depicted in Fig. 4.7.

Solution: The conduction mechanism is clearly displayed by the beautiful grill lines on the fish fillets. The hot grill conducts heat directly to the fish via direct contact.

One can realize the convection mechanism while standing close to the barbeque as Dr. JAS opens the cover. The highly buoyant plume of hot air rises up into the atmosphere. Closing the barbeque helps to contain the convective fluid motions dancing within the barbeque, augmenting the heat and cooking the fish fillets on all exposed surfaces.

Radiation from the fire is only effective when the burning is not clean, that is, when the fire is reddish because of the combustion of carbon particles. A clean propane-air flame is blue in color, and this happens when all the propane is completed burned and reacted with the oxygen in the air to form carbon dioxide and water vapor. A transparent and bluish propane-air flame signifies clean combustion, and the flame has a low value of emissivity, in the ballpark of 0.1 to 0.2. On the other hand, an unclean, reddish flame has an emissivity of around 0.9.

Have you ever wondered why the interior surfaces of a good barbeque are painted black? Camouflaging soots and dirts to presenting a cleaner barbeque is likely one reason. In terms of utilization and/or engineering, the black surface furnishes an emissivity value that approaches unity. Because of that, the hot surfaces, especially when the barbeque lid is closed, radiates thermal energy to the food that you are trying to cook.

Problems

4.1 Heat capacity at constant volume versus that at constant pressure.
An amount of 500 kJ of thermal energy is added to each of two sealed, insulated, 1-L containers filled with air at 20°C and 1 atm initially. The heat addition to Container A takes place at *constant volume* while that to Container B occurs at *constant pressure*. The final (equilibrium) temperature of
A) both containers will be lower than 20°C
B) both containers will be equal to 20°C
C) Container A will be higher than that of Container B
D) Container B will be higher than that of Container A

4.2 A perfect insulator.
A perfect insulator is a material across which the temperature
A) decreases   *B)* stays constant   *C)* increases

4.3 Thermals.
Many birds take advantage of thermals by gliding around them in circles without flapping their wings. The rising thermals, that often end up displaying as cumulus clouds when the water vapor condenses into tiny water droplets, depict which heat transfer mechanism in action?
A) conduction   *B)* convection   *C)* radiation

4.4 Ironing.

The first-generation of irons operate not by electricity but by burning charcoals inside a chamber. How does heat transfer from a hot first-generation iron to a shirt you are ironing? It is transferred via

A) conduction   *B*) convection   *C*) radiation

4.5 Cold tiles, warm carpet.

Why do your feet feel cold when walking on ceramic tiles and warm on carpet?

4.6 Double-pane window air gap.

The air gap of a typical double-pane window is based on the idea of minimizing heat transfer. As such, the engineer would attempt to keep the Nusselt number at its minimum. The smallest Nusselt number that we can have is

A) -70   *B*) -10   *B*) -7   *C*) -1   *D*) -0.7   *E*) 0
F) 0.7   *G*) 1   *H*) 7   *I*) 10   *J*) 70   *K*) 2300

This smallest possible Nusselt number implies that

A) conduction is negligible   *B*) radiation is insignificant
C) conduction = convection   *D*) conduction = radiation
E) there is no phase change   *F*) there is phase change
G) it is a perfect insulator   *H*) it is a perfect conductor

4.7 Keeping warm at an ice rink.

What type of heater is used at an ice rink to keep the spectators warm? Which heat transfer mechanism is exploited by this type of heater?

4.8 Polar bear heat transfer.

To thrive, a polar bear has to prevent excessive heat loss and maximize heat gain. Explain how polar bears manage heat, in terms of the three heat transfer mechanisms; see Shao et al. (2020).

4.9 Penguin thermal insulation.

Among many other tough creatures, penguins and polar bears have no problem braving frigid ice-cold water. Explore the intelligently designed insulation worn by penguins and polar bears; see Shao et al. (2020), August et al. (2019), and Metwally et al. (2019).

References

Advani, N.K., Parmesan, C., Singer, M.C., 2019. Takeoff temperatures in Melitea cinxia butterflies from latitudinal and elevational range limits: a potential adaptation to solar irradiance. Ecol. Entomol. 44 (3), 389–396.

August, A., Kneer, A., Reiter, A., Wirtz, M., Sarsour, J., Stegmaier, T., Barbe, S., Gresser, G.T., Nestler, B., 2019. A bionic approach for heat generation and latent heat storage inspired by the polar bear. Energy 168, 1017–1030.

Balaji, C., Srinivasan, B., Gedupudi, S., 2021. Heat Transfer Engineering: Fundamentals and Techniques. Elsevier, London.

Bayaztoglu, Y., Bayazitoglu, Y., Ozisik, M.N., 2012. A Textbook for Heat Transfer Fundamentals. Begell House, Danbury.

Bergman, T.L., Lavine, A.S., Incropera, F.P., DeWitt, D.P., Fundamentals of Heat and Mass Transfer, Eighth Edition, Wiley, Hoboken, NJ, 2018.

Cengel, Y., Ghajar, A., 2020. Heat and Mass Transfer: Fundamentals and Applications, Sixth Edition McGraw-Hill, New York, NY.

Holman, J.P., Transfer, H., 2010. Tenth Illustrated Reprint Edition. McGraw-Hill, New York, NY.

Kreith, F., Manglik, R.M., Bohn, M., 2010. Principles of Heat Transfer, SI Edition Cengage Learning, Boston, MA.

Lienhard IV, J.H., Lienhard V, J.H., 2019. A Heat Transfer Textbook, Fifth Edition Dover, Mineola.

Metwally, S., Comesaña, S.M., Zarzyka, M., Szewczyk, P.K., Karbowniczek, J.E., Stachewicz, U., 2019. Thermal insulation design bioinspired by microstructure study of penguin feather and polar bear hair. Acta Biomater. 91, 270–283.

Nellis, G., Klein, S., Heat Transfer, Cambridge University Press, Cambridge, 2009.

Rathakrishnan, E., 2012. Elements of Heat Transfer. CRC Press, Boca Raton, FL.

Rathore, M.M., Kapuno Jr., R.R.A., 2011. Engineering Heat Transfer, Second Edition Jones & Bartlett Learning, Burlington.

Sawhney, G.S., 2008. Heat and Mass Transfer. I.K. International Publishing House Pvt. Limited, New Delhi.

Shao, Z., Wang, Y., Bai, H., 2020. A superhydrophobic textile inspired by polar bear hair for both in air and underwater thermal insulation. Chem. Eng. J. 397, 125441.

Sukhatme, S.P., 2005. A Textbook on Heat Transfer, Fourth Edition Universities Press, Hyderabad.

Sullivan, M.V., Novelty Device, United States Patent Office, 2402463, Filed August 6, 1945, Serial No. 609114, 1946.

Sunden, B., 2012. Introduction to Heat Transfer. WIT Press, Ashurst.

Treehugger, https://www.treehugger.com/ask-pablo-electric-kettle-stove-or-microwave-oven-4858652, (accessed September 19, 2021) 2021.

Tsai, C-C., Childers, R.A., Shi, N.N., Ren, C., Pelaez, J.N., Bernard, G.D., Pierce, N.E., Yu, N., 2020. Physical and behavioral adaptations to prevent overheating of the living wings of butterflies. Nat. Commun. 11, 551.

Part 2

An Ecological View on Engineering Thermodynamics

Chapter 5

The four laws of ecology

"Earthworms are the intestines of the soil."

Aristotle

Chapter Objectives

- Understand what ecology is.
- Appreciate the importance of ecology in engineering.
- Learn how creatures survive via thermoregulation.
- Fathom and differentiate thermoregulation of endotherms and ectotherms.
- Recognize the four laws of ecology, and the parallels with the first and second laws of thermodynamics.
- Comprehend intelligent designs and mimic them in engineering applications.

Nomenclature

c_P	specific heat capacity at constant pressure
m	mass
Q	heat
T	temperature; ΔT is temperature difference

5.1 What is ecology?

Ecology is a subset of biology that is concerned with the interplay among organisms, including human beings, and their physical surroundings. Only the latter part, that is, the interaction between an organism and its environment is of interest here. We aim to learn about the thermofluid aspect of how an organism copes with its environment and thrives. It is primarily the thermoregulation that is of interest here. Let us follow Unger et al. (2020) and define some key terms in Table 5.1.

Thermoregulation is a mechanism by which creatures regulate body temperature within a desirable range; see, for example Ganslosser and Jann (2019). When the environment is warmer than desirable, creatures including humans, apes, monkeys, horses, and hippos sweat. Dogs do not sweat much, instead, they reject heat effectively via evaporative cooling, that is, panting. To stay warm when the environment is colder than thermally comfortable, mammals can increase

Thermofluids: From Nature to Engineering. DOI: https://doi.org/10.1016/B978-0-323-90626-5.00006-9

TABLE 5.1 Thermoregulation, endotherm, and ectotherm.

Term	Meaning	Illustrations
Thermoregulation	A mechanism by which creatures regulate body temperature within a desirable range	Humans shiver when it is cold, and we sweat when it is hot
Endotherm	Warm-blooded animals that maintain the body at a metabolically favorable temperature	Mammals, including whales, elephants, humans and chipmunks, and birds
Ectotherm	Creatures that rely on external heat sources to stay alive; their body temperatures vary with the surroundings	Amphibians, reptiles

their body temperature by developing goosebumps, creating added insulation to reduce heat loss.

Endotherms are warm-blooded creatures that maintain their body at a metabolically favorable temperature, regardless of the environment. This is achieved primarily by the use of heat released by their internal bodily functions.

Ectotherms are cold-blooded creatures that rely heavily on external heat sources to regulate their body temperature. The heat produced by internal physiological sources is relatively small. That being the case, their body temperature varies with the environment.

From the basic understanding of how animals interact with their natural environment, we can draw a parallel in engineering to facilitate our acquisition of thermofluids knowledge. Upon being fully fledged with natural ecological thermofluids, we can better engineer a nature-friendly tomorrow.

An everyday example of an ectotherm is a frog, as shown in Fig. 5.1. Frogs belong to the group of small vertebrates that survive in a moist environment. This group of animals is called amphibians, animals that can live both on land and in water, and they are cold blooded. Cold blooded implies that frogs adjust their body temperature to that of the environment that they are in. To put it another way, they do not warm themselves, like warm-blooded animals, by generating heat via metabolism. Since we are talking about frogs, let us take a few minutes to revel in the "Boiling Frog" fable, as illustrated in Fig. 5.2. The fable is credited to Edward Wheeler Scripture (1897), where he cited and utilized frog kinetics studies by Sedgwick (1883) and others to illustrate a concept in psychology. Namely, Scripture said that "a live frog can actually be boiled without a movement, if the water is heated slowly enough; in one experiment the temperature was raised at a rate of 0.002°C per second and the frog was found dead at the end of the 2.5 hours without having moved." The actual experiment or experimental claim that a frog stayed still as it was boiled to

FIGURE 5.1   A cold-blooded frog adjusting itself to its surroundings (created by S. Akhand). The left figure shows that the body temperature of the frog is low because it is in a cool autumn environment. The right figure exhibits that the body temperature of the frog is high because its surroundings is hot, during a hot summer day.

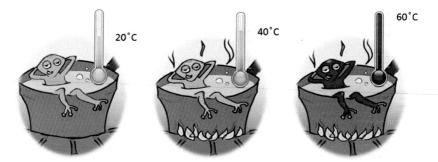

FIGURE 5.2   An illustration of the "Boiling Frog" fable (created by S. F. Zinati). If we put a live frog into a pot of hot water, it will jump out immediately. On the other hand, if we place it in a pot of room-temperature water, and heat it up slowly, the frog will adjust its body temperature to that of the slowly increasing water temperature. According to the "Boiling Frog" narrative, if the rise in water temperature is sufficiently slow, the frog will eventually be boiled to death without realizing the water is too hot for it to stay.

death is debatable. In spite of that, the moral behind the Boiling Frog typification remains. When there are changes in the environment that we are in, we should first take appropriate actions, including adjusting ourselves to the changes and adapting to the new environment. However, when the changes are heading in the destructive direction, we must jump out!

Coming back to cold-blooded frogs. The way to thrive in a changing environment is to adjust the body as close to the ambient temperature as possible. In terms of thermofluids in its simplest form, this relies on heat transfer and the first and second laws of thermodynamics. On the basis of the second law of thermodynamics, heat transfer occurs naturally from high to low temperature. The first law of thermodynamics states that the amount of heat lost by the higher temperature body is equal to that gained by the lower temperature one. Cold-blooded animals such as frogs let the natural flow of heat from high to low temperature happen until their body temperature is at equilibrium with the surroundings. They rely heavily on the warmer surroundings to transfer heat to warm and energize them. When the surroundings cools off, they adjust their body and behavior and drop their body temperature and, thus, the temperature gradient, to minimize heat loss. Some frogs have biological antifreeze and can survive frigid winters without breathing and their heart beating (Emmer, 1997). Warmer-than-comfortable-temperatures, on the other hand, can pose a challenge for cold-blooded creatures, especially terrestrial ectotherms. They thermoregulate the best they can according to the availability of shade and by altering their seasonal timing of activity and reproduction (Kearney et a., 2009).

5.2 The four laws of ecology

While many have contributed to the establishment of these and other informal laws, Barry Commoner (1971) is recognized as the one who stands out from his predecessors. The four classical laws of ecology are as follows:

1. Everything is connected to everything else.
2. Everything must go somewhere.
3. Nature knows best.
4. Nothing comes from nothing, that is, there is no such thing as a free lunch.

Leonardo da Vinci rightly put it, "Learn how to see. Realize that everything connects to everything else." The first of the four informal laws of ecology, everything is connected to everything else, indicates how interconnected we are to other creatures and living things, and the physical environment that all of us are in. To give an example, water flows in a closed cycle within the ecosystem we call earth. "All streams run to the sea, but the sea is not full; to the place where the streams flow, there they flow again." (Ecclesiastes 1: 7). Water, like energy, is conserved; though, for energy conservation, we need to include the entire universe. Polluting the water will harm everything along the cycle and, ultimately, ourselves. To put it positively, purifying water will lead to a chain of positive impacts. Tending the garden and caring for others and the environment is our responsibility, duty, and joy, and it is an essential part of caring for ourselves. Interested readers can refer to Swithinbank et al. (2019) to be comprehensively enlightened on this topic.

The second law of ecology states that everything must go somewhere. Following the water example, human discharge that goes into the sewage system becomes part of our drinking water, along with a little solid residual that can be made into sewage sludge fertilizer. This shows the need for water reclamation,

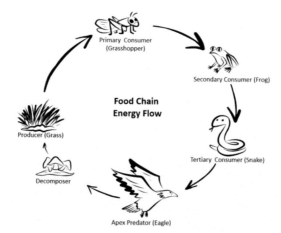

FIGURE 5.3   The simplified food chain (created by S. Akhand). A fallout of the second law of thermodynamics is that even the mighty eagle, the king of the sky, dies and decomposes. It takes the always-giving sun to transform the compost back into all kinds of hearty plants that provide high quality food and/or energy. Note that food web is a more accurate description that illustrates the complexity involved, as most organisms have a more diverse appetite than one food.

and we have much to learn from nature in furthering the engineering of water conservation, usage, and cleaning. An effective measure is to not pollute beyond what is absolutely necessary, such as nontoxic washroom visits. Mr. (Fred) Rogers reminds us of this when he said, "If you were a fish, you wouldn't want somebody dumping garbage into your home." We are reminded that polluting the fish, or the environment in general, is polluting our dinner plate.

In terms of energy, the second law of ecology is akin to the first law of thermodynamics, which states that energy cannot be created nor destroyed; it can only transform from one form to another. The second law of thermodynamics, however, dictates the direction of this energy transformation. Energy flows from higher quality to lower quality. In spite of the fact that the total energy in the universe is conserved, the energy, as a whole, is deteriorating in quality. In ecology, however, the system of interest is Earth. The sun is not considered a part of the system and, thus, makes it possible for the energy within the earth to return to the higher-quality state. Viewing the earth as an isolated system is an incomplete view that violates the second law of thermodynamics. Along the same line, considering solar energy as a renewable energy is also incorrect, strictly speaking. This is because the sun is continuously losing energy. Thankfully, because of its massive energy content, the rate of decrease is negligible when cast in a human timescale. The return to higher quality energy is illustrated in terms of the food chain, or food web, in Fig. 5.3. For instance, the death of the king of the sky, an eagle, is not the end but part of the cycle of life on earth. The corpse of the eagle is decomposed by nature's decomposers including fungi and detritivores, organisms that feed on dead plants and animals. The largest group of detritivore is probably earthworms. This is likely why Aristotle called earthworms the intestines of the soil. Gilbert White warned us to appreciate our slimly, wiggling little friends, "Earthworms, though in appearance a small and despicable link in the chain of nature, yet, if lost, would make a lamentable chasm."

The third law of ecology, nature knows best, can be observed everywhere in nature. To paraphrase this in terms of Proverbs 16: 4, everything has a purpose. For example, frogs play an important role in the food chain. Their babies, tadpoles, feed on algae and, in so doing, keep the water clean (Earth Rangers, 2013). Adult frogs feed on insects, keeping bug populations in check. In return, they serve as a tasty meal for fish, snakes, birds, and many others. The lessons we can learn from the third law of ecology include minimizing our disturbance of natural settings, engineer to blend into and coexist with our natural surroundings and habitats and learn from the multitude of intelligent designs in nature. The fact that breastfeeding is superior testifies that nature knows best. By way of illustration, Wood et al. (2021) show that the proportion of regulatory T cells for breastfed babies of three weeks or younger is about two times higher than their counterparts who are fed with formula.

The fourth informal law of ecology, nothing comes from nothing, confesses that the exploitation of nature always carries an ecological cost. In plain English, there is no free lunch. Parallel to the second law of thermodynamics, usage of resources inevitably converts them into a less-useful form. In using energy, we transform it into forms that are less readily available for work. As an illustration, combusting fuels to power engineering systems results in significant waste heat. As a matter of fact, all the energy we utilize in sustaining our standard of living ends up as heat. While it is impossible for us or other creatures to live without utilizing energy and/or food, some energy and food sources are considerably less taxing on the environment than others. There is much more room for us to harness more ecologically friendly energy and food. With the sun as literally the sole supplier, resources that are easily replenished by the sun are the preferred choices. An equally essential measure is conservation and waste reduction, without which renewable energy and eco-friendly food can only take us so far. Mahatma Gandhi nailed this, "There's enough on this planet for everyone's needs but not for everyone's greed."

Example 5.1 Engineering campus ecological makeover.
Given: Dr. JAS received the go-ahead from the administration of her university to make the campus eco-friendly. She decided to start with the kitchen that feeds a thousand students.
Find: Specific measures you would take as an ecological engineer-in-training working under the supervision of Dr. JAS.

Solution: An efficacious composter for breaking down organic waste from the kitchen is a good starting point. This will significantly reduce the amount of garbage going into a landfill. The compost can be bagged as organic, environmentally friendly fertilizer.

Related to the composter is a garden of selected common vegetables such as tomatoes and green, red, and orange peppers that are relatively easy to manage and are popular on the menu. The in- or next-to-kitchen garden will ensure the freshness of these nutritional vegetables. Dirt-free hydroponic and aeroponic gardening can also be considered, especially for growing salubrious herbs and spices. As fish is likely a popular cuisine, aquaponic gardening, where fish and vegetables are grown in complement, is also an option. If the Biology Department is active in research involving aquatic species other than fish, those species may also be bred. Biodiversity generally enhances the ecosystem, promoting the healthy growth of all organisms and plants involved.

Harvesting rainwater for irrigation is probably worthwhile. If treated, the water can also be used for dishwashing. Solar and/or kitchen waste heat can be harnessed to heat and sanitize the water.

Not explicitly ecological engineering measures that may be taken include harnessing the cooking oil and grease as biofuels. There is also a handsome amount of waste heat from the stove that can be recovered for space and hot water heating.

The ecological-footprint method is picking up pace in making university campuses more sustainable. When you get right down to it, if we want to sustain tomorrow, our educational institutions must lead by example. Genta et al. (2021) invoked this method when executing a comprehensive study of the Politecnico di Torino. Among the six main categories of consumption studied, transportation, followed by energy, topped the list, with 49.4% and 40.1%, respectively, of the total campus environmental impact. Food made up 5.7% and waste 3.7%. Isildar and Morsali (2020) compared the environmental footprint of numerous university campuses among developed and developing countries. It is no surprise to see some serious discrepancies from institution to institution. Interested readers can also refer to Genta et al. (2019) and Ortegon and Acosta (2019). As well, the ecological footprint and other sustainability data of nations around the world are tallied and presented by Data Footprint (2021).

5.3 Animal thermoregulation

There is much for us to learn from animals concerning thermoregulation. Thermoregulation, or temperature regulation, does not just apply to warm-blooded creatures to regulate our body temperature within the ideal range primarily via metabolism. Poikilotherms, creatures whose body temperature varies with ambient temperature, such as insects can also vary part of their body temperature away from that of their surroundings. Table 5.2 draws attention to temperature regulation for endotherms (warm-blooded creatures) and ectotherms (cold-blooded creatures). Warm-blooded creatures regulate their body temperature primarily

TABLE 5.2 Thermoregulation of endotherms and ectotherms.

Creature type	Examples	Thermoregulation mechanisms
Endotherms (warm-blooded)	Elephants, penguins	*Physiological*—metabolism is their primary control mechanism *Anatomical*—large ears for animals such as black-tailed jackrabbits and elephants to keep them cool; big and round with extra fat for emperor penguins to keep them warm *Behavioral*—elephants flap their ears and spray mud and water on their bodies to stay cool; emperor penguins huddle in a group to stay warm
Ectotherms (cold-blooded)	Lizards, butterflies	*Behavioral*—lizards hide in warmer soil when it is cold, stay in the shade when it is hot, and bask cozily in warm sun *Anatomical + Behavioral*—butterflies control the radiation by the extent their wings are spread out or the angle between their wings and the sun; in addition, the details on the wing scale structures adjust the amount and range of light spectrum and heat to be absorbed

by controlling their metabolic rate. Anatomically, elephants have large ears to compensate for their relatively small surface area with respect to volume ratio in rejecting metabolic heat to the ambient. Similarly, black-tailed jackrabbits have oversized ears with many blood vessels that maintain their ears at a higher than body temperature to dissipate heat to their warm surroundings. These and other animals also tap into a variety of thermal-regulating behaviors. For descriptive purposes, elephants flap their ears and spray muddy water on their bodies to stay cool when their surroundings are hot.

What about humans? How does our body regulate heat? On the issue of fertility, testicular thermoregulation is key to safeguarding the continuation of one's family line. Pham and Schultz (2021) found that conduction and blood perfusion have more influence on testicular temperature dynamics than metabolism. To highlight but one more human example, proper thermoregulation is critical for a person recovering from a heart attack; see, for example, Hoeyer-Nielson et al. (2021). Timely and appropriately targeted temperature management can make a life-and-death difference.

For ectotherms, or cold-blooded creatures, they behave well, or they die. Fig. 5.4 depicts that a lizard buries its body in the ground, away from the cold midnight-to-dawn air. Shade provides a comfortable temperature when direct sun rays are too hot. During later part of the morning and late afternoon, the warm sun is just right for basking. The goal is to balance conduction between the ground, direct solar radiation, convection heat exchange with the surrounding air, radiation to and from the sky and the environment, and evaporation cooling.

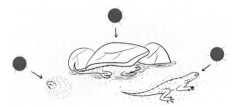

FIGURE 5.4 Ectotherms survive changes in their environment primarily via behavioral thermoregulation (created by S. Akhand). A lizard takes advantage of the ground, shade, and the sun appropriately to achieve a comfortable thermal condition.

Example 5.2 Motherly pythons.

Given: Pythons have a bad reputation because they are known to swallow animals and humans alive. In spite of that, mother pythons exhibit sacrificial love by the way they breed. According to Alexander (2018), the females do not eat during the 6-month-plus breeding cycle. Southern African pythons turn black and exercise "facultative melanism," an adaptation to increase heat absorption while basking. Pregnant and brooding females are 5°C, or more, warmer than non-reproductive females.

Find: The additional thermal energy a 50-kg mother python harnesses from the sun via facultative melanism, raising her body temperature by an extra 5°C.

Solution: Assume the specific heat capacity of a python, $c_P = 3$ kJ/(kg·°C). For a 5°C temperature difference between reproductive and non-reproductive female Southern African pythons, the thermal energy difference,

$$Q = mc_P \Delta T,$$

where m is the mass of the snake, and ΔT is the temperature difference. Substituting the values, we have

$$Q = (50)(3000)(5) = 750 \text{ kJ}.$$

We see that the intelligent adaptation mechanism, facultative melanism, makes an impressive difference!

Another fascinating ectotherm thermoregulation example is the morpho dragonfly, Zenithoptera lanei. According to Guillermo-Ferreira and Gorb (2021), this hot-weather dragonfly possesses a unique, complex system of thermoregulation that is not found in other dragonflies or insects. Specifically, the wings of the males are covered with wax nanocrystals that reflect ultraviolet light and infrared radiation. On top of that, the wing membrane is permeated by an intricate system of tracheae, which is hypothesized to be for thermoregulation, in addition to communication. Based on wing temperatures, Guillermo-Ferreira and Gorb (2021) suggested that the dorsal wing surface acts as a cooling system, whereas the ventral surface serves to elevate body temperature.

It is important to note that air humidity can have a significant effect on the way ectotherms thermoregulate. Le Galliard et al. (2021) conducted a controlled experiment on viviparous lizards, a dry-skinned ectotherm species. They found that short-term dehydration can slightly influence thermal preferences under some circumstances. In short, environmental humidity conditions can play a role in the way ectotherms thermoregulate. Along the same lines, Perez et al. (2021) explored the effects of food intake and hydration state on behavioral thermoregulation and locomotor activity in the western collared spiny lizards, Tropidurus catalanensis. They found that food consumption is the key predictor of preferred temperature, and the total distance traveled is marginally influenced by the hydration state.

The scope of this introductory textbook is to introduce Thermofluids: From Nature to Engineering. From here forward, we recommend interested readers to delve into specialized literatures on animal thermoregulation, such as McCafferty et al. (2018). For fish lovers, we recommend you start with Haesemeyer (2020). Those who are curious about how insects respond to extremely high temperature can refer to Ma et al. (2021).

5.4 Learning from intelligent designs

Intelligently designed butterfly wings are quite an inspiration for humans to imitate for products such as solar cells (Lou et al., 2012). Siddique et al. (2017) imitated some of the micro- and nano-structures to improve photovoltaic engineering. They copied efficient light in-coupling and light-trapping properties along with a high angle robustness to create thin-film photovoltaic absorbers of disordered nano-holes. They fabricated these absorbers using a scalable, self-assembly patterning technique based on the phase separation of a binary polymer mixture. They found the integrated absorption increased by up to 200%.

There is a wealth of intelligent designs on display in nature for us to appreciate and learn from. Let us look at a few more ideas while we are on the topic. We will focus on solar-related intelligent designs because the sun is the key source in sustaining life on earth. Furthermore, every added process leads to additional losses, thus, the most direct tapping of solar energy is the most efficient and should be sought.

5.4.1 Natural-convection-enabled air transport

Solar-powered natural convection plays an essential role in sustaining atmospheric circulation. Natural convection is also quintessential to many flying birds. A large flock of turkey vultures soaring in a circle is a direful proclamation of death on the ground. It is not the soul of the dead that is keeping the scavengers afloat but the equally invisible thermals, which are no more than rising volumes of hot air. More on this will be conveyed in a later chapter on natural convection.

Long-distant travelers such as peregrine falcons take advantage of rising thermals to gain altitude. After looping around an ever-shifting thermal to an appropriate height, the falcon leaves the thermal behind as it glides toward its destination (Ákos et al., 2010). The process of looping around a thermal to gain lift for gliding repeats as needed until the bird reaches its landing place. Lancaster (1885) is presumably the first researcher who formally documented the thermal-enabled soaring. Larger birds are particularly well designed to capitalize the lifts provided by natural convection to reduce the need of exerting their own energy in keeping their heavyweights up in the air. Pennycuick (1973) put forward that vultures are too big for their muscles and stamina beyond a few flaps at a time. The Andean condor, or vulture gryphus, is probably the heaviest soaring bird, whose three-dimensional traveling path has been tracked by researchers, such as Shepard et al. (2011). It appears that, when foraging, or hunting for food, the big birds soar primarily to ascend to a better position in terms of acquiring food on the ground. New vulture and condor movement ecology enthusiasts are encouraged to start with the review by Alarcón and Lambertucci (2018). To highlight, foraging has been quite extensively studied, but not the commuting and natal dispersal phases of these big birds.

Chimneys can be engineered to operate dynamically, sending exhaust to much higher up in the atmosphere via vortex rings. An excellent example of this is demonstrated by the Amager Bakker waste-to-energy plant just outside of Copenhagen, Denmark; see Big Vortex (2012). Below the atmospheric air, the display of the beautiful vortex rings is even more abundant. Novel underwater locomotives are taking advantage of vortex rings to excel; see, for example, Bi and Zhu (2018).

5.4.2 Wearing polar bear hair

If we want to fight the cold by wearing the proper gear, let us look no further than polar bears. Cui et al. (2018) mimicked the white and reflective hollow hairs of the polar bear. They fabricated synthetic fibers with an aligned porous structure similar to a shaft of a polar bear's fur and achieved excellent thermal insulation, breathability, and wearability. On top of that, doping the textile with electroheating materials can induce a fast thermal response and uniform electroheating. This group of researchers have furthered the technology into thermally insulating, protective clothing for hot environments (Wang et al., 2020). Imagine fire fighters appearing like the fluffy polar bear? The review paper by Eadie and Ghosh (2011) is also worth reading.

5.4.3 Ecological buildings

It is appalling to learn that both Americans and Canadians spend about 90% of their time indoors (Leech et al., 2002). This lack of outdoor time is a universal problem everywhere except in less developed and rural areas. Due to climate differences, Canadians savor summer outdoors more than their southern neighbors. This gain is reciprocated by the relatively longer Canadian winter, that is, the shorter winter south of Canada makes provision for the Americans to make up for their shorter outdoor stay in the summer. More importantly, each and every one of us are encouraged to spend more time outdoors because of the many health benefits; see, for example, Bowers et al. (2021) and Lee and Ho (2021). Over and above that, it is time to thoroughly include more healthy outdoor infrastructures in urban designs.

At the same time, we must accept that most of us will continue to spend the vast majority of our time indoors. With that in mind, efforts are needed to make healthy the indoor environment. In other words, when we cannot bring humans to appreciate nature, we bring nature for people to savor. There is much about nature that we can learn from and incorporate into our buildings. For instance, Fecheyr-Lippens and Bhiwapurkar (2017) learned from the African reed frog and the Hercules beetle and incorporated a hydrogel chamber and embedded a phase-change material, among other measures, into a building in Chicago. After performing a detailed energy analysis, they found a 66% reduction in the space-conditioning energy usage.

What about imitating a camel's nose strategy to manage thermal comfort, including water retention for desert buildings (Shahda et al., 2018)? The literature is becoming very rich concerning application of intelligent designs into sustainable buildings. Hershcovich et al. (2021) and Fu et al. (2020) provide two of the recent reviews on the topic.

5.4.4 Intelligent designs are complex and integrated

There is a multitude of possibilities for humans to advance simply by learning from nature. A word of caution is due at this point. The intelligent designs in nature always consider multiple factors simultaneously. The optimal design is meticulously fashioned for the unique settings and environments the creature is placed in. Bearing that in mind, we should be careful not to jump to conclusions after studying for but one or two aspects of an intelligent design. Due caution should also be exercised when copying the particular design to be employed in a scale and/or environment that is different from what was originally intended. For instance, Broeckhoven et al. (2017) discovered that there is a trade-off between strength and thermal capacity of dermal armor of girdled lizards. To be specific, a strong dermal armor is important in fending off enemies, but only to the extent that it does not hinder the lizard's thermoregulation ability.

Problems

5.1 Ecological problems.
List three of the top ecological problems that are caused by humans. List three serious ecological problems that are not caused by humans. Compare and contrast the man-caused ones with the natural-caused ones.

5.2 Wetlands and ecology.

Wetlands present the most known engineering solution that aims at restoring a healthy ecology. How do wetlands work? What are the outstanding challenges and limitations of wetlands?

5.3 Circular economy is ecology.

One of the best solutions to man-made environmental problems is to tackle the problems at the roots. A circular economy is an approach to make accountable the producer of the potential problem. Explain how a circular economy helps in reducing ecological damage.

5.4 Urban heat islands.

Urban heat islands are a real problem that deserves immediate action. New Orleans, Newark, and New York City are some of the cities in the United States that have the most serious heat island problem. Layout five measures as an ecological engineer to mitigate the problem.

5.5 Engineering coral reefs.

Coral reefs are dynamic and diverse ecosystems intelligently designed to play a paramount role in sustaining life on Earth. Delineate how a coral reef operates as an ecosystem. Conduct a literature review on man-made coral reefs.

5.6 Ecological transportation.

Man-made transport systems and traffic started escalating near the end of the twentieth century. Compile data from the literature to show the latest transportation trends. What are some ecological measures that we can feasibly implement to turn the tide around?

5.7 Plastic and ecology.

There is a movement trying to get rid of plastic. The challenge is obvious, that is, plastic is used in almost everything. It is used so extensively that even health care would cease to function if there were a sudden halt in plastic production. Devise a feasible engineering solution either to make plastic ecologically friendly or replace it with something that is ecologically friendly.

5.8 Overpopulation versus waste.

Some say that the world population is throwing our ecosystem off balance. Mother Teresa seemed to indicate that the culprit is waste, when she professed, "There must be a reason why some people can afford to live well. They must have worked for it. I only feel angry when I see waste. When I see people throwing away things we could use." Put this question into perspective by conducting some calculations on resources, urbanization, and waste.

References

Ákos, Z., Nagy, M., Leven, S., Vicsek, T., 2010. Thermal soaring flight of birds and unmanned aerial vehicles. Bioinspiration Biomimetics 5 (1), 015003.

Alarcón, P.A.E., Lambertucci, S.A., 2018. A three-decade review of telemetry studies on vultures and condors. Movement Ecol. 6 (13).

Alexander, G.J., 2018. Reproductive biology and maternal care of neonates in southen African python (Python natalensis). J. Zool. 305 (3), 141–148.

Bi, X., Zhu, Q., 2018. Numerical investigation of cephalopod-inspired locomotion with intermittent bursts. Bioinspiration Biomimetics 13 (5), 056005.

Big Vortex, https://www.youtube.com/watch?v=_GL3xAaIcvI, Reality: United, 2012, (accessed August 16, 2021).

Bowers, E.P., Larson, L.R., Parry, B.J., 2021. Nature as an ecological asset for positive youth development: empirical evidence from rural communities. Front. Psychol. 12, 688574.

Broeckhoven, C., du Plessis, A., Hui, C., 2017. Functional trade-off between strength and thermal capacity of dermal armor: insights from girdled lizards. J. Mech. Behav. Biomed. Mater. 74, 189–194.

Commoner, B., 1971. The Closing Circle: Nature, Man, and Technology. Alfred A. Knopf, New York, NY.

Cui, Y., Gong, H., Wang, Y., Li, D., H, B., 2018. A thermally insulating textile inspired by polar bear hair. Adv. Mater. 30, 1706807.

Data Footprint, https://data.footprintnetwork.org/#/, (accessed September 20, 2021), 2021.

Eadie, L., Ghosh, T.K., 2011. Biomimicry in textiles: past, present and potential—an overview. J. R. Soc. Interface 8 (59), 761–775.

Earth Rangers, "What's so great about frogs?" 2013. https://www.earthrangers.com/omg_animals/whats-so-great-about-frogs/ (accessed August 13, 2021).

Emmer, R., 1997. How do frogs survive winter? Why don't they freeze to death? Sci. Am. https://www.scientificamerican.com/article/how-do-frogs-survive-wint/. (accessed August 13, 2021).

Fecheyr-Lippens, D., Bhiwapurkar, P., 2017. Applying biomimicry to design building envelopes that lower energy consumption in a hot-humid climate. Architectural Sci. Rev. 60 (5), 360–370.

Fu, S.C., Zhong, X.L., Zhang, Y., Lai, T.W., Chan, K.C., Lee, K.Y., Chao, C.Y.H., 2020. Bio-inspired cooling technologies and the applications in buildings. Energy Build. 225, 110313.

Ganslosser, U., Jann, G., "Thermoregulation in animals: some fundamentals of thermal biology," Encyclopedia of Ecology, Second Edition, Volume 1, pages 328-336, 2019.

Genta, C., Favaro, S., Sonetti, G., Barioglio, C., Lombardi, P., 2019. Envisioning green solutions for reducing ecological footprint of a university campus. Int. J. Sustain. Higher Education 20 (3), 423–440.

Genta, C., Favaro, S., Sonetti, G., Fracastoro, G.V., Lombardi, P., 2021. Quantitative assessment of environmental impacts of the urban scale: the ecological footprint of a university campus. Environ., Development Sustain. doi:10.1007/s10668-021-01686-5.

Guillermo-Ferreira, R., Gorb, S.N., 2021. Heat-distribution in the body and wings of the morpho dragonfly Zenithoptera lanei (Anisoptera: Libellulidae) and a possible mechanism of thermoregulation. Biol. J. Linn. Soc. 133 (1), 179–186.

Haesemeyer, M., 2020. Thermoregulation in fish. Molecular Cellular Endocrinolol. 518, 110986.

Hershcovich, C., van Hout, R., Rinsky, V., Laufer, M., Grobman, Y.J., 2021. Thermal performance of sculptured tiles for building envelopes. Build. Environ. 197, 107809.

Hoeyer-Nielson, A.K., Holmberg, M.J., Christensen, E.F., Cocchi, M.N., Donnino, M.W., Grosses-treuer, A.V., 2021. Thermoregulation in post-cardiac arrest patients treated with targeted temperature management. Resuscitation 162, 63–69.

Isildar, G.Y., Morsali, S., 2020. Environmental footprint assessment of university campuses among developed and developing countries including a case from Turkey. Fresenius Environ. Bull. 29 (2), 1114–1120.

Kearney, M., Shine, R., Porter, W.P., 2009. The potential for behavioral thermoregulation to buffer "cold-blooded" animals against climate warming. In: Proceedings of the National Academy of Sciences of the United States of America, 106, pp. 3835–3840.

Landaster, I., 1885. The Problem of the Soaring Bird. University of Chicago Press, Chicago, IL.

Le Galliard, J-F., Rozen-Rechels, D., Lecomte, A., Demay, C., Dupoué, A., Meylan, S., 2021. Short-term changes in air humidity and water availability weakly constrain thermoregulation in a dry-skinned ectotherm. Public Library Sci. doi:10.1371/journal.pone.0247514.

Lee, J.L.C., Ho, R.T.H., 2021. Exercise spaces in parks for older adults: a qualitative investigation. J. Aging Phys. Act 29 (2), 233–241.

Leech, J.D., Nelson, W.C., Burnett, R.T., Aaron, S., Raizenne, M.E., 2002. It's about time: a comparison of Canadian and American time-activity patterns. J. Exposure Sci. Environ. Epidemiol. 12, 427–432.

Lou, S., Guo, X., Fan, T., Zhang, D., 2012. Butterflies: inspiration for solar cells and sunlight water-splitting catalysts. Energy Environ. Sci. 5, 9195–9216.

Ma, C-S., Ma, G., Pincebourde, S., 2021. Survive a warming climate: insect responses to extreme high temperatures. Annu. Rev. Entomol. 66, 163–184.

McCafferty, D.J., Pandraud, G., Gilles, J., Fabra-Puchol, M., Henry, P-Y., 2018. Animal thermoregulation: a review of insulation, physiology and behaviour relevant to temperature control in buildings. Bioinspiration Biomimetics 13, 011001.

Ortegon, K., Acosta, P., 2019. Ecological footprint: a tool for environmental management in educational institutions. Int. J. Sustain. Higher Education 20 (4), 675–690.

Pennycuick, C.J., 1973. The soaring flight of vultures. Sci. Am. 229 (6), 102–109.

Pham, S., Schiltz, J.S., 2021. Testicular thermoregulation with respect to spermatogenesis and contraception. J. Therm. Biol 99, 102954.

Perez, D.J.P., de Carvalho, J.E., Navas, C.A., 2021. Effects of food intake and hydration state on behavioral thermoregulation and locomotor activity in the tropidurid lizard Tropidurus caralanensis. J. Exp. Biol. 224 (6), jeb242199.

Scripture, E.W., 1897. The New Psychology. Charles Scribner's Sons, London, p. 300.

Sedgwick, W.T., 1883. On the variations of reflex-excitability in the frog, induced by changes of temperature. In: Martin, N., Brooks, W.K. (Eds.), Studies from the Biological Laboratory, II. N. Murray, John Hopkins University, Baltimore, pp. 385–410.

Shahda, M.M., Elhafeez, M.A.A., El Mokadem, A.A., 2018. Camel's nose strategy: new innovative architectural application for desert buildings. Sol. Energy 176, 725–741.

Shepard, E.L.C., Lambertucci, S.A., Vallmitjana, D., Wilson, R.P., 2011. Energy beyond food: foraging theory informs time spent in thermals by a large soaring bird. PLoS One 6 (11), e27375.

Siddique, R.H., Donie, Y.J., Gomard, G., Yalamanchili, S., Merdzhanova, T., Lemmer, U., Hölscher, H., 2017. Bioinspired phase-separated disordered nanostructures for thin photovoltaic absorbers. Sci. Adv. 3 (10), e1700232.

Swithinbank, H.J., Gower, R., Foxwood, N., 2019. Sustained by faith? The role of Christian belief and practice in living sustainably. In: Filho, W.L., McCrea, A.C. (Eds.), Sustainability and the Humanities. Springer International Publishing, Cham, pp. 375–391.

Unger, S.D., Rollins, M.A., Thompson, C.M., 2020. Hot- or cold-blooded? A laboratory activity that uses accessible technology to investigate thermoregulation in animals. Am. Biol. Teacher 82 (4), 227–233.

Wang, Y., Cui, Y., Shao, Z., Gao, W., Fan, W., Liu, T., Bai, H., 2020. Multifunctional polyimide aerogel textile inspired by polar bear hair for thermoregulation in extreme environments. Chem. Eng. J. 390, 124623.

Wood, H., Acharjee, A., Pearce, H., Quraishi, M.N., Powell, R., Rossiter, A., Beggs, A., Ewer, A., Moss, P., Toldi, G., 2021. Breatfeeding promotes early neonatal regulatory T-cell expansion and immune tolerance of non-inherited maternal antigens. Allergy 76 (8), 2447–2460.

Chapter 6

The first law of thermodynamics

"In all things of nature there is something of the marvelous."

Aristotle

Chapter Objectives
- Recognize energy and the various forms of energy.
- Differentiate open, closed, and isolated thermodynamic systems.
- Distinguish work transfer from heat transfer.
- Comprehend the first law of thermodynamics.
- Appreciate moving-boundary work.
- Fathom enthalpy and its relationship to internal energy.
- Learn what a thermodynamic cycle is.

Nomenclature

A	area or surface area
C	a constant
c	speed of light in a vacuum
c_P	specific heat capacity at constant pressure
c_V	specific heat capacity at constant volume
E	energy; E_{in} is the energy going into a system, E_{mech} is mechanical energy, E_{out} is the energy coming out of a system, E_{waste} is waste energy, ΔE is the path-independent change in energy, ΔE_{surr} is the energy change of the surroundings, ΔE_{sys} is the energy change of the system, ΔE_{turkey} is the energy change of the turkey, ΔE_{univ} is the energy change of the universe
e_{mech}	mechanical energy per unit mass
F	force
g	gravity
H	enthalpy; H_{in} is the enthalpy going into a system, H_{out} is the enthalpy coming out of a system
h	enthalpy per unit mass
k	exponent
KE	kinetic energy
m	mass
n	amount of a substance
P	pressure
PE	potential energy

Thermofluids: From Nature to Engineering. https://doi.org/10.1016/B978-0-323-90626-5.00019-7

Q   heat; Q' is heat transfer rate, δQ is the path-dependent heat transfer, Q_H is the heat transfer at T_H, Q_L is the heat transfer at T_L, $_1Q_2$ represents the specific path of heat addition from state 1 to state 2

r   radius, r_o is the outer radius

R_u the universal gas constant

S, s   displacement

T   temperature; ΔT is temperature difference or change, T_H is the higher temperature, T_L is the lower temperature

t   time

U   internal energy

V   velocity

W   work; W' is work transfer rate or power, $_1W_2$ represents the specific path of work addition from state 1 to state 2, δW_b is the path-dependent boundary work, W_A is work performed along Path A, W_B is work performed along Path B, W_C is work performed along Path C

x   distance

z   elevation

Greek and other symbols

Δ difference; path-independent change

ρ density

$\forall$ volume

6.1 Energy

Thermodynamics is concerned with energy, a quantitative property denoting the capacity for doing work or for heating. Energy can take place in many forms, as depicted in Fig. 6.1. For example, a frog uses mechanical energy to hop, swim, and catch bugs with its tongue. Fuel or electricity is converted into mechanical energy to move an automobile. Thermal energy is utilized for cooking yummy food. Radios and microwaves call upon electromagnetic energy to function. We depend on the power grid to supply electrical energy for powering light bulbs, computers, cell phones, kitchen appliances, heating and cooling systems, and many others. Chemical energy from the food we eat empowers us to go about our daily business. Above all, nuclear energy makes it possible for the sun to send forth light and heat, sustaining lives on Earth. Within the context of introductory thermofluids, energy in the forms of thermal, kinetic, potential, mechanical, and chemical are of particular interest. Thermal energy is the form of energy that increases with temperature. Kinetic energy and potential energy are the most common forms of macroscopic energy—macroscopic in the sense that the energy is related to the bulk of the material involved, that is, it can be seen with our naked eyes. An invisible fluid necessitates illumination so that we can see it moving, namely, its kinetic energy, or sitting above a reference altitude, that is, its potential energy.

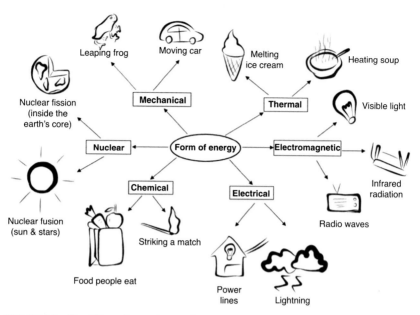

FIGURE 6.1   The different forms of energy include mechanical, thermal, electromagnetic, electrical, chemical, and nuclear (created by S. Akhand). Not included in the illustration are kinetic energy and potential energy.

Example 6.1 Water wall.
Given: To facilitate building-energy conservation research, Dr. JAS presents her vacation home near Charleston, South Carolina, as a living research facility. Due to her love for the water, she commissions the walls of her villa to be replaced with glass water columns. The building envelope is 25 m long, 20 m wide, and 12 m tall. Each wall is 1 m thick, with water enclosed in glass.

Find: The amount of solar thermal energy stored when the water temperature rises from 14°C to 17°C.

Solution: What we are dealing with is thermal energy; the change in energy of the water from 14°C to 17°C can be expressed as

$$\Delta E = mc_P \Delta T = \rho \forall c_P \Delta T = 1000 \text{ kg/m}^3 (1080 \text{ m}^3)(4180 \text{ J/kg·°C})(3°C)$$
$$= 1.35 \times 10^{10} \text{ J}.$$

Note that Δ signifies a path-independent change, that is, it is a change in a thermodynamic property (internal energy, in this case) that is explicitly defined by the state and not the process that gets it there.

This 1.35×10^{10} J stored by the water is an enormous amount of energy. A typical house that is ten times more modest (smaller) than Dr. JAS' villa would still amount to more than one giga-joule of energy storage!

Internal energy. At the microscopic end, we have the internal energy, U, which denotes the sum of all the microscopic forms of energy. Internal energy is the molecular energy of a system; it includes translational kinetic energy, vibrational and rotational kinetic energy, and potential energy from intermolecular forces. Internal energy is an extensive state property. State properties are independent of the path of the process. Extensive implies that it is size-dependent, that is, the larger the mass of material at a state, the larger the amount of internal energy. Sensible energy is the internal energy associated with the kinetic energy of the molecules. Latent energy is the internal energy associated with the phase, that is, the intermolecular forces. Chemical energy is the internal energy associated with the atomic bonds in a molecule, while nuclear energy is the internal energy associated with the bonds within the nucleus of the atom. The energy density or intensity increases from sensible to latent to chemical to nuclear. A little fanning can remove a little heat when the temperature is just above the thermally cool range. This is sensible energy (heat) transfer. Higher temperatures, however, force our body to resort to much more energy-intensive removal via sweating. This involves phase change and, hence, it is the transfer of latent energy. An average passenger car consumes about 40 m^3 of gasoline, with an energy density of 46 MJ/kg, over 20 years. The combustion of gasoline with air into heat to power an automobile demonstrates the working of the energy-intensive chemical energy to thermal and subsequently mechanical energy. We only need a peanut-size piece of uranium with an energy density of 3,900,000 MJ/kg, to do the same. Wow, how stupendously powerful nuclear energy is!

Molecules and atoms become more energetic with temperature and, thus, the internal energy increases. The thermal energy of a system increases with temperature. It follows that thermal energy is a part of the internal energy of a system. With the macroscopic kinetic and potential energies, the total energy of a system can be expressed as

$$E = U + KE + PE, \qquad (6.1)$$

where U is internal energy, KE is macroscopic kinetic energy, and PE is macroscopic potential energy. In Example 6.1, there are no changes in the macroscopic kinetic and potential energies and, thus, the change in the water-energy with temperature, $\Delta E = mc_P \Delta T$, is the change in internal energy, ΔU, of the volume of water.

Mechanical Energy. For a fluid in which the internal energy[1] is of little concern or does not change, mechanical energy is the energy of interest. Mechanical energy is the sum of the macroscopic kinetic and potential energies of a system. A water turbine is a mechanical work-producing system that extracts mechanical energy from a moving fluid by decreasing its pressure. A pump, on the other

---

1. And there are no magnetic, electrical, and other forms of energy involved.

hand, is a work-consuming system that requires mechanical energy input to raise the pressure of a fluid. On that account, the pressure of a fluid is associated with its mechanical energy. Let us take a closer look at the units of pressure, the pascal,

$$[Pa] = N/m^2 = Nm/m^3 = J/m^3 = \text{energy per unit volume.} \quad (6.2)$$

Multiplying pressure by volume, $\forall$, leads to

$$P\forall = [J/m^3][m^3] = J = \text{energy.} \quad (6.3)$$

This is flow energy, that is, the energy used for moving the fluid along the flow passage. Dividing it by mass, m, gives

$$P\forall/m = [J/m^3][m^3]/[kg] = J/kg = \text{energy per unit mass.} \quad (6.4)$$

We can view fluid pressure as flow energy, pushing the fluid against frictional and other resistance. With that in mind, the mechanical energy of a fluid per unit mass of fluid can be expressed as

$$E_{mech}/m = e_{mech} = P/\rho + \frac{1}{2}V^2 + gz, \quad (6.5)$$

where ρ is the density of the involved fluid, V is the velocity of the fluid, g is gravity, and z is elevation. The equation conveys that the mechanical energy of a fluid is the sum of flow or pressure energy, kinetic energy, and potential energy due to elevation under gravity. This mechanical energy can be expressed in rate form as

$$E'_{mech} = m'e_{mech} = m'\left(P/\rho + \frac{1}{2}V^2 + gz\right), \quad (6.6)$$

where m' is the mass flow rate of the fluid. Often, we are more interested in the change in mechanical energy from point 1 to point 2, that is, per unit mass of fluid,

$$\Delta e_{mech} = (P_2 - P_1)/\rho + \frac{1}{2}(V_2^2 - V_1^2) + g(z_2 - z_1). \quad (6.7)$$

To that end, we see that a flowing fluid has pressure or flow energy. This is in addition to the internal energy, kinetic energy, and potential energy that a nonflowing material has. Expressly, a more general equation for energy is

$$E = P\forall + U + KE + PE. \quad (6.8)$$

Whatever the form, our focus on energy is on its ability to do useful work.

Example 6.2 Mechanical to internal energy.
Given: Dr. JAS installed a wind turbine to power her cottage. The location is so abundant in wind energy that she decided to use the untapped wind energy to heat water. To do so, she harnessed the excess wind to spin a 500-W paddle wheel in a 1-m³ tank of water initially sitting at 18°C.
Find: The rise in water temperature after an hour.

Solution: We can express the first law of thermodynamics as

$$\Delta E_{sys} = E_{in} - E_{out},$$

that is the change in energy of the 1-m^3 tank of water is equal to the amount of energy entering the system, 1-m^3 tank of water, minus that leaving it. The system energy change consists of the change in its internal energy plus the change in kinetic energy and potential energy. Therefore, we can write

$$\Delta U + \Delta KE + \Delta PE = Q_{in} - Q_{out} + W_{in} - W_{out},$$

where, in the absence of mass transferring across the system boundary, energy can enter and exit the system as heat and work, Q and W, respectively. We note that there is no change in kinetic energy. In other words, we can look at the problem as the paddle is spinning at the same rate all the time or that the disturbance caused by the spinning paddle dissipates promptly into heat. There is also no change in potential energy, as the elevation of the tank is fixed. We can further assume that there is no heat transfer between the system of water and its surroundings, that is, the tank is insulated. Under those circumstances, we have

$$U = W_{in} = 500 \text{ W}(3600 \text{ s}) = 1.8 \times 10^6 \text{ J}.$$

The internal energy change can be expressed as $mc_P \Delta T$, that is,

$$mc_P \Delta T = 1000 \text{ kg/m}^3 (1 \text{ m}^3)(4186 \text{ J/kg} \cdot {}^\circ\text{C})\Delta T = 1.8 \times 10^6 \text{ J}.$$

This gives $\Delta T = 0.43{}^\circ$C.

We see that the temperature rise is small. This is partly because water has a high heat capacity, that is, it can take a lot of beating before it heats up. If one cubic meter of air is used instead of water, then we have

$$mc_P \Delta T = 1.2 \text{ kg/m}^3 (1 \text{ m}^3)(1000 \text{ J/kg} \cdot {}^\circ\text{C})\Delta T = 1.8 \times 10^6 \text{ J}.$$

This gives $\Delta T = 1500{}^\circ$C. This is enormous because there are two orders of magnitude of decrease in fluid density, along with a factor of four decrease in heat capacity, when changing from 1 m^3 of water to 1 m^3 of air. A fairer comparison is to keep the mass of the fluid the same, that is,

$$mc_P \Delta T = 1000 \text{ kg}(1000 \text{ J/kg} \cdot {}^\circ\text{C})\Delta T = 1.8 \times 10^6 \text{ J}.$$

For 1000 kg of air, the temperature rise, $\Delta T = 1.8{}^\circ$C. This is serviceable, that is, Dr. JAS can utilize the excess wind to spin a paddle inside her cottage. For safety reasons, the paddle should be placed away from occupants; it can be designed as a beautiful, functional, ceiling fan.

6.2 Thermodynamic systems

A system is an arbitrary or real region of interest. It is defined by a boundary, separating it from the surroundings. The boundary, arbitrary or real, is a surface that separates the system from the surroundings. Fig. 6.2 illustrates the three

FIGURE 6.2   Thermodynamic systems (A) open system, (B) closed system, and (C) isolated system (created by X. Wang). Both mass and energy can enter or leave an open system. Only energy can cross the boundary of a closed system. Nothing can enter or exit an isolated system.

basic thermodynamic systems—open, closed, and isolated. A cup of hot coffee without a lid is an open system. Mass tends to leave the cup in the form of steam or water vapor. When you cover the cup of coffee with a lid, it becomes a closed system, that is, no mass, neither coffee nor water vapor, is leaving or entering the system. Energy, however, can cross the boundary of a closed system. For the cup of hot coffee covered with a lid, thermal energy makes its way through the cup and lid into the cooler room. The heat transfer continues until the temperature of the coffee equals to that of its surroundings. Pouring hot coffee into to a thermos flask and seal it tight represents an isolated system across which neither mass nor energy transfer takes place. In reality, the insulation is not perfect and, as a consequence, the hot coffee eventually cools off because of heat transferring slowly across the thermos flask to the cooler ambient.

Open System. An open system is a system that both mass and energy can enter or leave. It is often referred to as a control volume with arbitrary boundaries. As discussed above, the cup of hot coffee in Fig. 6.2 depicts an open system, where the open top is an arbitrary boundary or surface. Heat is leaving the cup along with the mass transfer of evaporating steam across this imaginary boundary. Heat is also conducted across the wall of the cup. A water pump is an open system, where the inlet and outlet are arbitrary surfaces that allow water to flow through. Electric energy drives the pump, and it enters the open system as mechanical energy via a spinning impeller. This mechanical energy enters the fluid in terms of flow or pressure energy. In fact, almost all systems dealing with moving fluid are open systems that involve pressure or flow energy.

Example 6.3 Examination room energy change.
Given: The procedure for final examinations prohibits students from entering the
        examination room until the instructor is ready.
Find: What happens to the energy of the room before and after 250 students
        enter it.

Solution: As masses can enter the room, the room is an open system. Every
student carries a certain amount of mass and also a particular amount of energy.
Therefore, the energy of the room increases after the students enter it.

   Closed System. A closed system does not allow any mass to cross its boundary. Since it is closed, there is no fluid entering or exiting and, that being so, a
closed system does not involve flow energy. That said, energy can be transferred
into or out of a closed system. The cup with the top covered with a lid, as depicted
in Fig. 6.2, is a closed system. The lid prevents any steam (mass) from leaving
the cup. Energy in the form of heat, however, leaves the system via conduction
through the lid and also the other surfaces of the cup. A light bulb is another
common closed system. Electricity enters this closed system and leaves as light
and heat, even though no mass crosses the system boundary.
   Isolated System. An isolated system is a self-contained system of fixed mass
and energy. Neither mass nor energy can cross its boundary. A calorimeter is
an isolated system. It sets a reaction to proceed and measures the resulting heat
of combustion. For a constant-volume bomb calorimeter, the energy released
by the reaction goes completely into raising the temperature and pressure of
the calorimeter. As explained earlier, the thermos flask shown in Fig. 6.2 is an
isolated system. There is no mass leaving or entering a closed thermos flask.
A perfectly insulated thermos flask does not allow any heat transfer. Without
any interference such as magnetizing (magnetic energy), electricizing (electric
energy), or shaking (mechanical energy), etc., no energy enters or leaves the
isolated system. In the ideal case, a liter of mouth-watering, iced, homemade
honey lemonade sealed in a thermos flask will remain savory until eternity. All
real thermos flasks, on the other hand, have finite conductivity, permitting heat
to transfer across its boundary slowly. Within the time frame of a couple of
hours, a good thermos flask that is properly closed is a bona fide isolated system.
Technically speaking, the only definitive isolated thermodynamic system is the
universe that encompasses everything under creation.

6.3   Heat and work transfer

We can look at the energy transfer through a system as work and heat. Thermal energy in transition, heat, is transferred via a temperature difference. An
adiabatic system is a well-insulated system without any heat transfer. Work is
an energy interaction between a system and its surroundings. In this context,
the transfer of electricity and magnetic force is considered work (transfer).

In short, heat and work are energy transfer mechanisms between a system and its surroundings. Both are transient boundary phenomena, that is, systems possess energy but not heat or work. They are associated with a process, not a thermodynamic state and, for that reason, are path-dependent. As a metaphor, Montreal is akin to a thermodynamic state and so is Nashville. The corresponding latitude and longitude are akin to thermodynamic properties such as pressure, temperature, and volume, which are not path-dependent. The energy usage for driving from one city to the other is parallel to heat and energy, in the sense that it is path-dependent, that is, the energy usage varies depending on the route you take.

The most widespread sign convention is to consider heat transfer, Q, into a system as positive, while work, W, is positive when it is transferring out of the system. The sounder sign convention is to deem energy, whether heat or work, into the system as adding energy to the system and thus positive. This is adopted in chemistry but not in engineering. Unless specified, we use Q and W, in general, to signify the quantity of heat and work without attaching the energy flow direction. The change in energy of a system that exchanges heat and work with its surroundings can be expressed as

$$\Delta E_{sys} = Q + W. \tag{6.9}$$

Example 6.4 Heat or work?
Given: When autumn begins, the energy management facility has to choose an appropriate date to switch from cooling mode into heating mode, leaving some days in the building space colder than thermally comfortable for Dr. JAS. To take care of herself, she installs an electric heater in her office.
Find: If it is a transfer of heat or work into her office via the electric heater.

Solution: If the control volume of interest includes the electric heater, that is, if the heating coils are part of the system under study, then, it is the transfer of electric work into Dr. JAS' office. To put it another way, no heat transfers into the system, Dr. JAS' office, via the power wire in the ideal case. Only electrical energy flows along the wire, and only after entering the system that as it passes through the resistance coil it is transformed into thermal energy.

If we exclude the heating coils so that they are just outside the boundary of the system under consideration, then, it is heat transfer from the higher temperature heated coils into the office space.

6.4 Conservation of energy

According to the first law of thermodynamics, energy cannot be destroyed nor created. In plain English, the energy of the universe is constant. There can be energy exchange between a system under consideration and its surroundings, but the total energy remains the same. Specifically, the change in energy of the universe:

$$\Delta E_{univ} = \Delta E_{sys} + \Delta E_{surr} = 0, \tag{6.10}$$

where ΔE_{sys} is the change in energy of the system and ΔE_{surr} is the change in energy of the surroundings. Energy can also change from one form to another. An obvious example is the energy from the food we ate this morning, or yesterday, which is used to do useful work including pumping our heart and keeping us warm. Any excess amount of food we eat is not destroyed it is either saved or discarded. In short, the total amount of energy is conserved. To put it another way, the amount of energy in the food is equal to the sum used for doing work, generating heat, stored as fat, excreting via our washroom visits, and breathing[2]. Succinctly, energy in is equal to energy out. In the same token, we cannot perform work if we stop consuming hearty food for a few days. Simple as it may be, the first law of thermodynamics is essential and a necessary complement to second-law thermodynamics analysis. For example, Temiz and Dincer (2021) performed both the first- and second-law thermodynamics analyses of nuclear- and solar-based energy, food, fuel, and water production systems for an indigenous community. The obtained first-law energy efficiency complemented that of the exergy efficiency in describing the overall system performance. The first-law analysis outweighs that of the second law in studying nanofluid enhancement of a concentrator photovoltaic thermal system performed by Deymi-Dashtebayaz and Rezapour (2021). It is interesting to note that while the first law efficiency increases with increasing Reynolds number, the second law efficiency decreases. When it comes to monitoring the energy status in lithium-ion batteries, Chen et al. (2021) found that the first law alone is sufficient.

Example 6.5 Skateboarding the first law of thermodynamics.
Given: Dr. JAS brings her skateboarding skills to illustrate the first law of thermodynamics, as depicted in Fig. 6.3. She has equipped her latest skateboard with battery-operated small air jets, making her skateboard more like an off-the-ground hover-board to minimize friction. Dr. JAS positions herself completely rigidly and asks T-E Kaya to push her gentling from the top right platform, into the hemispheric cavity.
Find: How the first law of thermodynamics operates.

Solution: This is parallel to a pendulum. Dr. JAS starts at the highest level on the right platform, motionless. The amount of potential energy with respect to

---

2. The air we breathe out contains a fair amount of thermal energy. For this reason, a classical way to check if a person is dead is to place a mirror under the nose. The formation of fog, due to partial condensation of water vapor in the warm air being cooled by the cold mirror, indicates the person is breathing and, thus, alive. Solid scat contains more than thermal energy and, thus, dung beetles, rabbits, chimps, dogs, and some fish take advantage of its nutrients to thrive.

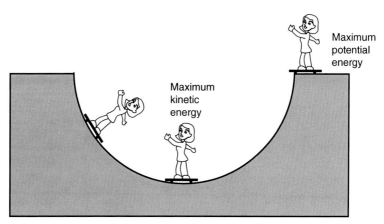

FIGURE 6.3   The conservation of energy between potential and kinetic energy of a frictionless skateboarder is analogous to that of a pendulum (created by Y. Yang). In the ideal frictionless situation, the sum of potential energy and kinetic energy is a constant.

the lowest point in the middle of the cavity is

$$PE = mgz,$$

where m is mass and z is elevation. As she slides down the curve, she gains kinetic energy at the cost of potential energy. More importantly, the total energy, the sum of internal energy, potential energy, and kinetic energy, remains the same, as long as there are no losses, that is,

$$E = U + PE + KE = \text{constant}.$$

We have not included the flow energy, as this is not a flowing-fluid problem. Without any change in temperature, the internal energy, U, stays the same and, thus,

$$E = PE + KE = mgz + \frac{1}{2}mV^2 = \text{constant}.$$

As elevation or height, z, decreases, Dr. JAS' speed, V, increases. The moment she zooms through the middle of the cavity at the highest speed, her potential energy with respect to the physical datum is zero. Beyond the lowest point, she moves upward to the left, losing speed but gaining potential energy. In the ideal situation without friction and drag, Dr. JAS would swing back and forth like a pendulum, illustrating the first law of thermodynamics until eternity.

Practical thermofluid systems can be divided into closed systems and open systems. As discussed above, energy can transfer across both closed and open systems, but only an open system allows mass to enter or exit. According to the first law of thermodynamics, the heat, δQ, added to a closed system in an infinitesimal process can increase the system's internal energy by dU and/or perform δW amount of work. Specifically,

$$\delta Q = dU + \delta W. \tag{6.11}$$

We are reminded that δ denotes that the corresponding value, of heat and work, is path-dependent, whereas system (internal) energy is a state variable, that is, path-independent. Alternatively, we can express the first law of thermodynamics for the change in energy from state 1 to 2 as

$$\Delta E_{1-2} = {}_1Q_2 - {}_1W_2. \tag{6.12}$$

Here, ${}_1Q_2$ represents the specific path of heat addition and ${}_1W_2$ indicates the path-dependent work output. Let us walk through a couple of everyday examples to illustrate the first law of thermodynamics for closed and open systems.

Example 6.6 A wild Turkey and Kaya's new oven.
Given: On teacher's day, T-E Kaya invites Dr. JAS and a few friends over for a home-cooked dinner. T-E Kaya exercises his culinary skills with a 3.5-kg wild turkey using his new oven. He accurately times the required time to roast the wild turkey at 180°C in his new 1000-W oven. He is fully aware that turkey meat is cooked when it reaches approximately 75°C.
Find: The energy required if it takes 2.5 h to roast the turkey into perfection. The required energy to bring the turkey from 20°C to 75°C. The waste heat that is not used for cooking the turkey.

Solution: Ignoring the initial warm-up, the energy usage is

$$E_{in} = E'_{in}\Delta t = 1000 \text{ W}(2.5 \text{ hr} \times 60 \text{ min/hr} \times 60 \text{ s/min}) = 9 \times 10^6 \text{ J},$$

where E_{in}' is the electric power input, and Δt is the time duration.

The amount of energy required to bring the 3.5-kg turkey at 20°C to 75°C is

$$\Delta E_{turkey} = mc_P\Delta T = 3.5 \text{ kg}(2810 \text{ J/kg} \cdot °C)(75°C - 20°C) = 5.4 \times 10^5 \text{ J},$$

where m is the mass, c_P is the specific heat at constant pressure, and ΔT is the temperature change. This is for the ideal case that the entire turkey is homogeneous in terms of temperature, that is, it rises from 20°C to 75°C. The first law of thermodynamics can be expressed as energy supplied by the oven is equal to the energy increase of the turkey plus the unused energy that is wasted. Specifically,

$$E_{in} = \Delta E_{turkey} + E_{waste}$$

which can be rearranged into

$$E_{waste} = E_{in} - \Delta E_{turkey} = 9 \times 10^6 \text{ J} - 5.4 \times 10^5 \text{ J} = 8.5 \times 10^6 \text{ J}$$

In other words, about 94% of the energy is wasted.

In reality, however, when the internal temperature of the 3.5-kg turkey reaches 75°C, the external temperature is at the same temperature as the oven, that is, 180°C. Let us assume a linear spatial distribution of temperature into the turkey, which is approximated as a sphere. With a density that is around 1000 kg/m^3, the volume of the 3.5-kg spherical turkey is

$$\forall = (4/3)\pi r^3 = m/\rho = 3.5/1000 = 0.0035 \text{ m}^3.$$

From which, the radius, r = 0.094 m. The temperature in degrees of centigrade as a function of the radius can be expressed as

$$T(r) = 75 + (180 - 75) \, r/0.094 = 75 + 1115 \, r.$$

Multiplying both sides with differentiable radius, dr, gives

$$T(r)dr = (75 + 1115 \, r) \, dr.$$

We can integrate from r = 0 to 0.094 m, that is,

$$\int T(r)dr = \int (75 + 1115 \, r) \, dr.$$

Note that the average temperature, $T_{avg} = \int T(r) \, dr / \int dr$, and, therefore, we have

$$T_{avg} \int dr = \left(75r + 557 \, r^2\right) \text{ from } r = 0 \text{ to } 0.094 \text{ m}.$$

Solving, we get

$$T_{avg} = 127.5°C.$$

This average does not account for the increasing weighting due to increasing (thermal) mass with increasing radius. Consider an infinitely small spherical shell element of thickness dr, the volume and thus the (thermal) mass of this shell increases as we move farther out from the center of the sphere.

To account for the increasing thermal mass with radius effect, we have

$$T_{avg} = \int T(r)4\pi r^2 \, dr/\left[(4/3)\pi r_o^3\right],$$

where $4\pi r^2$ dr is the surface area of the sphere at radius r multiplied by the thickness of the shell, giving the volume of the spherical shell. With constant density, this volume corresponds also to the mass of the spherical shell at radius r, that is, mass is equal to the volume times the density. The denominator, $(4/3) \pi r_o^3$, is the volume of sphere with radius r_o. Substituting for temperature variation with respect to radius, $T(r) = 75+1115r$, and canceling terms in the numerator with similar terms in the denominator, we get

$$T_{avg} = \int (75 + 1115 \, r) \, r^2 dr/\left(r_o^3/3\right).$$

Integrating from r = 0 to r_o, we obtain

$$T_{avg} = \left[75r_o^3/3 + 1115r_o^4/4\right]/\left(r_o^3/3\right) = 3(75/3 + 1115r_o/4) = 153.75°C.$$

The average body temperature of a cooked turkey is 153.75°C. The amount of energy required to bring the 3.5-kg turkey at 20°C to 153.75°C is

$$\Delta E_{turkey} = mc_P \Delta T = 3.5 \text{ kg} (2810 \text{ J/kg} \cdot {}^\circ C)(153.75^\circ C - 20^\circ C)$$
$$= 1.32 \times 10^6 \text{ J}$$

The unused energy,

$$E_{waste} = E_{in} - \Delta E_{turkey} = 9 \times 10^6 \text{ J} - 1.32 \times 10^6 \text{ J} = 7.68 \times 10^6 \text{ J}$$

That is, 85% of the energy input is not capitalized for roasting the yummy turkey.

A much more efficient way to cook is to use a microwave, but one would have to be extremely hungry to cram down a piece of microwaved turkey.

Concerning conservation, it is worthwhile to pause and be enlightened by a couple of sayings put forward by wise teachers. British physicist, Sir Oliver Joseph Lodge promulgated, "There is a conservation of matter and of energy, there may be a conservation of life; or if not of life, of something which transcends life." Earlier in the first century, Roman philosopher Seneca asserted, "If you live in harmony with nature you will never be poor; if you live according what others think, you will never be rich." Not to over-exploit or waste energy, food, and other resources will carry us a long way forward, leaving enough for future generations to treasure and enjoy.

Example 6.7 Einstein's $E = mc^2$ and the first law of thermodynamics.
Given: According to Albert Einstein's Theory of Relativity, the energy content of matter is given by

$$E = mc^2$$

where m is the resting mass of the matter and c is the speed of light in a vacuum. Imagine that Dr. JAS successfully realized a biological Einstein reactor that can convert matter into energy and energy into matter, and she implanted this into the body of Turbulence and Energy Lab student T-E Kaya.

Find: The amount of durian T-E Kaya needs, without and with the biological Einstein reactor, over a period of 50 years. Does Einstein's Theory of Relativity violate the first law of thermodynamics?

Solution: According to EatRight (2021), an adult female needs between 1600 and 2200 calories per day, depending on factors such as age, work, and activities. With relatively larger body mass and higher metabolism, an adult male requires 2000 to 3200 calories every day. Assume 2500 calories per day is just right for the moderately active student, T-E Kaya. Over a 50-year period, T-E Kaya would require

$$E = 2,500 \text{ calories/day} \times 365 \text{ days} \times 50 \text{ years} = 4.56 \times 10^7 \text{calories}.$$

This is a lot of calories. If we convert it into joules, we get

$$E = 4.56 \times 10^7 \text{ calories} \times 4.184 \times 10^3 \text{ J/calorie} = 1.909 \times 10^{11} \text{ J}.$$

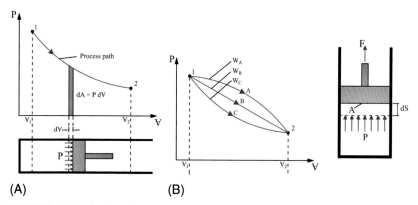

FIGURE 6.4   Moving-boundary work of a piston-cylinder closed system (created by F. Kermanshahi). Note that work is a path-dependent transient process, where a different path leads to a different amount of work. Expansion along Path A leads to more work than along Path B and Path C.

According to Mtakverbia (2021), the average 1-kg durian contains 1350 calories. It follows that 4.56×10^7 calories can be provided by

kilograms of durians $= 4.56 \times 10^7$ calories/1350 calories/kg $= 3.38 \times 10^4$ kg.

In other words, if T-E Kaya feeds on a diet of only durian, without the biological Einstein reactor, he would be consuming over 33 thousand 1-kg durians, over a 50-year period.

With the biological Einstein reactor, the required energy can be obtained from a tiny mass of durian, that is,

$$E = 1.909 \times 10^{11} \text{ J} = mc^2 = m(3 \times 10^8 \text{ m/s})^2.$$

From this, we get m $= 2 \times 10^{-6}$ kg. This tiny amount is enough to convince durian-loving T-E Kaya to never entertain a biological Einstein reactor.

Concerning whether Einstein's Theory of Relativity violates the first law of thermodynamics, the answer is no. In plain English, energy is always conserved, that is, the first law of thermodynamics is valid. The explanation lies with how we look at energy and mass. In a nutshell, Einstein's Theory of Relativity as expressed by $E = mc^2$, says that the quantity of energy and matter in the Universe is conserved. To rephrase it, mass is a form of energy, and the total energy is always conserved. Interested readers are recommended to take a close look at Moskowitz (2014).

6.5   Moving boundary work

Consider a closed system of a freely moving piston with no mass transfer in or out of the cylinder, as shown in Fig. 6.4. The first law of thermodynamics states that energy into the system, E_{in}, minus energy out of the system, E_{out}, is equal to the net change in the energy of the system. Namely:

$$\Delta E_{system} = E_{in} - E_{out}. \tag{6.13}$$

The moving boundary work is the expansion or compression work. Work is needed to compress the fluid (gas) inside the cylinder, leading to a gain in energy of the system. Expansion is work done by the system on the surroundings and, thus, the system loses energy. The moving boundary work is the area under the pressure-versus-volume plot. For an infinitely small change in the displacement of the piston, the differential work done:

$$\delta W_b = Fds = PAds = Pd\forall, \tag{6.14}$$

where F is the force, s is the displacement, and A is the surface area of the piston. Note that δ, instead of Δ, is used for work because work is energy in transition, and it is path-dependent. We can integrate Eq. (6.14) to obtain the total boundary work over the span of the piston movement, that is,

$$W_b = \int Pd\forall. \tag{6.15}$$

As mentioned earlier, this is path-dependent, that is, the area under the pressure-versus-volume plot depends on how pressure varies with volume. There are an infinite number of paths this can take place. Let us look at some of the more common ones.

Example 6.8 Isothermal compression of an ideal gas.

Given: The standard way to experimentally illustrate moving-boundary work is to employ a piston–cylinder system similar to that depicted in Fig. 6.4. For a piston–cylinder system initially containing 0.2 m³ of air at 101 kPa and 20°C.

Find: The required work to compress it to (a) 0.1 m³, and (b) 0.05 m³, isothermally.

Solution: Isothermal denotes constant temperature. We know from experience when pumping air into a ball or a tire, the temperature increases with compression. Therefore, isothermal compression necessitates continuous heat removal in the specific manner that the temperature remains fixed.

The given pressure and temperature conditions are way below the critical pressure and way above the critical temperature. As such, they are favorable for the ideal gas assumption, where

$$P\forall = nR_uT.$$

Here, n is the amount of substance, air in this case, which is a constant for a closed system, R_u is the universal gas constant (8314.3 J/kmol·K or 8.3143 J/mol·K), another constant, and for an isothermal process, T is also a constant. The ideal gas equation can thus be simplified into

$$P\forall = C,$$

where C is a constant.

The boundary work,

$$W_b = \int P d\forall = \int (C/\forall) d\forall = C \ln(\forall_2/\forall_1) = P_1 \forall_1 \ln(\forall_2/\forall_1)$$

(a) compress the 0.2 m^3 of air to 0.1 m^3

Substituting the corresponding values, we get

$$W_b = P_1 \forall_1 \ln (\forall_2/\forall_1) = (101,000)(0.2) \ln (0.1/0.2) = -14 \text{ kJ}.$$

It is negative because it is work done on the system, the standard sign convention in engineering.

(b) compress the 0.2 m^3 of air to 0.05 m^3

Substituting the respective values, we get

$$W_b = P_1 \forall_1 \ln (\forall_2/\forall_1) = (101,000)(0.2) \ln (0.05/0.2) = -28 \text{ kJ}.$$

In other words, it takes another 14 kJ to compress it isothermally from 0.1 m^3 to 0.05 m^3. We see that every halving of the volume requires 14 kJ of boundary work.

The isothermal process is but one of the polytropic processes, that is, $P\forall^k$ = constant. For the more general polytropic process, the boundary work,

$$W_b = \int P d\forall = \int (C\forall^{-k}) d\forall = C(\forall_2^{-k+1} - \forall_1^{-k+1})/(-k+1)$$
$$= (P_2\forall_2 - P_1\forall_1)/(1-k), \tag{6.16}$$

for $k \neq 1$. By invoking the ideal gas law, this can also be expressed as

$$W_b = nR_u(T_2 - T_1)/(1-k). \tag{6.17}$$

Note that when $k = 0, P\forall^0 = P =$ constant, that is, it is an isobaric or constant-pressure process.

If $k = 1$, we have

$$W_b = \int P d\forall = \int (C/\forall) d\forall = C \ln (\forall_2/\forall_1) = P_1 \forall_1 \ln (\forall_2/\forall_1), \tag{6.18}$$

the isothermal process.

When $k = c_P/c_V =$ specific heat at constant pressure with respect to that at constant volume, then, for an ideal gas, it is an adiabatic process.

If $k = \infty$, it is an isochoric or constant-volume process.

6.6 Enthalpy

From chemistry, we learned that enthalpy, H, is the amount of heat released by a chemical reaction such as one mole of methane, CH_4, reacting with two moles of oxygen, O_2, to form one mole of carbon dioxide, CO_2, and two moles of water, H_2O, under a constant-pressure process. We can envision this reaction in a cylindrical chamber with a free-moving piston. The amount of heat extracted by bringing the temperature of the products back down to the initial temperature of the reactants is the enthalpy change from reactants to products. To put it another way, enthalpy is the heat content of a system at constant pressure, and it can be defined as

$$H = U + P\forall. \tag{6.19}$$

Note that U, P, and $\forall$ arc all state functions in the sense that they are not path-dependent. It follows that enthalpy, H, which is explicitly defined by three state functions, is also a state function. With that in mind, we only need to deal with the change of enthalpy from one state to another without worrying about the particular path the process proceeds from one state to the other. Namely:

$$\Delta H = \Delta U + \Delta(P\forall) \tag{6.20}$$

or

$$\Delta H = \Delta U + P\Delta\forall, \tag{6.21}$$

as it is a constant-pressure process.

According to the first law of thermodynamics, energy, $E = P\forall + U + KE + PE$. In the absence of kinetic and potential energy, we have

$$\Delta U = \Delta E - P\Delta\forall = Q + W. \tag{6.22}$$

We note that work is done on the system to compress it, that is, reduce its volume. If the process occurs in a constant-volume manner, such as setting the reaction to take place in a sealed chamber, then $\Delta\forall = 0$ and

$$\Delta U = \Delta E = Q. \tag{6.23}$$

Example 6.9 Enthalpy analysis of a steam turbine.
Given: A steam turbine produces 800 kW. The steam is moving at 2 kg/s with a specific enthalpy at the inlet of 3000 kJ/kg and leaves the turbine with a specific enthalpy of 2500 kJ/kg.
Find: The rate at which the turbine loses heat.

Solution: The first law of thermodynamics can be expressed as

$$\Delta E_{sys} = \Delta U + P\Delta\forall = \Delta H = Q + W.$$

For this case, we can write

$$H_{out} - H_{in} = Q + W.$$

Substituting the values, we get

$$2\,kg/s\,(2500 - 3000)\,kJ/kg = Q + (-800\,kW),$$

noting that work is produced by the turbine and, thus, energy is leaving the system. Rearranging, we get the heat transfer rate, $Q = $ -200 kW. The negative sign denotes that this is the rate at which heat leaves the turbine.

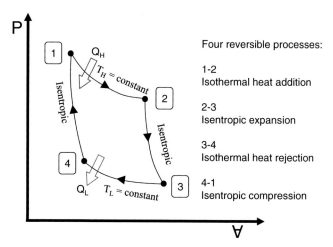

FIGURE 6.5   The Carnot engine cycle (created by Y. Yang). Isothermal heat addition of amount Q_H at the higher temperature, T_H. Adiabatic (Q=0) expansion, decreasing pressure as the volume increases. Isothermal heat rejection of amount Q_L at the lower temperature, T_L. Adiabatic compression process back to its initial state.

6.7   Thermodynamic cycle

A thermodynamic cycle consists of a sequence of heat- and work-transfer processes leading to variation in temperature, pressure, and other state variables, and the sequence returns to the initial state. Fig. 6.5 shows the ideal Carnot engine cycle along with the pressure versus volume, P-∀, plot from which work can be deduced. Process 1 is isothermal heat addition at the high temperature, T_H; the amount of heat added to the system is denoted as Q_H. This is followed by adiabatic (Q=0) expansion, Process 2. Subsequently, in Process 3, Q_L amount of heat is rejected to the surroundings isothermally at the lower temperature, T_L. Process 4 is an adiabatic compression process, bringing the system back to its initial state. Interested readers can refer to Borgnakke and Sonntag (2019), Moran et al. (2018), Reynolds and Colonna (2018), Çengel and Boles (2015), and Balmer (2011).

Example 6.10 Thermodynamic cycle and the first law of thermodynamics.
Given: A closed system undergoes a cycle consisting of four processes.
　　　From state 1 to 2: 4 kJ of heat is added to the system and 8 kJ of work is done by the system.
　　　From state 2 to 3: The system energy increases by 25 kJ adiabatically.
　　　From state 3 to 4: 6 kJ of work is done on the system and the system gains 20 kJ of energy.
　　　From state 4 to 1: The system does 32 kJ of work.
Find:　The respective heat and work entering or leaving the system and the change in the system energy. In other words, fill in the blanks in Table 6.1.

TABLE 6.1 A closed system undergoing a four-process cycle.

Process	Q [kJ]	W [kJ]	ΔE [kJ]
State 1 to 2	4		
State 2 to 3			25
State 3 to 4			
State 4 to 1			
Total			

Solution: Let us adopt the notation that heat or work transfer that leads to an increase in the system energy is positive.

State 1 to 2:

With $Q = 4$ kJ and $W = -8$ kJ, the change in total energy,

$$\Delta E = Q + W = 4 - 8 = -4 \text{ kJ}.$$

In other words, the system loses a net of 4 kJ of energy from State 1 to State 2. Specifically, $_1Q_2 = 4$ kJ, $_1W_2 = -8$ kJ, $\Delta E_{1\text{-}2} = -4$ kJ.

State 2 to 3:

This is an adiabatic and, hence, $Q = 0$.
From the first law of thermodynamics, the change in the total energy,

$$\Delta E = Q + W = 25 \text{ kJ}.$$

In other words, $W = 25$ kJ, that is, 25 kJ of work is added to the system. Namely, $_2Q_3 = 0$, $_2W_3 = 25$ kJ, $\Delta E_{2\text{-}3} = 25$ kJ.

State 3 to 4:

The work done on the system is the energy transferred into the system and for this case, $W = 6$ kJ.
According to the first law of thermodynamics, $\Delta E = Q + W = Q + 6$ kJ $= 20$ kJ. Accordingly, $Q = 14$ kJ.
In short, $_3Q_4 = 14$ kJ, $_3W_4 = 6$ kJ, $\Delta E_{3\text{-}4} = 20$ kJ.

State 4 to 1:

Work done by the system is energy transfer from the system, $W = -32$ kJ.
It follows from the first law of thermodynamics that $\Delta E = Q + W = Q - 32$ kJ.
In short, $_4Q_1 = ?$ kJ, $_4W_1 = -32$ kJ, $\Delta E_{4\text{-}1} = ?$ kJ.

As we close the cycle, returning to state 1, the energy level should return to its original level. In other words, the sum of the changes in energy should be zero, that is,

$$\Delta E = \Delta E_{1-2} + \Delta E_{2-3} + \Delta E_{3-4} + \Delta E_{4-1} = 0.$$

Substituting the values, we have

$$\Delta E = (-4kJ) + (25kJ) + (20kJ) + \Delta E_{4-1} = 0.$$

This gives $\Delta E_{4\text{-}1} = -41$ kJ.

For state 4 to 1,

$$\Delta E_{4-1} = {_4}Q_1 + {_4}W_1 = {_4}Q_1 + (-32 \text{ kJ}) = -41 \text{ kJ}.$$

This results in ${_4}Q_1 = $ -9 kJ.

We see that there is a net heat of 9 kJ entering the system, and a net work of 9 kJ out of the system, in one cycle. If the entire 9 kJ of work were in the net work output, then the system would be a heat engine, which will be discussed in the next chapter, running at a first law efficiency:

$$\eta = W_{out,net}/Q_{in} = 9 \text{ kJ}/9 \text{ kJ} = 100\%.$$

This, by all means, does not happen in reality because of various losses and, of course, according to the second law of thermodynamics, not all the heat input into a heat engine can be converted into work some heat must exit the system as waste heat. These interesting details are left to the next chapter.

Problems

6.1 Time to boil 2 L of water.

How long does it take a 1-kW electric kettle to bring 2 L of water from 20°C to 100°C? Suppose, in reality, it takes 2 min more than the calculated. How much heat is lost?

6.2 Balancing building heat loss with lively souls.

The power grid went out during a winter day and shut down all electric-powered devices, including furnaces. Dr. JAS' resident loses heat at 5 kW. How many students does she have to invite to produce enough heat to make up for the heat lost to the outdoors? Assume each student generates 90 W of heat.

6.3 Harnessing water-jet kinetic energy.

A water jet with a velocity of 20 m/s has a mass flow rate of 70 kg/s. What is the power generation potential in watts?

6.4 River running down a ramp.

A river with a volumetric flow rate of 485 m³/s runs down a ramp with a 68 m elevation difference at 2.7 m/s. How much power can be harnessed?

6.5 Compressed air energy storage.
Unused renewal energy is stored as 100 kg of compressed air at five bars and 20°C. What is the available energy that can be harnessed when the need arises?

6.6 Isentropic compression of air.
A volume of air at 27°C and 101 kPa is compressed isentropically into 1/7th its initial volume. Determine the work done per unit mass of air. Calculate the temperature and pressure after compression.

6.7 Heat flow rate of a boiler.
Water with specific enthalpy of 160 kJ/kg enters a boiler at 2000 kg/h and exits as steam with specific enthalpy of 2400 kJ/kg. What is the required heat input rate?

6.8 Constant-volume versus constant-pressure heat addition.
How much heat is required to bring 2 kg of air from 20°C and 101 kPa to 45°C via a constant-volume process versus a constant-pressure process?

6.9 Cylinder with piston and spring.
A cylinder containing 0.001 m^3 of an ideal gas initially at 150 kPa has a piston on top. Just above the piston is a linear spring restricting upward expansion. The spring touches the piston but exerts no force at the initial condition. The gas is heated until the volume is tripled, and the pressure is 1000 kPa.
Part I) Draw the P-∀ diagram.
Part II) Calculate the work done by the gas. What is the work done against the piston? What is the work done against the spring?
Hint: Pressure, $P = P_{piston} + P_{spring} = P_{piston} + k_{spring}x/A$, where P_{piston} is the pressure exerted by the piston, P_{spring} is the pressure caused by the spring, k_{spring} is the spring constant, A is the area of the piston, and the displacement of the spring, $x = (∀ − ∀_{initial})/A$. In short, $P = P_{piston} + (k_{spring}/A^2) (∀ − ∀_{initial})$.

References

Balmer, R.T., 2011. Modern Engineering Thermodynamics. Academic Press, Oxford.
Borgnakke, C., Sonntag, R.E., 2019. Fundamentals of Thermodynamics, Tenth Edition Wiley, Hoboken, NJ.
Cengel, Y.A., Boles, M.A., 2015. Thermodynamics: An Engineering Approach, 8th ed. McGraw-Hill, New York, NY.
Chen, Y., Yang, X., Luo, D., Wen, R., 2021. Remaining available energy prediction for lithium-ion batteries considering electrothermal effect and energy conversion efficiency. J. Energy Storage 40, 102728.
Deymi-Dashtebayaz, M., Rezapour, M., 2021. The effect of using nanofluid flow into a porous channel in the CPVT under transient solar heat flux based on energy and exergy analysis. J. Therm. Anal. Calorim. 145 (2), 507–521.
EatRight, https://www.eatright.org/food/nutrition/dietary-guidelines-and-myplate/how-many-calories-do-adults-need, (accessed September 24, 2021).

Moran, M.J., Shapiro, H.N., Boettner, D.D., Bailey, M.B., 2018. Fundamentals of Engineering Thermodynamics, Ninth Edition Wiley, Hoboken, NJ.

Moskowitz, C., 2014. Space & physics – fact or fiction?: Energy can neither be created nor destroyed. Is energy always conserved, even in the case of the expanding universe? Sci. Am. https://www.scientificamerican.com/article/energy-can-neither-be-created-nor-destroyed/. (accessed September 24, 2021).

Mtalvernia, https://mtalvernia.sg/education/durians-are-they-good-for-me/, (accessed September 24, 2021).

Reynolds, W.C., Colonna, P., 2018. Thermodynamics: Fundamentals and Engineering Applications. Cambridge University Press, Cambridge.

Temiz, M., Dincer, I., 2021. Design and analysis of nuclear and solar-based energy, food, fuel, and water production system for an indigenous community. J. Cleaner Prod. 314, 127890.

Chapter 7

The second law of thermodynamics

"If someone points out to you that your pet theory of the universe is in disagreement with Maxwell's equations – then so much the worse for Maxwell's equations. If it is found to be contradicted by observation – well, these experimentalists do bungle things sometimes. But if your theory is found to be against the second law of thermodynamics I can give you no hope: there is nothing for it but to collapse in deepest humiliation."

Sir Arthur Stanley Eddington

Chapter Objectives

- Be acquainted with the three laws of thermodynamics.
- Catch on to the notion that energy has both quantity and quality.
- Appreciate the fact that energy naturally flows in the decreasing-quality direction.
- Understand entropy in terms of disorder and that all processes generate entropy.
- Make sense of thermal reservoirs as heat source and heat sink.
- Know what a heat engine is and comprehend the theoretically best Carnot heat engine.
- Recognize and differentiate first-law and second-law efficiencies.
- Be aware of reverse heat engines such as refrigerators, air conditioners, and heat pumps.

Nomenclature

COP   coefficient of performance; $COP_{Carnot,HP}$ is the Carnot coefficient of performance of a heat pump, $COP_{Carnot,Ref}$ is the Carnot coefficient of performance of a refrigerator, COP_{HP} is the coefficient of performance of a heat pump, COP_{Ref} is the coefficient of performance of a refrigerator

E   energy
g   gravity
h   height
m   mass
n   number of moles
P   pressure

Fluid Mechanics from Nature to Engineering. DOI: https://doi.org/10.1016/B978-0-12-900765-3.00007-1
109

PE   potential energy

Q   heat; Q' is the heat transfer rate, Q_H is the heat transfer to or from the high-temperature reservoir, Q_L is the heat transfer to or from the low-temperature reservoir

S   entropy; ΔS is the entropy change, ΔS_H is the entropy change on the higher temperature process, ΔS_L is the entropy change on the lower temperature process, ΔS_{tot} is the total entropy change

T   temperature; ΔT is the temperature change, T_H is the higher temperature, T_L is the lower temperature

W   work; W' is the work transfer rate or power, W_{in} is the work input, W_{net} is the net work output

Greek and other symbols

Δ difference

η efficiency; η_I is the first-law efficiency, η_{II} is the second-law efficiency, $\eta_{thermal}$ is the thermal efficiency, $\eta_{thermal,Carnot}$ is the Carnot thermal efficiency

ρ density

$\forall$ volume

7.1 Introduction

Let us beef up our cognizance of thermodynamics. To recap, thermodynamics is the science of energy and entropy, and the study of energy in its various forms and transformation of energy from one form to another. Other than its quantity and conservation, thermodynamics also deals with the quality of energy. More elementary than these fundamentals is the zeroth law of thermodynamics, see Fig. 7.1.

The zeroth law of thermodynamics. If two bodies are in thermal equilibrium with a third body, they also are in thermal equilibrium with each other. To put it another way, two bodies are in thermal equilibrium if both have the same temperature reading, even if they are not in contact with each other. This enables a physician to use the same temperature sensor (thermometer) for multiple patients. All patients who reach thermal equilibrium with the thermometer at 37°C, plus or minus approximately 0.5°C to account for uncertainties in measurements and allowable hale and hearty variation, are at thermal equilibrium with each other, and achieving this equilibrium is considered a healthy sign.

Consider the three bodies, Body A, Body B, and Body C, as depicted in Fig. 7.1. If a human (Body A) is in thermal equilibrium with a polar bear (Body C) and, in like manner, a panda (Body B) is in thermal equilibrium with the polar bear[1], then the human body is also in thermal equilibrium with the panda. Namely, there will be no heat exchange between the human and the panda when they hug each other.

---

1. A healthy human, polar bear (Whiteman et al., 2015), panda (Panda, 2021), and sea lion (Melero et al., 2015) all have a body temperature of approximately 37°C.

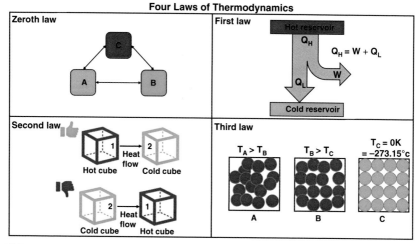

FIGURE 7.1 The four laws of thermodynamics illustrated (created by O. Imafidon). The zeroth law deals with thermal equilibrium. The first law is about the conservation of energy. The second law is concerned with the natural flow direction of energy. According to the third law, the absolute zero temperature corresponds to zero entropy for a perfect crystal.

This principle is more basic than the conservation of energy that constitutes the first law of thermodynamics. On the grounds of this, it is designated as the zeroth law of thermodynamics, in spite of the fact that it was recognized after the first law was discovered and established.

The first law of thermodynamics. Energy cannot be created nor destroyed; it can only transform from one form to another. An approximate everyday analogy is that calorie intake is equal to the extent of useful work performed by the consumer plus the amount of heat generated, assuming there is negligible energy left in the excreted body waste. This is equivalent to a work-producing heat engine, as illustrated in Fig. 7.1. The energy input is heat; the goal of a heat engine is to produce useful work. Due to the second law of thermodynamics, there is always undesirable waste heat. According to the first law of thermodynamics, the quantity of heat in from the hot, or high-temperature, reservoir,

$$|Q_H| = |W| + |Q_L| \tag{7.1}$$

where W is the work produced and Q_L is the waste heat dissipated to the lower temperature environment. Absolute or magnitude signs are employed to emphasize that we are talking about the quantity or amount, as by typical engineering convention, work produced by a system is positive, while the heat produced by the same system is negative. This conventional engineering energy sign designation is confusing, to say the least. For this reason, some suggest "HIP to WIN" to connote "Heat In Positive, Work In Negative" to ease memorization,

Nevertheless, a better convention is to specify all forms of energy added to a system as positive.

The second law of thermodynamics. Actual processes occur in the direction of decreasing the quality of energy. An everyday example is thermal energy flows in the direction of decreasing temperature. Thermal energy from the hot cube shown in Fig. 7.1 moves to the cooler environment, and not the other way round. According to the first law of thermodynamics, the amount of energy loss from the hot cube is equal to that gained by the environment. However, the quality of the energy in the environment is of lower quality. For example, placing the hot side of a thermoelectric generator in a cup of hot coffee can generate some electricity. No electricity is generated by placing the thermoelectric generator in the environment.

Another illustration for the second law of thermodynamics is the movement of air from a balloon to the surroundings when we open the aperture, and not the other way round. The rushing air through the open aperture can spin a turbine, that is, produce work and, thus, is of higher quality than the same air in the surroundings, where no work can be produced.

The third law of thermodynamics. The third law of thermodynamics deals with order or its antonym, disorder, which is quantified in terms of entropy, with respect to order at absolute zero temperature. It may be worded in this manner: the entropy of a pure crystal is zero at absolute zero temperature. To put it another way, a pure crystal is in perfect order at 0 K.

This chapter is dedicated to the second law of thermodynamics. We are going to look at this fascinating law from different angles.

7.2 One-way energy flow

Energy can flow in the decreasing quality direction only; it cannot traverse back in the reverse direction on its own. An everyday parallel is our temporal movement. We can only travel forward in time; that being the case, we do not waste time looking back. Without a functioning time-traveling machine, the best that we can pretend as far as time traveling is concerned is to play a movie backward. We are much better off to embrace the bright reality of the future as asserted by Lyndon B. Johnson, "Yesterday is not ours to recover, but tomorrow is ours to win or lose."

Example 7.1 Dr. JAS' ice cube in a classroom.
Given: At the beginning of her 3-h class, Dr. JAS places an ice cube, at 0°C, on a table in a classroom full of keen minds. The room is at 20°C.
Find: What happens to the ice cube at the end of the class.

Solution: Part of the ice cube would melt by the end of the 3-h class. This is because the students and the classroom are at a higher temperature than the 0°C ice cube. The thermal energy from the students, with a body temperature of 37°C, transfers to the air in the classroom, and part of the thermal energy of the air transfers to the significantly cooler ice cube.

We are reminded that the 0°C ice cube still contains a lot of thermal energy, as it is 273.15 K above absolute zero (0 K). To harness this thermal energy in Dr. JAS' 0°C ice cube, say, to melt another ice cube, it would require work input. This can be realized in practice by placing Dr. JAS' ice cube in a freezer and the other ice cube next to the radiator of the freezer. The point is that electricity is required to power the freezer to transfer some thermal energy from Dr. JAS' ice cube and dissipate it via the radiator onto the other ice cube.

In short, the second law of thermodynamics limits the natural or spontaneous flow of thermal energy from high to low temperature.

One of the most taken-for-granted one-way-energy-flow realities is solar energy. Earth has been blessed by solar energy for more than a few years, yet we have never returned the blessing. Keep in mind that all actual processes occur in the direction of decreasing the quality of energy. This is the second law of thermodynamics. How valid is it? Seth Lloyd assured that "Nothing in life is certain except death, taxes, and the second law of thermodynamics." If you can time travel, maybe you can cheat death and avoid paying taxes; even then, it is unlikely that you are outside the domain encompassed by the second law of thermodynamics. Let us go through another everyday example illustrating the truthfulness of the second law of thermodynamics.

Example 7.2 Rambunctious class versus focused Dr. JAS.

Given: On Valentine's Day, a number of the 350 students in Dr. JAS' class have a little too much sugar and are a little overexcited. Finding the rambunctious students distracting those who are trying to grasp the second law of thermodynamics, Dr. JAS stops her lecture, and poses a challenge to those who are high on sugar. She effortlessly lifts up her 1-kg thermodynamics textbook to 0.5-m above the table and drops it back onto the table. She then asks the hyperactive students to harness the noise they are making to lift up the book.

Find: The amount energy needed to lift the 1-kg book 0.5-m above the table.

Solution: The potential energy of the 1-kg book at 0.5-m above the table is

$$PE = mgh = (1\,kg)(9.81\,m/s^2)(0.5\,m) = 4.9\,kg \cdot m^2/s^2, \text{ or } 4.9\,J$$

In reality, somewhat more energy than this is needed to account for losses such as drag when lifting up the book. The students can generate on the order of 120 dB or 1 W of sound power. Even if they try really hard for an entire minute, they will not be able to lift the book by 1 cm. In 1 min of 1 W sound, there is 60 J of energy. This is more than ten times the focused (high quality) energy Dr. JAS utilizes to lift the book. Thus, it is clear that the random noise is of a much lower quality of energy than the well-organized energy Dr. JAS uses to lift the book.

FIGURE 7.2 Decreasing order and, thus, increasing entropy from solid to liquid to gaseous phase (created by X. Wang). In the solid phase, the H_2O molecules are rigidly confined within the ice cubes. They are free to move around within the spread of the water in the pan when the ice cubes melt. Boiling the water into steam grants full freedom to the H_2O molecules, allowing them to roam everywhere into the environment.

We can also start from the state when the book is at 0.5 m above the table. That 4.9 J of potential energy is of higher quality than the resulting vibrations and noise generated upon the book hitting the table. As such, the energy does not flow in the reverse direction, that is, from noise and vibrations to lifting the book back up to 0.5 m above the table. A machine can be engineered to produce vibrations and noise specifically aimed to bounce the book up to 0.5 m off the table. However, considerably more energy than the 4.9 J potential energy is required, even with the best book-lifting machine based on vibrations and noise. The second law of thermodynamics wins again.

The second law of thermodynamics asserts that processes occur in a certain direction and that energy has quality in addition to quantity. A process will not occur unless it satisfies both the first and second laws of thermodynamics. The second law of thermodynamics states that all processes proceed in the decreasing-energy-quality direction. The direction of decreasing energy quality is marked with increasing entropy. But what is entropy?

7.3 Entropy

Entropy is a thermodynamic property that signifies (molecular) disorder. We can state the second law of thermodynamics in terms of entropy. In an isolated system, natural processes proceed spontaneously when they lead to an increase in entropy of the entire[2] isolated system. Concerning order, Fig. 7.2 shows

---

2. It is important to highlight "the entire isolated system" because the entropy of a subsystem can decrease at the expense of a larger increase in entropy outside the subsystem. All players must be included in the system under study.

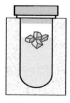

FIGURE 7.3   An isolated system of ice cubes in hot water (created by S. Zinati). The entropy of the entire system increases as the hot water loses thermal energy to melt the ice cubes.

10 cubes of ice placed in a frying pan. In the solid phase, these ice cubes are quite orderly. All the H_2O molecules are held rigidly together within the ice cubes, they do not move around. These molecules start to move as the ice cubes melt, but they stay within the spread of liquid water in the pan. To that end, the H_2O molecules in the liquid phase are less orderly, compared to their initial rigid solid-ice phase. Turning on the burner will lead to progressively more disordered water. The liquid-phase water reaches its maximum-disorder stage when it starts to boil off, at which point liquid water transforms into the substantially more disordered vapor phase.

The underlying thermodynamic property for the ice-to-water-to-steam transition is temperature. The higher the temperature, the more disordered the molecules are[3]. Conversely, the lower the temperature, the more orderly the molecules are. The third law of thermodynamics states that, as the temperature approaches absolute zero, the entropy of a pure substance approaches zero. A pure crystal is perfectly ordered, with an entropy of zero, at absolute zero, that is, zero degrees Kelvin.

In terms of thermal energy, the second law constrains the natural flow of heat from higher to lower temperature. This is interesting, as a higher temperature implies more energetic molecules and, hence, indicates more disorder. Manifestly, the natural direction of heat flows from the more energetic, and apparently disordered, region to the less energetic, more ordered, one. At the same time, all processes proceed from order to disorder, according to the second law of thermodynamics. While heat flow may seem to contradict the second law in terms of apparently moving from disorder to order, this is not the case. The system under consideration, as portrayed in Fig. 7.3, consists of a container of hot water with ice cubes. The total entropy of this system increases as the heat is transferred from the hot water to the ice cubes to melt the ice cubes. When all the ice cubes are melted and become homogeneous with the water, the system is more disordered, as a whole, because all the molecules are moving

---

3. Confinement also plays a role, as far as disorder is concerned. Enclosing the (gas) molecules in a small volume would limit them from moving far apart from each other. As such, confinement limits or reduces disorder.

around, as compared to the initial stage where ice cubes are orderly in their positions, with relatively little room left for the molecules of the hot water to roam around. This is true notwithstanding the fact that the molecules of the initial portion of hot water were more disordered due to their higher initial temperature. Note that the initial liquid volume is smaller and, more importantly, the initially ordered ice cubes become substantially more disordered after melting. As a result, the entropy of the hot water-ice cubes system increases as the ice cubes melt.

Example 7.3 Entropy in a nutshell.

Given: Entropy is a thermodynamic property that measures the amount of molecular disorder.

Find: *Part I.* Relative entropy value for a) a volume of gas at a high temperature, b) a solid at a low temperature, and c) molecules arranged in perfect order, in general terms.

Part II. If entropy can be destroyed.

Part III. If the entropy of a system can be increased and decreased.

Solution:

Part I.

A volume of gas at a high temperature has a high degree of molecular disorder. Therefore, it has a high entropy value.

A solid at a low temperature has a low degree of molecular disorder. That being so, a solid at a low temperature has a low entropy value.

Molecules arranged in perfect order signify an ideal crystal at 0 K, which has an entropy value of zero.

Part II.

Entropy can only be produced. This is the reason why some scientists theorize that the universe started from a singularity, in other words, perfect order, and it has since been expanding, that is, becoming more disordered.

Part III.

All processes result in entropy generation or increase. Transporting some entropy out of a system can lead to a decrease in the entropy of the system at a cost of an increase in entropy outside the system that is larger than the decrease inside the system. To come to the point, the answer is yes, we can increase or decrease the entropy of a system, but the total entropy, when considering all players involved, is always increasing.

How can we reduce the entropy of a system? Consider a completely sealed pressure cooker containing boiling water. Since it is completely sealed, no molecules escape the pressure cooker; neither can any molecules enter it. Heat, on the other hand, can escape through the wall of the pressure cooker. In doing so, the temperature inside the pressure cooker and, thus, also the disorder, decreases as some steam condenses to liquid water. That being the case, the entropy of the pressure cooker system decreases as it cools down. This decrease is due to the removal of entropy with heat transferring out of the pressure cooker. Under those circumstances, the entropy of the system, the pressure cooker, decreases at the expense of an increase in entropy outside the pressure cooker. Over and above that, the increase in entropy in the surroundings is larger than the decrease in the pressure cooker. In plain English, the total entropy increases in the process of decreasing the entropy of the pressure cooker. Let us go through one more example on this to illustrate, beyond a shadow of doubt, that entropy always increases. Sir Arthur Stanley Eddington rightly put it, "The law that entropy always increases—the Second Law of Thermodynamics—holds, I think, the supreme position among the laws of Nature. If someone points out to you that your pet theory of the universe is in disagreement with Maxwell's equations— then so much the worse for Maxwell's equations. If it is found to be contradicted by observation—well, these experimentalists do bungle things sometimes. But if your theory is found to be against the second law of thermodynamics I can give you no hope; there is nothing for it but to collapse in deepest humiliation."

Example 7.4 The entropy of T–E Kaya's room decreases.

Given: T–E Kaya is all pumped up after Dr. JAS' lecture on entropy. He goes home cantillating Stephen Hawking's quote, "You may see a cup of tea fall off a table and break into pieces on the floor… But you will never see the cup gather itself back together and jump back on the table. The increase of disorder, or entropy, is what distinguishes the past from the future, giving a direction to time." T–E Kaya is pleasantly surprised when stepping into his room and seeing that everything is clean and organized. He takes a few photos and finds some older photos showing the previously messy room of his and he is ready to prove to Dr. JAS that she is wrong concerning the one-way street of entropy of increasing entropy.

Find: If T–E Kaya is right.

Solution: It is true that the entropy of T–E Kaya's room has decreased. This is only because his mother spent the entire day cleaning and organizing. The decrease in entropy of T–E Kaya's room is realized at the cost of a larger increase in entropy executed by his mother. The organized, or orderly, almonds, cashews, hazenuts, and oats that T–E Kaya's mother consumed for breakfast are broken down and become disordered to provide the energy for her to clean the messy room. The resulting increase in entropy is larger than the amount of entropy decrease in T–E Kaya's room. The total, or net, entropy has gone up. It takes honesty and humility to accept that we can never win when it comes to entropy!

7.4 Heat source and sink

Before we proceed further, let us define the terms heat source and heat sink. These are essential concepts and components for most engineering and natural processes. Heat sources and heat sinks are thermal energy reservoirs. A thermal energy reservoir is a body with a large thermal energy capacity (mass times specific heat) that can supply or receive large amounts of heat without undergoing any change in temperature.

A heat source is a thermal energy reservoir at a high temperature that supplies heat without decreasing its temperature. The sun is a thermal energy reservoir. The magma chamber, a large pool of liquid rock beneath the surface of the Earth, is a thermal energy reservoir. As a heat source, the magma chamber continuously supplies heat to the many hot springs in Iceland, for example.

A heat sink is a thermal energy reservoir that receives heat without increasing its temperature. Atmospheric air is possibly the most familiar heat sink, where waste heat from engineering processes such as power plants and computers is dumped. It is also the heat sink where all metabolizing creatures dissipate their heat. Other common heat sinks include oceans, lakes, or rivers. These bodies of water are excellent heat sinks for nuclear power plants. When it comes to the second law of thermodynamics, a heat sink is the most common dumpster for entropy.

7.5   Heat engine

Work can be converted to heat directly and completely. An everyday example is turning on an electric heater, resulting in every joule, including losses in the wire, ending up as heat. On the other hand, converting heat to work requires the use of some special devices called heat engines. The most familiar heat engine is possibly an internal combustion engine. The heat generated from the chemical reaction is used to power the nonelectric cars we drive.

Example 7.5 Worked-up into heat.
Given: Dr. JAS gives a test on the second law of thermodynamics in a 7-m long by 5-m wide by 3-m high classroom. Many students find the topic and the test challenging. One student, who lacks self-control, throws a temper tantrum for a good 5 min.
Find: What happens to the worked-up work of the hotheaded student. Where the heat goes.

Solution: As there is no useful work produced when we lose our cool, all the worked-up energy is converted to waste heat directly and completely.

Where does the heat go? It goes into the classroom (air). To raise the temperature of the room by 0.5°C so that some of the more thermally sensitive individuals can sense it would require

$$E = mc_P\Delta T = \rho \forall c_P \Delta T = 1.2(7)(5)(3)(1000)(0.5) = 63{,}000\,J,$$

where E is energy, m is the mass of air in the room, c_P is the specific heat at a constant pressure of air, ΔT is the temperature change, ρ is the density of air, and $\forall$ is the volume of air.

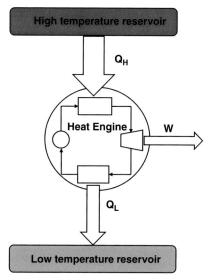

FIGURE 7.4 A heat engine (created by O. Imafidon). A portion of the heat from the high-temperature reservoir, Q_H, is converted into useful work, W, while the remaining unused heat is dissipated into the low-temperature reservoir as waste heat, Q_L.

A typical adult can generate over 300 W when undergoing very intense activity. If the hothead is in a frenzy for 5 min, the person can generate 300 W times 5 min time 60 s/min, that is, 90,000 J of heat. This will raise the room temperature by 0.7°C. To put it another way, the calculations indicate that, without ventilation and air conditioning, the heat dissipated by just one hothead can be felt by the rest of the class.

This is a real-life example illustrating work can be converted directly and completely into waste heat. Note that the other students also generate about 100 W or 30,000 J of heat during those 5 min, while harnessing most of their metabolic energy (power) for constructive work.

A heat engine is an engineering system that converts heat to work. As depicted in Fig. 7.4, a good heat engine receives heat from a high-temperature source and converts as much of the thermal energy to work as possible. The remaining thermal energy that is not converted to work is rejected as waste heat to a low-temperature sink.

A steam power plant, as illustrated in Fig. 7.5, is a heat engine. The high-temperature source is the combustion chamber, which is not part of the portrayed system. The heat produced by the combustion process is supplied at a rate Q_H' to the heat engine indicated by the dashed line. The heat-receiving component of the heat engine, the boiler, receives this heat, producing steam to spin the work-producing component, the turbine, which produces useful power, W_{net}'. The unused thermal energy is dissipated at a rate Q_L' into the atmosphere via

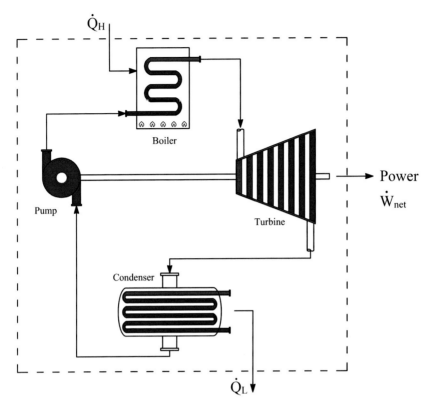

FIGURE 7.5 A steam power plant is a heat engine (created by F. Kermanshahi). High-temperature heat from a high-temperature reservoir enters the system (heat engine) at Q_H'. Steam produced by the boiler spins the turbine, leading to a net useful power output of W_{net}'; this is the net power after a portion of it is employed to operate the pump. The remaining unused heat is dissipated into the low-temperature reservoir, at a rate Q_L', via the condenser.

the heat-rejecting component, the condenser. It is clear that it is better if less of the received energy goes into waste heat. To better illustrate this point, let us first define the thermal efficiency of a heat engine.

Example 7.6 Thermoelectric generation heat engine.

Given: Thermoelectric generators such as thermoelectric fans are some of the simplest heat engines. Consider the thermoelectric fan shown in Fig. 7.6. Thermal energy from the higher-temperature heat source at T_H enters the heat engine as Q_H. Part of the energy is harnessed to perform work, W, by turning the fan. The remaining thermal energy leaves the heat engine at Q_L, it is dissipated into the lower-temperature heat sink at T_L.

Find: Illustrate a) the first law of thermodynamics and b) the second law of thermodynamics in terms of increasing entropy and one-way energy flow, that is, heat flows from high to low temperature.

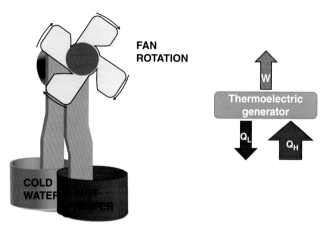

FIGURE 7.6 A thermoelectric generator as a heat engine (created by O. Imafidon). Thermal energy enters the system from the heat source at T_H via the hot leg. Part of this energy is harnessed to rotate the fan. The remaining untapped thermal energy leaves the system, via the cold leg, into the heat sink at T_L.

Solution: According to the first law of thermodynamics, the amount of energy entering the heat engine is equal to that exiting it, under steady-state operation. For the thermoelectric generator depicted in Fig. 7.6, the conservation of energy can be expressed as

$$Q_H = W + Q_L,$$

which is equally valid when expressed in rate form, $Q_H' = W' + Q_L'$. Since work, W, is positive or greater than zero, that is, $W > 0$, for spinning the fan, it follows that

$$Q_H > Q_L.$$

According to the second law of thermodynamics, the total entropy always increases in any process. For the thermoelectric generator shown in Fig. 7.6, the change in total entropy:

$$\Delta S_{tot} = \Delta S_H + \Delta S_L \geq 0,$$

where the total change in entropy on the hot and cold sides is equal to zero in the non-existing, ideal reversible process. To put it another way, the total change in entropy, ΔS_{tot}, is always greater than zero in a real process such as the thermoelectric fan under consideration. It is also worth mentioning that work is entropy-free energy, as entropy is a measure of how much energy is not available to do work. Substituting the respective thermal energy and temperature into the total entropy equation, we get

$$\Delta S_{tot} = \Delta S_H + \Delta S_L = Q_H/T_H - Q_L/T_L > 0.$$

Implicitly, we have assumed Q_H/T_H is entering, while Q_L/T_L is leaving the system (heat engine) and, therefore, Q_L/T_L is negative. We can rearrange the above equation into

$$Q_H/T_H > Q_L/T_L$$

or

$$Q_H/Q_L > T_H/T_L.$$

From the first law analysis, we have $Q_H > Q_L$, thereby, $T_H > T_L$. As you can see, it follows that heat goes from high to low temperature.

Thermal efficiency. Efficiency is the desirable output divided by the required input. For a heat engine, a measure of its performance is the thermal efficiency. The thermal efficiency is the amount of desirable power produced, W', with respect to the required thermal energy input rate, Q_H'. Namely,

$$\eta_{thermal} = W'/Q_H' = (Q_H' - Q_L')/Q_H' = 1 - Q_L'/Q_H'. \qquad (7.2)$$

It is clear that the smaller the waste heat or the waste heat rejection rate, Q_L', the better the heat engine. We have been enlightened that, as a consequent of the second law of thermodynamics, waste heat cannot be zero. The question then is what is the maximum thermal efficiency that we can possibly strive for? Before we answer that question, let us illustrate the thermal efficiency using a numerical example.

Example 7.7 Internal combustion engine thermal efficiency.
Given: An internal combustion engine can be modeled as a heat engine receiving heat from a high-temperature reservoir at 200 kW and rejecting waste heat at 130 kW.
Find: The thermal efficiency.

Solution: The thermal efficiency,

$$\eta_{thermal} = 1 - Q_L'/Q_H' = 1 - 130/200 = 35\%.$$

We see that around 65% of the energy associated with the fuel goes into waste heat. For this reason, much research effort continues into improving the thermal efficiency and recovering part of the waste heat for useful work. In cold weather, a portion of the waste heat is used to keep the driver and passengers from freezing.

Carnot heat engine. Nicolas Leonard Sadi Carnot [1796–1832] is credited with the discovery of the ideal heat engine, called Carnot heat engine. Imagine trying to capture all the kinetic energy that approaches a wind turbine. To do so, we would like to suck out every single drop of kinetic energy. There is, however, a problem with this; for, without any kinetic energy left, the air cannot leave the turbine. Keeping that in mind, we see that it is impossible for all the kinetic energy to be harnessed; some kinetic energy must leave the system as moving air. Then, what is the theoretical best? For a heat engine, this theoretical best is

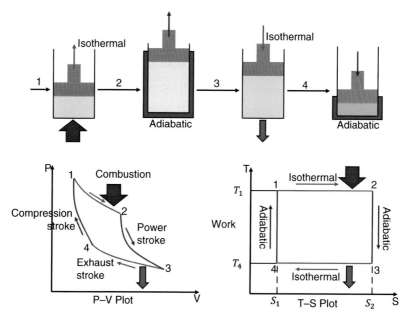

FIGURE 7.7   A Carnot heat engine (created by O. Imafidon). It is an ideal, or theoretically most-efficient, engine, setting the limit for real engines to strive for.

the Carnot engine cycle, as depicted in Fig. 7.7. The Carnot engine cycle is a thermodynamic cycle. A (closed) thermodynamic cycle consists of a sequence of thermodynamic processes involving transfer of heat and work into and out of the system. The working fluid undergoes changes in temperature and pressure, and the cycle ends when the system returns to its initial state.

The Carnot cycle shown in Fig. 7.7 is a reversible cycle, with absolutely no losses due to friction, etc. In this ideal case, the working fluid is an ideal gas, which can be described mathematically as

$$P\forall = nRT, \tag{7.3}$$

where P is the pressure, $\forall$ is the volume, n is the number of molecules, R is the gas constant, and T is the temperature in the absolute scale. For a given number of molecules or amount of a given gas such as air, we see that

$$P\forall/T = nR = \text{constant.} \tag{7.4}$$

This can be expressed as

$$P_1\forall_1/T_1 = P_2\forall_2/T_2, \tag{7.5}$$

where subscripts 1 and 2 denote State 1 and State 2, respectively.

The four reversible processes of the Carnot cycle are:

Processes 1–2: Reversible Isothermal Expansion. Heat from the high-temperature source is added into the system at a constant temperature T_H. As the engine cylinder receives the heat, the gas expands at a constant temperature. With temperature fixed at T_H, the ideal gas equation, as described by Eq. (7.5), becomes $P_1 \forall_1 = P_2 \forall_2$. The volume of the cylinder increases as the gas expands and, it follows that, the in-cylinder pressure decreases. As a consequence of that, the system produces work and the amount of work done by the system is equal to the area under the P-$\forall$ curve. According to the first law of thermodynamics, the change in internal energy is equal to the amount of heat added to the system minus the amount of work done by the system. Specifically,

$$\Delta U = Q - W_{1-2}. \qquad (7.6)$$

For the constant-temperature process involving an ideal gas, the kinetic energy and, thus, the internal energy remains unchanged. That being the case, we have the work done by the system, W_{1-2}, is equal to the heat transferred into the system, Q. Furthermore, from the temperature versus entropy (T–S) plot, the entropy of the system increases because entropy from the high-temperature source enters the system via heat transfer and the expanded cylinder gives room for the gas molecules to become more disorderly or spread out.

Processes 2–3: Isentropic (Reversible and Adiabatic) Expansion. The gas expands without any heat exchange with the surroundings. The volume increases and, according to $P_2 \forall_2 = P_3 \forall_3$, the cylinder pressure decreases. The corresponding work produced is the area under this segment of the P-$\forall$ curve. Note from the temperature versus entropy plot that this process amounts to no net entropy change. While the expansion makes room for the gas molecules to roam, energy is lost via work output and, hence, the molecules become less energetic. This reversible adiabatic process proceeds along the path of constant entropy or zero net entropy change.

Processed 3–4: Reversible Isothermal Compression. Heat transfers out of the system to the low-temperature sink at a constant temperature T_L. This process is opposite to Processes 1–2, but it occurs at a lower temperature and pressure. The work performed on the system is described by the area under the corresponding P-$\forall$ curve. Similar to Processed 1–2, this work is equal to the amount of heat removed from the system. From the temperature versus entropy plot, the entropy of the system decreases because some entropy leaves the system with thermal energy into the low-temperature sink. The compressed cylinder keeps the less energetic gas molecules less spread out, that is, more orderly. The amount of entropy lost to the low-temperature sink is equal to that gained from the high-temperature source in Processes 2–3. As a result, the net change in entropy is zero.

Processes 4–1: Isentropic (Reversible Adiabatic) Compression. Work is done on the system by pushing the piston down, compressing the air and, thus, causing its temperature to rise back to T_H. While the rising temperature causes the gas molecules to be more energetic and chaotic, the decreasing volume counters this to keep the process along a constant entropy path.

So, what is the maximum thermal efficiency? The Carnot efficiency is the highest efficiency. For a heat engine operating between the high and low-temperature reservoirs at T_H and T_L, we have

$$\eta_{thermal,Carnot} = 1 - Q_L/Q_H = 1 - T_L/T_H, \tag{7.7}$$

where the temperatures, T_L and T_H, are in the absolute scale, i.e., in degrees Kelvin. The second law efficiency can be defined as the actual thermal efficiency over the corresponding Carnot efficiency.

Example 7.8 Two heat engines of equal thermal efficiency.
Given: Heat Engine I receives thermal energy from a high-temperature reservoir
       at 1000 K and dumps waste heat to a low-temperature reservoir at 300 K.
       Heat Engine II acquires heat from a 2000-K heat source and dissipates
       waste heat into the same 300-K heat sink. The thermal efficiency of both
       engines is 35%.
Find: If there is any difference in performance between the two heat engines.

Solution: The first-law thermal efficiency can be expressed in terms of the thermal energy gained from the high-temperature heat source, Q_H, and that lost to the low-temperature heat sink, Q_L,

$$\eta_{thermal} = 1 - Q_L/Q_H = 1 - Q_L'/Q_H'.$$

The last two terms represent the corresponding heat transfer rates, as in practice we typically deal with power in watts rather than energy in joules.

The highest possible efficiency is the Carnot efficiency, and it can be expressed in terms of the temperature, in absolute scale, of the high-temperature heat source, T_H, and the low-temperature heat sink, T_L, as

$$\eta_{thermal,Carnot} = 1 - T_L/T_H.$$

For Heat Engine I, the thermal efficiency,

$$\eta_{thermal} = 1 - Q_L/Q_H = 35\%.$$

The corresponding Carnot, or reversible, efficiency,

$$\eta_{thermal,Carnot} = 1 - T_L/T_H = 1 - 300/1000 = 70\%.$$

The second-law efficiency is thus

$$\eta_{thermal}/\eta_{thermal,Carnot} = 35\%/70\% = 50\%.$$

For Heat Engine II, the thermal efficiency,

$$\eta_{thermal} = 1 - Q_L/Q_H = 35\%.$$

The corresponding Carnot, or reversible, efficiency,

$$\eta_{thermal,Carnot} = 1 - T_L/T_H = 1 - 300/2000 = 85\%.$$

Accordingly, the second-law efficiency is

$$\eta_{thermal}/\eta_{thermal,Carnot} = 35\%/85\% = 41\%.$$

We see that the performance of Heat Engine I is better than that of Heat Engine II. This is because there is more and/or better resource in terms of thermal energy available for Heat Engine II, but it is not being tapped into as well as Heat Engine I. This example illustrates that the first law efficiency alone does not tell the whole story.

Another way to delineate the difference between the first-law efficiency and the second-law efficiency is to look at the delivery of two able graduates. Suppose both T–E Kaya and T–E Rajin went through rigorous entrepreneurship training under the able guidance of Dr. JAS. Upon completion, T–E Kaya convinced Dr. JAS to invest three million dollars in his startup company while T–E Rajin only asked for one million. Over the course of twelve years after the two companies are founded, Dr. JAS reaped six hundred thousand dollars in returns from each company. How do we define and differentiate the efficiencies of these two graduates? Let us use Example 7.9 to elucidate this.

Example 7.9 Two entrepreneurs making an equal annual return.
Given: T–E Kaya starts his company with $3,000,000 capital. Dr. JAS reaps $600,000 over the course of 12 years or, on average, $50,000 per year. T–E Rajin starts his company with $1,000,000 capital. Dr. JAS makes $600,000 over the course of 12 years or, on average, $50,000 per year.
Find: The analogous first and second law efficiencies.

Solution: The first-law efficiency can be viewed as looking at these handsome returns that Dr. JAS received from the two companies. Both T–E Kaya and T–E Rajin have the same first-law efficiency, that is,

$$\eta_I = [\$50,000/year](12\ years) = \$600,000.$$

We could probably use $2,000,000 valuation for both companies as a base to normalize this, that is,

$$\eta_I = \$600,000/\$2,000,000 = 30\%.$$

The second-law efficiency is "To whom much is given, much will be required" [Luke 12: 48]. T–E Kaya's second law efficiency is

$$\eta_{II} = \$50,000/year\ (12\ years)/\$3,000,000 = 20\%.$$

For T–E Rajin, the second law efficiency is

$$\eta_{II} = \$50,000/year(12\ years)/\$1,000,000 = 60\%.$$

The point is that we cannot just look at the output, delivery, or the first-law efficiency alone. The first-law efficiency is essential, but it does not furnish the full picture. We must also evaluate the desired output with respect to the resource available or given.

7.6 Reverse heat engines

A reverse heat engine is a heat engine that operates in the reverse direction, thermodynamically speaking, as depicted in Fig. 7.8. Heat pumps, refrigerators, and air conditioners are common reverse heat engines. For a heat engine, the desired output is work and, thus, heat engines are work-producing devices. Reverse heat engines require work input and, hence, are work-consuming devices.

Refrigerators and air conditioners. The desired output for refrigerators and air conditioners is the heat removal from the lower-temperature thermal space, Q_L. The heat from a cooler space, along with the required work input, is dumped into the warmer surroundings, the high-temperature reservoir. For refrigerators, the lower-temperature region is commonly a compartment keeping dead things frozen. The lower-temperature space associated with air conditioning is generally much milder, keeping living beings thermally comfortable. With the change in the desired output and the required input with respect to that of a heat engine, we need a more suitable performance indicator. This performance indicator is the coefficient of performance:

$$COP_{Ref} = Q_L/W_{in} = Q_L{'}/W_{in}{'}, \tag{7.8}$$

where W_{in} is the required work input to operate the reverse heat engine, that is, the refrigerator or air conditioner. A Carnot refrigerator operating reversibly would give the ideal, or highest, COP, that is,

$$COP_{Carnot,Ref} = Q_L/W_{in} = T_L/(T_H - T_L). \tag{7.9}$$

We are reminded that the temperatures must be in the absolute scale, that is, in degrees Kelvin.

Heat pump. The desired output of a heat pump is Q_H, the heat transferred into the higher-temperature space. This heat is the sum of heat from a lower-temperature reservoir and the required work input to transfer heat from lower to higher temperature. The corresponding coefficient of performance for the heat pump is

$$COP_{HP} = Q_H/W_{in} = Q_H{'}/W_{in}{'}. \tag{7.10}$$

The corresponding Carnot heat pump provides the limit for best performance, where the ideal coefficient of performance can be expressed as

$$COP_{Carnot,HP} = Q_H/W_{in} = T_H/(T_H - T_L). \tag{7.11}$$

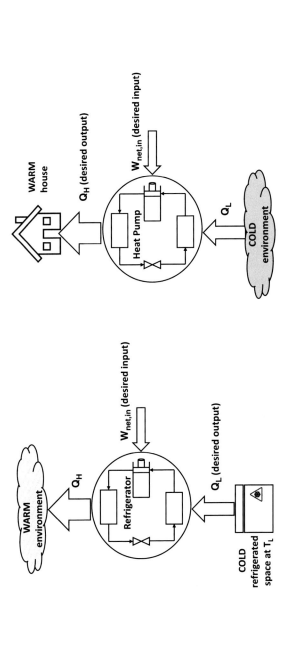

FIGURE 7.8 A reverse heat engine (created by O. Imafidon). Work input, W_{in}, is required to move heat from a lower-temperature region at T_L to a higher-temperature zone at T_H. For heat pumps, the desired output is Q_H, the heat for warming up a zone that is at a higher temperature than the region where the thermal energy is extracted from. On the other hand, the thermal energy removal from a cooler region Q_L is the desirable output for refrigerators and air conditioners, and the dumping of this heat into a higher-temperature reservoir is made possible via a reverse heat engine with appropriate work input.

Note that the coefficient of heat pump is equal to that of a refrigerator or air conditioner plus one, that is,

$$COP_{HP} = COP_{Ref} + 1. \tag{7.12}$$

Let us look at an example concerning a heat pump. As it is for heating purposes, we will cast it in comparison to the familiar electric-resistance heater.

Example 7.10 Resistance heater versus heat pump.

Given: Dr. JAS bought a cottage in West Virginia that is not equipped with a heating system. Wishing to spend some quality time at the cottage with her family during the Christmas-New Year holiday season, she called a couple of her former students who are in space-heating business. T–E Kaya enthusiastically recommends their top-of-the-line 100% electric-resistance heater. After checking out the weather for the location, T–E Rajin suggests a heat pump, claiming that it is superior to the best electric-resistance heater. The average temperature at the location during the Christmas-New Year time is $-5°C$, and Dr. JAS would like the indoor temperature to be around $20°C$.

Find: The appropriate heating system.

Solution: The first-law efficiency of the electric-resistance heater is 100%. In plain English, all electricity input into the heater ends up as heat inside the cottage. In terms of coefficient of performance, it is

$$COP = Q_H/W_{in} = 1.$$

For a heat pump operating between low temperature T_L and high temperature T_H, the ideal Carnot efficiency is

$$COP_{Carnot,HP} = Q_H/W_{in} = T_H/(T_H - T_L)$$
$$= 293.15/(293.15 - 268.15) = 11.7$$

A real heat pump working at a realistic 70% of Carnot efficiency would result in a COP of 8.2. This is more than eight times what the best resistance heater can offer!

To bring this example into context with the second law of thermodynamics, the second-law efficiency of the resistance heater can be defined as the actual coefficient of performance with respect to the ideal Carnot coefficient of performance. Specifically,

$$\eta_{II} = COP/COP_{Carnot} = 1/11.7 = 0.085 = 8.5\%.$$

In plain English, the best electric-resistance heater can perform less than 9% that of the ideal Carnot heat pump at the given conditions, that is, at the particular high and low temperatures.

If the outdoor temperature drops to $-30°C$, the corresponding Carnot heat pump coefficient of performance becomes

$$COP_{Carnot,HP} = Q_H/W_{in} = T_H/(T_H - T_L) = 293.15/(293.15 - 243.15) = 5.0$$

For a real heat pump with 70% Carnot efficiency, the coefficient of performance is 4.1. We see that the same heat pump transferring thermal energy from T_L of $-30°C$ is only 50% as efficient as when it is moving thermal energy from T_L of $-5°C$. To put it another way, the performance of heat pump deteriorates significantly as the temperature difference between the heating space and the lower-temperature thermal energy source increases. For this reason, heat pumps are not economically appealing in colder regions such as Canada as compared to her southern neighbor with more moderate winter.

We have introduced the second law of thermodynamics in this chapter. Its importance has been illustrated with familiar examples. Interested readers can continue this enlightenment by referring to standard thermodynamic textbooks such as Borgnakke and Sonntag (2019), Moran et al. (2018), Çengel and Boles (2015), and Balmer (2011). Note that entropy is a critical parameter when it comes to mitigating environmental damage. While everything we do, including every breath we take, contributes to disorder, we can make a big difference when we exercise entropy minimization; see, for example, Ziya Sogut (2021). Earthquakes can also be better understood from the second law of thermodynamics perspective (Posadas et al., 2021).

Problems

7.1 Heat engine and first law of thermodynamics.
For a heat engine, which of the following describes the first law of thermodynamics? Recall that Q_H is the heat input to the heat engine from the high-temperature reservoir, Q_L is the waste heat dumped into the low-temperature reservoir, and W is work output.
A) $|Q_H| = |Q_L| + |W|$
B) $|Q_H| + |Q_L| = |W|$
C) $|Q_L| > 0$
D) $|Q_L| < 0$

7.2 Carnot cycle.
The ideal Carnot cycle consists of
A) two isothermal and two adiabatic processes
B) two constant pressure and two adiabatic processes
C) two constant volume and two adiabatic processes
D) one constant pressure, one constant volume and two adiabatic processes

7.3 Total entropy change.
During a process, a system undergoes zero entropy change. The corresponding change in the total entropy of the system and the surroundings is
A) zero
B) greater than or equal to zero
C) less than or equal to zero

7.4 Heat engine and second law of thermodynamics.
Which of the following describes the second law of thermodynamics concerning a heat engine? Note that Q_H is the thermal energy supplied by

the high-temperature reservoir, Q_L is the waste heat dumped into the low-temperature reservoir, and W is work output.

A) $|Q_H| = |Q_L| + |W|$

B) $|Q_H| + |Q_L| = |W|$

C) $|Q_L| > 0$

D) $|Q_L| < 0$

7.5 Reverse heat engine and the second law of thermodynamics.

Which of the following describes the second law thermodynamics concerning a reverse heat engine?

A) $|Q_H| > 0$

B) $|Q_L| > 0$

C) $|W| > 0$

7.6 Heat source temperature and the second law of thermodynamics.

Both Engines A and B have the same first-law efficiency, say 37%. The high-temperature source associated with Engine A is at a higher temperature. What can be said about their second-law efficiency, if they have a common low-temperature heat sink?

A) Engine A has the same second-law efficiency as Engine B

B) Engine A has a lower second-law efficiency than Engine B

C) Engine A has a higher second-law efficiency than Engine B

7.7 Heat sink temperature and the second law of thermodynamics.

Both Engines A and B have the same first-law efficiency, say 37%. The environment (low-temperature heat sink) that Engine A is in is at a lower temperature. What can be said about their second-law efficiency, if they have a common high-temperature source?

A) Engine A has the same second-law efficiency as Engine B

B) Engine A has a lower second-law efficiency than Engine B

C) Engine A has a higher second-law efficiency than Engine B

7.8 Transferring heat from low to high temperature.

What does the second law of thermodynamics say about moving heat from a cooler environment to a warmer space?

A) The amount of heat removed from the cooler environment must be equal to the amount transferred into the warmer space.

B) The amount of heat removed from the cooler environment must be greater than the amount transferred into the warmer space.

C) The amount of heat transferred into the warmer space must be greater than the amount of heat removed from the cooler environment.

7.9 Thermoelectric fan.

One leg of a thermoelectric fan receives heat at Q_H' from a reservoir of hot water. The other leg dissipates heat into cold water at Q_L'. The power driving the fan is W'.

Part I)

Which of the following cases are possible?

A) $Q_H' = 10W, Q_L' = 4W, P = 6W$

B) $Q_H' = 2W, Q_L' = 5W, P = 3W$

C) $Q_H' = 4W, Q_L' = 0, P = 4W$

D) $Q_H' = 0, Q_L' = 3W, P = 3W$

E) $Q_H' = 7W, Q_L' = 7W, P = 0$

Part II)

The thermal, or first-law, efficiency for the case where $Q_H' = 6$ W, $Q_L' = 4$ W, P = 2 W

is $\eta_I =$ (expression)

The numerical value is   (value).

References

Balmer, R.T., 2011. Modern Engineering Thermodynamics. Academic Press, Amsterdam.

Borgnakke, C., Sonntag, R.E., 2019. Fundamentals of Thermodynamics, Tenth Edition Wiley, Hoboken, NJ.

Çengel, Y.A., Boles, M.A., 2015. Thermodynamics: An Engineering Approach, Eighth Edition McGraw-Hill, New York, NY.

Melero, M., Rodríguez-Prieto, V., Rubio-García, A., García-Párrage, D., Sánchez-Vizcaíno, J.M., 2015. Thermal reference points as an index for monitoring body temperature in marine mammals. BioMed Central Research Notes 8 (1), 411.

Moran, M.J., Shapiro, H.N., Boettner, D.D., Bailey, M.B., 2018. Fundamentals of Engineering Thermodynamics, Ninth Edition Wiley, Hoboken, NJ.

Panda, https://www.chinadaily.com.cn/regional/2012-09/21/content_15774766.htm, (accessed September 25, 2021), 2021.

Posadas, A., Morales, J., Ibanez, J.M., Posadas-Garzon, A., 2021. Shaking earth: non-linear seismic processes and the second law of thermodynamics: a case study from Canterbury (New Zealand) earthquakes. Chaos Solitons Fractals 151, 111243.

Sogut, M.Z, 2021. New approach for assessment of environmental effects based on entropy optimization of jet engine. Energy 234, 121250.

Whiteman, J.P., Harlow, H.J., Durner, G.M., Anderson-Sprecher, R., Albeke, S.E., Regehr, E.V., Amstrup, S.C., Ben-David, M., 2015. Summer declines in activity and body temperature offer polar bears limited energy savings. Science 349 (6245), 295–298.

Part 3

Environmental and Engineering Fluid Mechanics

Chapter 8

Fluid statics

"The virtue of a man ought to be measured, not by his extraordinary exertions, but by his everyday conduct."

Blaise Pascal

Chapter Objectives

- Comprehend the pressure distribution in a fluid at rest.
- Understand the working principles of manometers and barometers.
- Learn to calculate the hydrostatic pressure force on a surface.
- Appreciate Archimedes' principle, buoyancy, and stability of floating objects.

Nomenclature

A	area; A_b is the area of the bottom surface
a	acceleration; a_x is the acceleration in the x direction, a_z is the acceleration in the z direction
B	center of gravity
C_B	center of gravity
C_G	geometric center
C_M	meta-center
C_P	center of pressure
F	force; F_B or F_b is the buoyancy force, F_H is the horizontal component of the resultant force, F_{net} is the net force, F_R is the resultant force, F_{Rx} is the resultant force component in the x direction, F_{Ry} is the resultant force component in the y direction, F_{Rz} is the resultant force component in the z direction, F_V is the vertical component of the resultant force, F_x is the force in the x direction, F_z is the force in the z direction
G	center of gravity
g	gravity
H	height
h	height; h_C is the height of the centroid
HVAC	heating, ventilation, air conditioning
i	unit vector in the x direction
I_x	second moment of area (area moment of inertia) about x axis
$I_{x,C}$	second moments of area about the axes passing through the centroid of the area
I_{xy}	second moment of area about y axis
$I_{xy,C}$	second moments of area about y axis passing through the centroid of the area
j	unit vector in the y direction

Thermofluids: From Nature to Engineering. DOI: https://doi.org/10.1016/B978-0-323-90626-5.00012-6

k	unit vector in the z direction
M	meta-center
m	mass
P	pressure; P_{atm} is the atmospheric pressure, P_{avg} is the average pressure, P_{bottom} is the pressure at the bottom, P_C is the centroid pressure, P_n is the pressure in the normal direction, P_{top} is the pressure at the top, P_x is the pressure in the x direction, P_z is the pressure in the z direction
R	radius
s	slope, distance along, or length of, the slope
SI	International System of Units
STAP	standard temperature and atmospheric pressure
T	temperature
W	width, or weight; W_{air} is weight in air, W_{water} is weight in water
x	x direction or distance in the x direction, x_C is the x coordinate of the centroid, x_R is the x coordinate of the resultant force
y	y direction or distance in the y direction, y_C is the y coordinate of the centroid, y_R is the y coordinate of the resultant force
z	z direction or distance in the z direction

Greek and other symbols

Δ	difference
θ	angle
ρ	density; ρ_{duck} is the density of a typical duck, ρ_G is density of gas, ρ_{human} is the density of a typical human, ρ_L is the density of liquid, ρ_{water} is the density of water
$\forall$	volume; $\forall_{crown}$ is the volume of a crown, $\forall_{duck}$ is the volume of a duck, $\forall_{human}$ is the volume of a human, $\forall_{submerged}$ is the submerged or displaced volume

8.1   What is pressure?

Pressure is force per unit area. The SI (International System of Units) unit for pressure is named after child prodigy and French mathematician, inventor, and theologian, Blaise Pascal. One pascal is one newton per square meter. We can envision 1 Pa of pressure by placing a 0.102-kg apple on a weighing scale with a 1-m by 1-m surface, as shown in Fig. 8.1. The 0.102-kg apple under the normal gravity of 9.81 m/s² furnishes a 1-N force. This 1-N force distributed over the 1-m by 1-m surface gives a pressure of 1 N/m², that is, 1 Pa.

Another common unit for quantifying pressure is bars. One bar is 10^5 Pa, that is, 100 kPa. This is very close to standard atmospheric pressure,

$$1 \text{ atm} = 101.325 \text{ kPa} = 1.01325 \text{ bars}. \tag{8.1}$$

Even though the bar is not part of the International System of Units, it is a metric unit, that is, it is used in multiples of tens, for example, a millibar is a bar divided by 10^3. Both the bar and the millibar were introduced by Vilhelm Bjernes, a Norwegian meteorologist who founded the modern practice of weather forecasting (Marvin, 1918). To hold water, a bucket, a pipe, or a plumbing system must be able to withstand the working pressure.

FIGURE 8.1 Pressure is force per unit area (created by Y. Yang). A 0.102-kg apple under normal gravity produces $0.102 \times 9.81 = 1$ kg·m/s^2 = 1 N of force or weight. The right figure illustrates that this 1 N force is exerted on Newton's hat and, thus, head. If this 1-N force is distributed over a surface area of 1 m^2, then the corresponding pressure on the surface is 1 Pa.

8.2 Fluid statics

In fluid statics, pressure is the normal force exerted by a fluid per unit area. The term hydrostatic is commonly used in engineering when we deal with stagnant water. Underwater engineering such as submarines can only go as deep as the hydrostatic pressure the submarine can withstand. It is important to note that hydrostatic applies to other fluids, including air. In atmospheric air, we are typically more concerned with the negative, or suction, pressure. For example, airplanes can fly as high as they can retain approximately one bar of pressure inside the cabin so that the passengers can breathe properly to stay alive. We have seen in movies that an opening of a door or a puncture on the body of a flying airplane can lead to a dangerous suction, dragging even passengers out of the airplane.

In a static, homogeneous fluid, a fluid particle retains its identity and fluid elements do not deform. This is also true in a fluid undergoing rigid-body motion, such as a cup of bubble tea sitting on a tray in an airplane at cruising speed. Changes in the speed, or more correctly velocity, can disturb the cup of bubble tea such that it splashes and spills. For this reason, it is not a good idea for the flight attendants to serve drinks when the airplane is going through rough air.

Consider the wedge-shaped fluid element of unit width depicted in Fig. 8.2, where the width is into the page. Recall that pressure is force per unit area. In a static, homogeneous fluid, there is no shear; that is, there are only normal and body forces acting on the fluid element. Newton's second law of motion states that the sum of forces is equal to mass times acceleration. For a static fluid, the

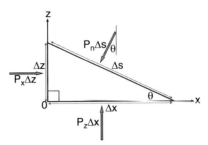

FIGURE 8.2   Forces acting on a small wedge-shaped fluid element of a fluid at rest (created by D. Ting). The infinitesimal fluid element has a unit width into the page. The z direction is pointing vertically up and, hence, gravity is acting in the negative z direction.

acceleration is zero and, thus,

$$\sum \mathbf{F} = \mathbf{ma} = 0, \tag{8.2}$$

where bold variables signify a vector, F is force, m is mass, and a is acceleration. This equation states that the net force in any direction is equal to mass times the acceleration in that direction and, when the fluid is not moving, this net force is zero. As an illustration, the sum of forces in the horizontal or x direction, which is equal to mass times the acceleration in the x direction, a_x, is zero. Mathematically, this can be expressed as

$$\sum F_x = ma_x = 0 : P_x \Delta z - P_n \Delta s \sin \theta = 0. \tag{8.3}$$

Here, as can be seen in Fig. 8.2, P_x is the pressure in the x direction, Δz is the z dimension of the wedge-shaped fluid element under consideration, P_n is the normal pressure, Δs is the dimension of the inclined surface in the x-z plane, and θ is the inclination angle. What is not shown is the unit width in the y direction that is pointing into the page. Similarly, in the vertical or z direction,

$$\sum F_z = ma_z = 0 = P_z \Delta x - P_n \Delta s \cos \theta - \frac{1}{2}\rho g \, \Delta x \, \Delta z = 0, \tag{8.4}$$

where P_z is the pressure in the z direction, Δx is the dimension in the x direction, ρ is the density of the fluid, and g is gravity. The last force term in the z direction is the weight of the fluid element, noting that the fluid element has a unit width into the page, in the y direction. We see from the figure that $\Delta z = \Delta s \sin \theta$ and $\Delta x = \Delta s \cos \theta$, where s is the length of the hypotenuse of the right-angled triangle. With this, we can rewrite the two force balance equations as

$$P_x(\Delta s \sin \theta) - P_n \, \Delta s \sin \theta = 0, \tag{8.3a}$$

and

$$P_z(\Delta s \cos \theta) - P_n \Delta s \cos \theta - \frac{1}{2}\rho \, g \, (\Delta s \cos \theta)(\Delta s \sin \theta) = 0. \tag{8.4a}$$

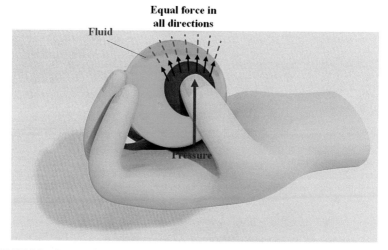

FIGURE 8.3 An everyday workings of Pascal's law, a pressure applied at a point in an enclosed fluid is distributed equally throughout the fluid in all directions (created by X. Wang). The volume of fluid is relatively small, with no appreciable variation in hydrostatic pressure, and, therefore, it can be considered as a point for all practical purposes.

These two equations can be simplified into

$$P_x = P_n, \tag{8.3b}$$

and

$$P_z - P_n - \frac{1}{2}\rho g \, (\Delta s \sin \theta) = 0. \tag{8.4b}$$

The wedge-shaped fluid element represents an infinitely small fluid element with dimensions Δx, Δz, and Δs approaching zero. That being the case, the third term in Eq. 8.4b, $\frac{1}{2}\rho g \, (\Delta s \sin \theta)$, can be neglected. As a result, we have

$$P_x = P_z = P_n = P. \tag{8.5}$$

This says that the pressure at a point in a fluid at rest is not a function of direction. To put it another way, the pressure at any point in a resting fluid is the same in all directions.

The discovery that the pressure at a point in a static fluid is the same in all directions is also credited to child prodigy Blaise Pascal. Pascal's principle or Pascal's law can be worded as pressure applied at any point in a contained incompressible fluid is transmitted equally in all directions throughout the entire enclosed fluid. An everyday example is a water balloon punctured with identical tiny holes throughout its surface; see Fig. 8.3. Applying a pressure by squeezing the balloon with your fingers will lead to water leaking uniformly through all the holes.

A common engineering illustration is the working of a car jack, as illustrated in Fig. 8.4. A relatively small force, F_1, is exerted via the car jack handle. This

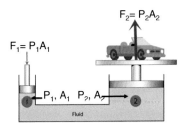

FIGURE 8.4   An engineering application of Pascal's law, where a pressure applied at a point in an enclosed fluid is distributed equally throughout the fluid in all directions (created by F. Fashami). For a car jack, the pressure throughout the enclosed fluid is the same, other than a possible small difference in hydrostatic pressure due to differing height, if any, between Point 1 and Point 2. Without any height difference between Point 1 and Point 2, we have $F_1 = F_2\, A_1/A_2$. This says that a relatively small, applied force, F_1, can result in a large F_2 for lifting a car, when A_2 is much larger than A_1.

induces a pressure, $P_1 = F_1/A_1$, where A_1 is the surface area of the small piston associated with the car jack handle. Because the surface area, A_1, is very small, therefore, the resulting pressure, P_1, is high. According to Pascal's law, P_2 is equal to P_1, that is, F_2/A_2 is the same as F_1/A_1, or $F_2 = F_1\, A_2/A_1$. We see that the force, F_2, is larger than F_1 by the area ratio A_2/A_1. To force the car upward, we employ a small A_1 and a large A_2. Note that we have assumed a negligible difference in the height of the fluid between Point 1 and Point 2. This is important; for otherwise, we have to account for the hydrostatic pressure (discussed in the next section), due to the difference in height.

8.3   Hydrostatic pressure

A water tower is a common landmark when entering a city. Water towers are tall and big and, thus, the perfect billboards for proclaiming the name of the town. Contained inside these large towers is essentially static fluid for providing the required pressure to distribute the water. The higher up the column of water, the higher the pressure at the base for overcoming pressure losses in the piping network, readily delivering water throughout the municipality. This increasing hydrostatic pressure with respect to the height of the fluid column above the point of interest is illustrated in Fig. 8.5. it is clear that the lower the opening from the free surface, the stronger the jet of water spouting through the opening. The hydrostatic pressure varies with respect the atmospheric pressure on the free surface and the height of the water column as

$$P = P_{atm} + \rho g h, \qquad (8.6)$$

where P_{atm} is the atmospheric pressure acting (pushing) on the free surface, ρ is the density of the fluid (water), g is gravity, and h is the height of water above the point of interest. We note that in the case shown, only the (net)

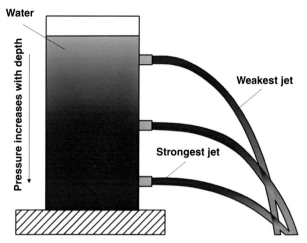

FIGURE 8.5   Hydrostatic pressure illustrated (created by X. Wang). The pressure increases linearly with depth, or height, h. The higher the pressure, the stronger the jet.

hydrostatic pressure, ρgh, dictates the strength of the water jet. This is because the surroundings where the water jet is issuing into is at the atmospheric pressure, P_{atm}, and, thus, this term cancels out. With the density, ρ, and the gravity, g, fixed, the hydrostatic pressure varies linearly with the water depth, or height, h. Fig. 8.5 clearly illustrates that the closer the opening to the free surface, the smaller the relative height, h, of the water column with respect to the opening and hence the lower the hydrostatic pressure.

Example 8.1 The shrinking of a balloon underwater.

Given: A balloon at standard temperature and atmospheric pressure (STAP) has a volume of 0.001 m³. It is immersed in water at room temperature, as shown in Fig. 8.6.

Find: The volume at a) 1-m and b) 10-m water depth.

*Solution:*Assume that the air inside the balloon behaves like an ideal gas, that is,

$$P_1 \Psi_1 / T_1 = P_2 \Psi_2 / T_2,$$

where Ψ is the volume and T is the temperature in the absolute scale. Since the temperature is not changing, we can simplify the equation into

$$\Psi_2 = P_1 \Psi_1 / P_2.$$

It is convenient to choose State 1 as the room condition, at STAP, where $P_1 = P_{atm} = 101325$ Pa and, it follows that the volume of the balloon at STAP, $\Psi_1 = 0.001$ m³.

FIGURE 8.6   Shrinking a balloon using hydrostatic pressure (created by X. Wang). An interesting experiment is to use a long balloon and position it vertically in a water column. The increasing hydrostatic pressure with depth will result in a shrinking cross section with water depth.

The hydrostatic pressure at 1 m underwater, Point 2, is the sum of the atmospheric pressure acting on the free surface plus that imposed by the column of water above it, that is,

$$P_2 = P_{atm} + \rho gh = 101325 + (1000)(9.81)(1) = 111135 \text{ Pa}.$$

Therefore, the volume of the balloon at Point 2,

$$\forall_2 = (101325)(0.001)/(111135) = 0.00091 \text{ m}^3.$$

The balloon shrinks by approximately 9% at 1 m below water.

The hydrostatic pressure at 10 m underwater is

$$P_2 = P_{atm} + \rho gh = 101325 + (1000)(9.81)(10) = 199425 \text{ Pa}.$$

Accordingly, we have

$$\forall_2 = (101325)(0.001)/(199425) = 0.00051 \text{ m}^3.$$

The balloon shrinks by approximately 49% at 10 m below water.

While our skull is much stronger than a balloon, the crushing of hydrostatic pressure is real. For this reason, advanced divers can go no more than about 40 m below water. A submarine can go to a few hundred meters in depth. Beaked whales, on the other hand, can dive 2000 m below the ocean's surface (Tyack et al., 2006). In fact, beaked whales have been tagged to spend over two hundred minutes at a depth of 2992 m (van Aalderink, 2021); also see Quick et al. (2020). On this account, the intelligent design of beaked whales is something for Dr. JAS to consider mimicking when designing her super dive suit.

Let us solidify an important point, that is, the hydrostatic pressure is not a function of the shape of the fluid-containing container. This can be inferred from Eq. (8.6), where the only parameters at play are atmospheric pressure that acts

FIGURE 8.7   The hydrostatic pressure depends on the height of the fluid column and not the shape of the container or the surface area on which the pressure acts (created by X. Wang). It goes without saying that the hydrostatic pressure is also a function of the fluid density, the acting gravity, and the surrounding pressure that acts on the free surface (but not the area of the free surface).

on the free surface, P_{atm}, the density of the fluid, ρ, the gravity, g, and the height of the fluid column, h. This is illustrated in Fig. 8.7, where, for the same fluid under the same atmospheric pressure and gravity, the hydrostatic pressure is solely a function of the height of the fluid column. In other words, the shape of the container does not play a role as far as the hydrostatic pressure is concerned.

We are less aware of the hydrostatic pressure exerted by the much-lighter fluid, air, that surrounds us. The decrease in the hydrostatic pressure of air is what causes our ears to pop when the airplane that we are in climbs. For those who have sensitive ears, the descent is much more painful, that is, increasing air pressure pushes in on our eardrums. One way to relieve the pain is to try blowing air out of our nose while clamping it tight with our fingers. This will force the air to rush into the inner side of our eardrums and, thus, balance the pressure on the two sides. The even more common everyday experience is riding the elevator up or down a tall building. Typically, the increase in pressure when descending is more noticeable because of the inward-pressing pressure on our eardrums.

Example 8.2   Determine the height of a building from pressure difference.
Given: Dr. JAS takes her class to visit a tall building. To give her students an assignment, she makes sure the height of the building is not revealed to them. She measures the pressure on the ground to be 99.9 kPa. At the top of the building is an open balcony where Dr. JAS measures the pressure to be 95.7 kPa.
Find: The height of the building.

Solution: The hydrostatic pressure at the base of the building is equal to that at the top of the building plus the pressure exerted by the column of invisible air

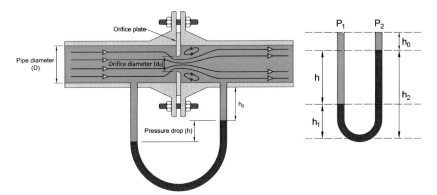

FIGURE 8.8 Measuring the static pressure difference using a manometer (created by F. Kermanshahi). In general, the density of the moving fluid in the conduit is orders of magnitude less than that of the manometer fluid in the U tube and, hence, $P_1 - P_2 = \rho_L gh$, where ρ_L is the density of the manometer liquid, and h is the height difference of the manometer liquid in the two arms of the manometer.

between the top of the building and the base of the building. Specifically,

$$P_{bottom} = P_{top} + \rho gh = 95700 + (1.2)(9.81)(h) = 99900 \text{ Pa},$$

where the density of air, ρ, has been assumed to be 1.2 kg/m^3. Solving the equation gives h = 356.8 m. While this is quite a bit shy of the 828-m tall Burj Khalifa, it is within the top one hundred tallest man-made structures in the world.

Note that it is important for Dr. JAS to measure the pressure at the base and at the garret in the open atmosphere. This is because hot air rises and this tends to pressurize the top floor of an enclosed building the most, preventing the determination of the building height from the indoor pressure.

8.4 Measuring pressure

We have learned from the above that a column of fluid (water, air) of height h gives rise to a pressure-specific weight ratio $P/(\rho g)$. To put it another way, a change of Δh in the elevation or depth of fluid causes a change in the pressure of the amount $\rho g \Delta h$. On account of this, a fluid column can be utilized to measure pressure and pressure differences. Fig. 8.8 shows such a case where an orifice flow meter, as an example, is employed to gauge the volumetric flow rate of a fluid moving in a pipe. More on orifice and other flow-measurement devices will be discussed in the next chapter. For now, let us focus on the deduction of the static pressure drop across the orifice, from which the volumetric flow rate of the moving fluid can be determined. The height difference of the heavier manometer fluid between the two arms of the U tube indicates the static pressure difference between the two locations in the flow conduit. It should be emphasized that the manometer fluid must be heavier than the moving fluid in the pipe for the pressure measurement system to function.

Let us follow the upstream static pressure at the corresponding port along the U tube to the downstream port according to the picture on the right-hand side of Fig. 8.8. Right at the upstream static port the pressure is P_1. Moving from this upstream static port down into the upstream arm of the U tube leads to increasing pressure, from P_1, at the uppermost horizontal dashed line at the upstream port, to P_1 plus $\rho_G g h_0$, at the second most elevated dashed horizontal line. Here, ρ_G denotes the density of the moving fluid in the pipe, which is assumed to be a gaseous fluid and, thus, the subscript G. Farther down into the upstream arm of the U tube, the pressure reaches P_1 plus $\rho_G g(h_0+h)$ at the second-lowest dashed horizontal line. Right at the bottom of the U tube, as indicated by the lowest dashed horizontal line, the pressure is P_1 plus $\rho_G g(h_0+h)$ plus $\rho_L g h_1$, where ρ_L is the density of the much denser manometer fluid, with subscript L signifying liquid. We note that the meniscus of the denser manometer fluid along the upstream arm is at h_1 from the bottom of the U tube. Coming up from the bottom, on the downstream arm of the U tube, the pressure at the second-lowest horizontal dashed line is P_1 plus $\rho_G g(h_0+h)$ plus $\rho_L g h_1$ minus $\rho_L g h_1$. Continue on the upward path along the right arm to the second-highest dashed horizontal line the pressure becomes P_1 plus $\rho_G g(h_0+h)$ plus $\rho_L g h_1$ minus $\rho_L g(h_1+h)$, noting that (h_1+h) is equal to h_2. Finally, the pressure at the downstream static port is P_1 plus $\rho_G g(h_0+h)$ plus $\rho_L g h_1$ minus $\rho_L g(h_1+h)$ minus $\rho_G g h_0$. We note that the meniscus of the manometer fluid along the downstream arm is at h_2 from the bottom of the U tube or h_0 from the top of the U tube. The pressure at the downstream static port is, of course, equal to the static pressure P_2. Mathematically, we can express this as

$$P_1 + \rho_G g\,(h_0 + h) + \rho_L g h_1 - \rho_L g(h_1 + h) - \rho_G g h_0 = P_2. \tag{8.7}$$

Canceling terms, we get

$$P_1 + \rho_G g h - \rho_L g h = P_2. \tag{8.7a}$$

The equation can be rearranged into

$$P_1 - P_2 = (\rho_L - \rho_G)g h. \tag{8.7b}$$

Since ρ_L is generally much larger than ρ_G, negligible error is introduced when dropping ρ_G from the expression, that is,

$$P_1 - P_2 \approx \rho_L g h. \tag{8.7c}$$

In other words, the decrease in pressure across the orifice plate or, more correctly, from the upstream static pressure port and the downstream static pressure port, is equal to the hydrostatic pressure drop in the U tube.

What about the pressure exerted by atmospheric air? Thanks to the good coaching of Galileo, his student Evangelista Torricelli [1608–1647], an Italian physicist and mathematician, invented the barometer, as shown in Fig. 8.9

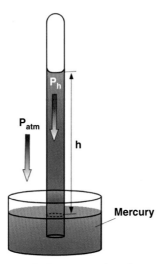

FIGURE 8.9 Measuring the atmospheric pressure using a barometer (created by X. Wang). The weight of the atmospheric air exerts a force on the free surface of the barometric fluid, such as mercury, in the reservoir at the base and the resulting pressure keeps the column of fluid standing inside the column at the corresponding height. Note that the pressure acting on the free surface at the top of the column of the barometric fluid in the tube is virtually zero. To put it another way, the only pressure acting on the column of barometric fluid inside the tube is the hydrostatic pressure induced by the weight of the fluid column, and this is counter balanced by the atmospheric pressure acting on the free surface of the barometric fluid in the reservoir.

Traditionally, liquid mercury, with a density of 13,593 kg/m³, is employed as the working fluid. With negligible mercury vapor pressure trapped at the top of the inverted tube, the pressure there is virtually zero, and the pressure at the free surface level of the reservoir at the base is

$$P_{atm} = \rho_L gh, \tag{8.8}$$

where ρ_L is the density of the liquid mercury at room temperature. In plain English, the atmospheric air exerts a pressure that keeps the column of mercury standing tall inside the inverted tube. For standard atmospheric pressure of 101,325 Pa and, gravity, g, of 9.81 m/s², the height of the mercury, h, is equal to 0.760 m, a convenient height for a typical person to gage. If we use water as the working fluid instead, this height will become 10.33 m. This would require a very tall ladder for the user to make the reading. For air, with a density of 1.2 kg/m³, the column height is 8.6 km! Tall as it is, this height is within the lower atmospheric layer called the troposphere. Even within this lowest atmospheric layer, the troposphere, the air temperature decreases approximately linearly, the pressure lessens somewhat parabolically, and the air becomes thinner, with increasing altitude. Furthermore, there is also a small reduction in gravity with elevation. To put it another way, the 8.6 km air column strictly applies only under the ideal condition that the entire column of air is homogeneous, with a constant density of 1.2 kg/m^3. For typical terra-firma applications, the hydrostatic pressure described by Eq. (8.8) is accurate. A couple of points worth highlighting are illustrated in Fig. 8.9. First and foremost, for a barometer fluid of density, ρ_L, in standard gravity, g, it is only the height, h, of the barometric fluid that determines the pressure. Neither the size, shape, height, or inclination of the tube comes into play.

Absolute pressure is the pressure with respect to an absolute vacuum. The barometer, as discussed above, measures the absolute pressure of the local atmosphere, as the vapor pressure above the column of the working fluid (mercury) inside the tube is negligible, that is, approaching an absolute vacuum. According to Sciencing (Mentzer, 2021), the highest air pressure recorded was 108.4 kPa in Siberia and the lowest air pressure of 87.0 kPa was recorded in a typhoon in the Pacific Ocean. In everyday living, however, gage pressure is more commonly used. Gage pressure is the pressure with respect to the local atmospheric pressure. It is extensively employed because we are typically more concerned with the pressure difference. For instance, in heating, ventilation, and air conditioning (HVAC) applications, HVAC engineers have to make sure no indoor space is over- or under-pressurized with respect to the outdoors. This is because over-pressurization will lead to conditioned air escaping, that is, exfiltrating from the higher pressure inside the building to the lower pressure ambient, leading to significant energy wastage. On the other hand, under-pressurization will result in infiltration of unconditioned atmospheric air into the building, deviating the condition of the indoor space from the intended thermally comfortable state.

8.5 Hydrostatic force on a surface

In engineering, the amount of force acting on a fluid-containing system, such as the cylindrical water tank depicted in Fig 8.10, needs to be calculated when designing the tank. In this way, we can ensure that the tank is strong enough to hold the water. Of interest is the point of application of the force, called the center of pressure. For the bottom surface of the cylindrical tank portrayed in Fig. 8.10, the pressure is uniform because every point on the bottom surface is holding a column of fluid of the same height. The center of pressure is at the center of the circular base plate. Therefore, as far as stability is concerned, the support structure for shouldering the tank at a desirable height should circulate the circumference of the base. At the same time, we need to make sure that the base of the tank is adequately strong, otherwise, the support structure should extend over the entire base plate of the tank.

We note that atmospheric pressure acts downward on the free surface, and this is counterbalanced, that is, canceled out, by the same atmospheric force acting upward from the bottom of the base plate. That being the case, we need only concern ourselves with the gage pressure, this is, the net pressure exerted by the fluid (water) in the tank. Specifically, the resultant force acting

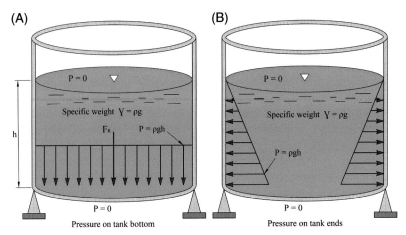

FIGURE 8.10 Hydrostatic force on the flat surfaces of a cylindrical tank (created by F. Fashami). Here, the gage pressure, P, is taken as the pressure with respect to the atmospheric pressure, that is, P = 0 indicates that the pressure is atmospheric, and the vertically upward acting atmospheric pressure cancels out the downward acting pressure. The bottom surface must be able to withstand a net force of ρghA or a pressure of ρgh, in order for the tank to hold water.

on the bottom surface is equal to the upward (and, hence, positive) force due to atmospheric pressure minus the hydrostatic force due to both the atmospheric pressure acting downward on the water surface and the column of water in the tank. Mathematically, we have

$$F_R = P_{atm}A_b - (P_{atm} + \rho gh)A_b = -\rho ghA = -\gamma hA. \tag{8.9}$$

Here, A is the area of the bottom surface and γ is the specific weight, that is, the product of fluid density and gravity. The negative sign implies that the resulting force is acting in the negative z direction, where z is positive when pointing upward. To put it succinctly, the magnitude, the direction, and the line of action of the force are all important. To hold the water, the bottom surface must be strong enough to withstand this force.

The sidewall of the cylindrical tank shown in Fig. 8.10 is subjected to a linearly varying pressure in the vertical direction. This is because the hydrostatic pressure is ρgh, and with constant density and gravity, this hydrostatic pressure increases linearly with the water depth, h. Let us first acquaint ourselves with the relevant fundamentals, including centroid and center of pressure, for a generic piece of flat surface inclined at an angle θ, as illustrated in Fig. 8.11. The centroid or geometric center of the flat surface, C_G, is (x_C, y_C). For a homogeneous flat plate, the density of the plate is uniform and, thus, the centroid is also the center of mass. The depth of this centroid below the free surface is h_C. The center of pressure, C_P, is at (x_R, y_R); this is where the resultant force, F_R, acts. This produces an equivalent force and, thus, moment, due to the hydrostatic pressure of the static fluid. The net force on an infinitesimal element, defined by dx by

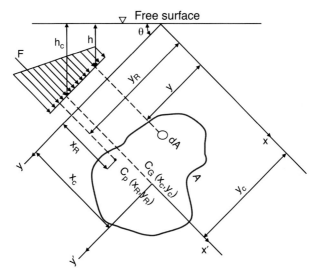

FIGURE 8.11 Hydrostatic force acting on a flat surface submerged in a fluid (created by O. Imafidon). The centroid, C_G, at (x_C, y_C), is the geometric center of the plate and the center of pressure, C_P, at (x_R, y_R), is the point where the resultant force acts.

dy dimensions, due to the gage pressure (tank pressure on the interior surface minus the ambient air pressure acting on the exterior surface), is

$$dF_{net} = \rho g h \, dx \, dy. \tag{8.10}$$

From Fig. 8.11, we see that the depth, h, is equal to y sin θ, and the product of dx by dy is the infinitesimal surface area dA. Accordingly, the infinitesimal net force on an infinitesimal element of the flat surface can be expressed as

$$dF_{net} = \rho g \, y \sin \theta \, dA. \tag{8.10a}$$

The total resultant force on the flat surface can be determined from the sum of the elemental force over the entire area. Mathematically, the summation can be realized exactly via integration, that is,

$$F_R = \int \rho g y \sin \theta dA = \rho g \sin \theta \int y \, dA. \tag{8.11}$$

We note that the y coordinate of the centroid:

$$y_C = 1/A \int y \, dA. \tag{8.12}$$

It follows that the resultant force can be expressed

$$F_R = \rho g \sin \theta \, y_C A. \tag{8.11a}$$

But $y_C \sin \theta$ is h_C, where h_C is the depth of water at the centroid, as shown in Fig. 8.11, and, thence,

$$F_R \quad \rho g \, h_C A. \tag{8.11b}$$

At the same time, $\rho g h_C$ is the pressure at the centroid P_C, and that being the case, we have

$$F_R = P_C A. \tag{8.11c}$$

We note that the centroid pressure, P_C, is equivalent to the average pressure. To put it another way, the magnitude of the resultant force is simply the average pressure multiplied by the surface area, that is,

$$F_R = P_C A = P_{avg} A. \tag{8.11d}$$

Note that the centroid of the area in the y direction, y_C, is the first moment of the area, $\int y dA$. This is a measure of the distribution of the area in relation to an axis normal to the y coordinate. The centroid is typically not the point through which the resultant force acts; the horizontal bottom plate is an exception rather than the norm. Let us deduce the location where the resultant force, F_R, acts, that is, (x_R, y_R). The moment of F_R about the x axis is equal to the moment due to the distributed hydrostatic pressure force. Namely:

$$F_R y_R = \int y dF = \int y (P_{atm} + \rho g y \sin \theta) dA = P_{atm} \int y dA + \rho g \sin \theta \int y^2 dA. \tag{8.13}$$

We include P_{atm} to leave the equation in its general form. The first integral is the first moment of the area, $\int y \, dA = y_C A$, and the integral $\int y^2 \, dA$ is the second moment of area (area moment of inertia) about the x axis, I_x. But $F_R = (P_{atm} + \rho g \sin \theta \, y_C)A$ and, thereupon, we have

$$(P_{atm} + \rho g \sin \theta \, y_C) A y_R = P_{atm} y_C A + \rho g \sin \theta \, I_x. \tag{8.14}$$

This can be rearranged into

$$y_R = P_{atm} y_C / (P_{atm} + \rho g \sin \theta \, y_C) + \rho g y \sin \theta \, I_x / (P_{atm} A + \rho g \sin \theta \, y_C A) \tag{8.14a}$$

In general, the atmospheric pressure cancels out, or is relatively small, and we are left with

$$y_R = I_x / (y_C A). \tag{8.14b}$$

The second moment of area about the axes passing through the centroid of the area, $I_{x,C}$, for many shapes are readily available in engineering handbooks. The second moment of area about the x axis, I_x, can be determined from $I_{x,C}$ via the parallel-axis theorem. Namely:

$$I_x = I_{x,C} + y_C^2 A. \tag{8.15}$$

With this, we can express the y location where the resulting force acts as

$$y_R = I_{x,C} / (y_C A) + y_C. \tag{8.14c}$$

In a similar fashion, the moment of F_R about the y axis is equal to the moment due to the distributed hydrostatic pressure force. Neglecting the atmospheric

TABLE 8.1 Moments of inertia for some common two-dimensional shapes.

Shape	Area, A	$I_{x,C}$	$I_{y,C}$	$I_{xy,C}$
Ellipse with width W height H	$\pi WH/4$	$\pi WH^3/64$	$\pi W^3 H/64$	0
Circle with radius R	πR^2	$\pi R^4/4$	$\pi R^4/4$	0
Semicircle with radius R	$\pi R^2/2$	$0.1098R^4$	$\pi R^4/8$	0
Quarter circle with radius R	$\pi R^2/4$	$0.05488R^4$	$0.05488R^4$	$-0.01647R^4$
Rectangle with width W & height H	WH	$WH^3/12$	$W^3 H/4$	0

Width, W, is in the x direction and height, H, is the y direction. Because of symmetry, the centroids for an ellipse, circle, and rectangle are at the middle. For a semicircle, $y_C = 4R/(3\pi)$ and, for a quarter circle, $x_C = 4R/(3\pi)$, and $y_C = 4R/(3\pi)$, from the center of the corresponding full circle.

pressure, we have

$$F_R x_R = \int x dF = \int x(\rho g y \sin \theta) dA = \rho g \sin \theta \int xy dA = \rho g \sin \theta \, I_{xy}.$$
(8.16)

Substituting for $F_R = \rho g \sin \theta \, y_C$, we end up with

$$(\rho g \sin \theta \, y_C) x_R = \rho g \sin \theta \, I_{xy}.$$
(8.17)

We can solve for x_R:

$$x_R = I_{xy}/y_C.$$
(8.17a)

The second moment of area about the y axis, I_{xy}, can be determined from $I_{xy,C}$ via the parallel-axis theorem. Doing so, we get

$$x_R = I_{xy,C}/(y_C A) + x_C.$$
(8.18)

Keep in mind that the center of pressure, P_C, is the point through which the resultant force, F_R, acts and this point is (x_R, y_R). Table 8.1 provides the formulas for some of the common shapes of uniform thickness and density.

Example 8.3 Underwater rescue of 007.

Given: Dr. JAS brings her students to the filming of a new James Bond episode. In one scene, 007 flies his car off a bridge and goes into a 7-m deep river. The director asks Dr. JAS about the required force for James Bond to push open the door and escape.

Find: The hydrostatic pressure on a 1-m wide and 1.5-m high flat surface representing the door, the pressure center, the required force to push open the door.

Solution: The average pressure acting on the 1-m wide by 1.5-m high vertical door is

$$P_{avg} = P_C = \rho g h_C = 1000(9.81)(6) = 58860 \text{ Pa}.$$

Here, we assume the door is 0.25 m above the riverbed, because of the tires, that is,

$$h_C = 7 - 0.75 - 0.25 = 6 \text{ m}.$$

The resultant force imposed by the water crashing in on the door is

$$F_R = P_{avg}A = 58860(1)(1.5) = 88290 \text{ N}.$$

The line of force action is

$$y_R = I_{x,C}/(y_C A) + y_C = WH^3/12/[(6)(1)(1.5)] + 6 = 6.031 \text{ m},$$

where the width, W = 1 m, and the height, H = 1.5 m, for the vertical car door.

To exert a force close to 88 kN, 007 has to be able to push 88290 N/9.81 m/s^2 = 9000 kg force. Being a very strong human, James Bond may be able to kick-push about 10% of this. In reality, in order to escape, 007 should let water into the car to counter much of the pressing-in force. Another option that may work, especially for a relatively tall vertical door, is to push out on the upper portion of the door, but not on the fragile glass window, as this will break the glass letting water gush into the car with broken pieces of glass. The idea is to take advantage of the higher hydrostatic pressure on the lower portion of the door as a counterforce. To put it another way, pushing out on the upper portion of the door while letting the deeper water push in at the lower portion creates a pivot at the middle line of the door and, hence, will flip open the door for 007 to escape.

8.5.1 Curved two-dimensional surfaces

For completeness, let us take a look at curved surfaces, as many surfaces subjected to hydrostatic forces are not flat. It should be pointed out that a good appreciation of the flat surface is the essential first step in understanding the more complex curved two-dimensional and three-dimensional surfaces. Three-dimensional surfaces are beyond the scope of an introductory textbook such as this.

Consider the generic curved two-dimensional surface shown in Fig. 8.12, where infinitesimal area elements point in varying directions. The resultant force vector can be obtained by integrating over the elemental area, d**A**, where the bold font is used to denote a vector. In the Cartesian coordinate system, the resultant force,

$$\mathbf{F_R} = F_{Rx}\mathbf{i} + F_{Ry}\mathbf{j} + F_{Rz}\mathbf{k} = -\int P d\mathbf{A}, \tag{8.19}$$

where F_{Rx}, F_{Ry}, and F_{Rz} are the resultant force components in the x, y, and z directions, respectively, and **i**, **j**, and **k** are unit vectors in the x, y, and z directions, respectively. For the two-dimensional curve in the x (horizontal) and z (vertical) directions, the horizontal and vertical force components are

$$F_H = P_C A, \quad F_V = \rho g \forall \tag{8.20}$$

The horizontal component is the familiar hydrostatic pressure that acts on a vertical flat surface, the projected vertical flat surface. The vertical force is

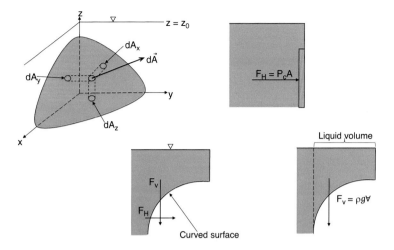

FIGURE 8.12 Hydrostatic force on a two-dimensional curved surface (created by O. Imafidon). The force acting in the horizontal direction is from the hydrostatic pressure that increases linearly with depth below the free surface. The force acting in the vertical direction is that due to the weight of the volume of fluid sitting on top of the surface.

that due to the weight of the volume of liquid sitting on top of the curve; see Fig. 8.12. The line of action of this vertical force passes through the center of gravity of volume $\forall$. The magnitude of the resultant force,

$$F_R = \sqrt{\left[(F_H)^2 + (F_V)^2 \right]}. \tag{8.21}$$

This resultant force acts at $\tan^{-1}(F_V/F_H)$ with respect to the horizontal axis.

8.6 Buoyancy

Archimedes would probably be playing with a rubber ducky, if he were born after Ernie, of Sesame Street, singing "Rubber Duckie" to his best bath buddy in 1970 (Rubber Duck, 2021). The buoyancy force that is keeping the rubber ducky in Fig. 8.13 afloat is equal to the net vertical force due to fluid pressure. Namely,

$$F_B = \rho g \forall_{submerged}, \tag{8.22}$$

where $\forall_{submerged}$ is the displaced or submerged fluid volume. Archimedes' principle states that the buoyancy force has a magnitude equal to the weight of the fluid displaced by the body and it acts upward at the center of the mass of the displaced volume. His eureka moment can be enunciated in terms of Archimedes' principle. A bathtub, initially filled up to the brim, overspills a volume of water when placing Archimedes into the bathtub gently. The weight of the displaced volume of water describes the magnitude of the buoyancy force acting on Archimedes, keeping him afloat.

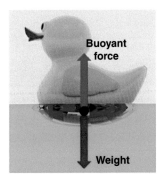

FIGURE 8.13   At equilibrium in the vertical direction, the downward weight of the rubber ducky is equal to the upward buoyant force (created by X. Wang). This buoyant force is equal to the weight of the displaced fluid (water) by the rubber ducky.

Example 8.4 Archimedes' principle for ducks.

Given: Ducks float so well on water that they are called "floating feathers." Their water-repelling, air-trapping feathers, along with hollow bones, and lungs and air sacs furnish an overall density that is much lighter than water; see, for example, Horton (2021).

Find: The Archimedes' principle expression in terms of the density of the duck, ρ_{duck}, density of water, ρ_{water}, volume of the duck, $\forall_{duck}$, volume of the water displaced by the duck, $\forall_{submerged}$, and gravity, g. If the density of a duck, $\rho_{duck} = 250$ kg/m^3, density of the fluid (water), $\rho_{water} = 1000$ kg/m^3, and gravity, g $= 9.81$ m/s^2, what is the volume fraction of the duck in water?

Solution: According to the Archimedes' principle, the weight of the duck is equal to the weight of the volume of fluid displaced by the duck, that is,

$$\rho_{duck}\, g\forall_{duck} = \rho_{water}\, g\forall_{submerged}.$$

This can be simplified and rearranged into

$$\forall_{submerged} = \rho_{duck}\forall_{duck}/\rho_{water} = (250/1000)\forall_{duck} = 0.25\forall_{duck}$$

In other words, for a duck with a density of 25% the density of water, 25% of its volume will be submerged in the water, with the remaining 75% above the free surface.

For a human with a density of approximately 985 kg/m^3,

$$\forall_{submerged} = \rho_{human}\forall_{human}/\rho_{water} = (985/1000)\forall_{human} = 0.985\forall_{human}$$

We see that a typical human can only keep approximately 1.5% of his or her body above water. Because of our dense body, we cannot float nearly as well as ducks.

The water in the Dead Sea has a density of 1238 kg/m^3 (Munwes et al., 2020). For a human with a density of 985 kg/m^3 in this much denser water, the

FIGURE 8.14 Neutrally buoyant rubber ducky (created by X. Wang). When the center of gravity of the rubber ducky, C_G, is below the center of buoyancy, C_B, the rubber ducky is stable. It will restore to its upright position when it is tilted, that is, the gravity will weigh the heavier base downward while the buoyancy will lift the lighter top upward.

submerged volume,

$$\forall_{\text{submerged}} = \rho_{\text{human}} \forall_{\text{human}}/\rho_{\text{water}} = (985/1238)\forall_{\text{human}} = 0.796\forall_{\text{human}}$$

In other words, we can keep more than 20% of our body above water in the Dead Sea. Because of this, there is a saying that it is impossible to drown in the Dead Sea. On the contrary, the Dead Sea has been named the second most dangerous place to swim in Israel (Gizmodo, 2021). It has been repeatedly proven that "it is impossible to drown in the Dead Sea" is erroneous (Staff, 2018). Drowning in the Dead Sea is unlike drowning in typical deep waters. When a person flips over and faces the salty water in the Dead Sea, the highly buoyant water hinders the individual's feet from reaching the ground to turn back over.

8.6.1 Immersed bodies

Have you ever wondered why typical fish swim bladders consist of two gas sacs? The two-gas-sac bladder allows for pitching stability. Assume we add a weight to the base of a rubber ducky so that it is neutrally buoyant, that is, it has the same density as the water and, thus, it is completely immersed, but does not sink to the bottom of the bathtub. This is illustrated in Fig. 8.14. Under those circumstances, the rubber ducky is stable when the center of gravity, C_G, is below the center of buoyancy, C_B, noting that the center of buoyancy, C_B, is the centroid of the displaced volume. The center of gravity will restore the rubber ducky to its upright

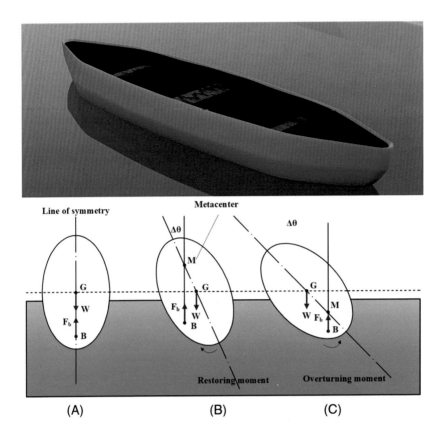

FIGURE 8.15 A canoeist in a canoe, where the center of gravity is above the center of gravity (created by X. Wang). It is (A) neutrally stable when the meta-center, M, coincides with the center of gravity, G, (B) stable when the meta-center is above the center of gravity, and (C) unstable when the meta-center is below the center of gravity.

position when it is wobbled. The rubber ducky is neutrally stable when its center of gravity coincides with the center of buoyancy. The immersed rubber ducky is unstable when the center of gravity is above the center of buoyancy. This can be achieved by adding the required weight to immerse the rubber ducky on the upper part of its body. Small disturbances will tip it over, turning it upside down.

8.6.2 Floating bodies

We see that keeping the center of gravity below the center of buoyancy furnishes stability. Because of that, ships, the most common man-made floating objects, are designed and built such that the center of gravity is below the center of buoyancy. This is where the keel of a ship contributes to its stability, in addition to preventing the ship from being blown sideways by the wind (Discover Boating, 2021). For canoes, however, the center of gravity tends to be above the center of buoyancy; see Fig. 8.15. The meta-center point, C_M or M, is the point where the new line of buoyancy intersects with the initial line of buoyancy at its neutrally stable orientation. To put it another way, the meta-center, M or C_M, coincides with the center of gravity, G or C_G, initially in the neutrally stable alignment. Tilting the canoe sideway by a little, assuming the canoeist remains upright with respect to the normal of the canoe, results in shifting the meta-center above the center of gravity. As a consequence, the restoring moment adjusts the canoe into its neutrally stable position. Too much tilting, on the other hand, can move the meta-center below the center of gravity, leading to capsizing.

Problems

8.1 Deep-diving leatherback turtle.

While most turtles have a hard shell, leatherback turtles have a soft, leather-like shell. They are not only the largest sea turtle species but also one of the most migratory turtles (World Wildlife, 2021). Leatherback turtles can stay underwater for more than one hour. Furthermore, they can dive to a depth of 1200 m when foraging (Oceana, 2021). Unlike hawksbill, olive ridley, and black sea turtles, leatherback turtles have a relatively shorter, wider, and larger volume nasal cavity (Kitayama et al, 2021). It is suggested that these unique nasal features help to lower water pressure and suppress heat dissipation when the leatherback turtles are in deep and cold waters (Kitayama et al., 2021). If a leatherback turtle swimming at the surface of a body of water dives to 1000 m below the free surface, what is the corresponding change in pressure?

8.2 Barometric fluid.

The barometer is the classical device for measuring atmospheric pressure using mercury as the working fluid. A new liquid is used in a barometer, and it gives a height of 2.1 m when the atmospheric pressure, $P_{atm} = 101.325$ kPa. What is the density of the new liquid?

8.3 Floating or sinking egg.

An egg has a small air pocket inside that is presumably for the chick to breathe before hatching. Even so, the average density of a fresh egg is 1031 kg/m³. With time, some of the water permeates through the shell and evaporates, lowering the average density of the egg. It takes about two weeks for the larger end of an egg, where the air pocket is located, to rise off the bottom of a glass of water. If a freshly laid egg has a 5%-volume air pocket, what is this volume after two weeks?

Readers who wish to learn more about the age of bird eggs can refer to, for example, Rush et al. (2007).

8.4 King Hiero's crown.

While the details vary, Archimedes was summoned by King Hiero to determine if the goldsmith had replaced a portion of the gold for the king's crown with some silver. An.Iiim.il.a ncoded to perform thio took without damaging the crown in the investigation process (Britannica, 2021, Live Science, 2021). Archimedes presumably solved this based on the Archimedes principle. Specifically, placing the crown in a tub filled to the brim with water and measuring the amount of water displaced or overflowed out of the tub gave Archimedes the volume of the crown, $\forall_{crown}$. This volume, along with the weight of the crown in air, W_{air}, and in water, W_{water}, provided the needed information for Archimedes to determine the density of the crown and, thus, if it is of pure gold or not.

Part I)

Express the density of the crown in terms of gravity, g, volume of the crown, $\forall_{crown}$, W_{air} and W_{water}.

Part II)

Find the specific gravity of the crown if $W_{air} = 4.2$ N and $W_{water} = 4.0$ N.

8.5 Archimedes junior.

After being flabbergasted by Archimedes, T–E Kaya decides to determine the density of a baby hippopotamus under his care. When submerged, the 1820-N (in air) baby hippo weighs 1022 N in water. What is its density?

8.6 Lighter bubbling water.

The T&E Lifeboat is 3 m long and has a cross-section of an inverted isosceles triangle (1.5 m wide and 2.8 m high). It is floating at 0.4 m above seawater with a density of 1050 kg/m^3. What is the weight of the T&E Lifeboat? Dr. JAS took her crew on a Bermuda-Triangle exploration. She suspects that the mysterious disappearance of ships in the region is due to gas bubbling from the seabed. The gas bubbling can reduce the density of the fluid significantly. At what fluid density would the T&E Lifeboat start to sink?

Keen minds will enjoy learning more from Denardo et al. (2001), Verschoof et al. (2016), and Huang et al. (2021). Note that the work by Verschoof et al. (2016) has to do with generating the appropriate bubbles underneath a ship's hull to reduce friction.

8.7 Salvaging the titanic.

Dr. JAS brings your class on a mission to lift the Titanic up to the surface of the ocean via ocean salvage bags. The depth is estimated to be approximately 12,000 ft and the corresponding hydrostatic pressure around 6,000 psi. What are the two most outstanding challenges? Note that an upward moving buoyant body is akin to a free-falling body in the opposite direction. Keep in mind that the hydrostatic pressure decreases as you move upward from the depth of the waters. What measures can you take to overcome or mitigate these challenges?

References

Britannica, https://www.britannica.com/biography/Archimedes, (accessed October 1, 2021) 2021.
Denardo, B., Pringle, L., DeGrace, C., 2001. When do bubbles cause a floating body to sink? Am. J. Phys. 69 (10), 1064.

Discover Boating, https://www.discoverboating.ca/resources/article.aspx?id=251, (accessed October 1, 2021), 2021.

Gizmodo, https://gizmodo.com/why-so-many-people-drown-in-the-dead-sea-5798844, (accessed October 4, 2021) 2021.

Horton, J., "How do ducks float?" https://animals.howstuffworks.com/birds/duck-float.htm, (accessed July 19), 2021

Huang, W., Yun, H., Huang, W., Zhang, B., Lyu, X., 2021. On the influences of air bubbles on water flow in a two-dimensional channel. Math. Probl. Eng. 6818673. doi:10.1155/2021/6818673.

Kitayama, C., Ueda, K., Omata, M., Tomita, T., Fukada, S., Murakami, S., Tanaka, Y., Kaji, A., Kondo, S., Suganuma, H., Aiko, Y., Fujimoto, A., Kawai, Y.K., Yanagawa, M., Kondoh, D., 2021. Morphological features of the nasal cavities of hawksbill, olive ridley, and black sea turtles: comparative studies with green, loggerhead and leatherback sea turtles. Public Library Sci. ONE 16 (4), e0250873.

Live Science, https://www.livescience.com/58839-archimedes-principle.html, (accessed October 1, 2021), 2021.

Marvin, C.F., 1918. Nomenclature of the unit of absolute pressure. Monthly Weather Rev. 46, 73–75.

Mentzer, A.P., "What is the range of barometric pressure?" Sciencing, https://sciencing.com/barometer-5047250.html, (accessed July 15, 2021), 2021.

Munwes, Y.Y., Geyer, S., Katoshevski, D., Ionescu, D., Licha, T., Lott, C., Laronne, J.B., Siebert, C., 2020. Discharge estimation of submarine springs in the Dead Sea based on velocity or density measurements in proximity to the water surface. Hydrol. Processes 34, 455–472.

Oceana, https://oceana.org/marine-life/sea-turtles-reptiles/leatherback-turtle, (accessed October 1, 2021), 2021.

Quick, N.J., Cioffi, W.R., Shearer, J.M., Fahlman, A., Read, A.J., 2020. Extreme diving in mammals: first estimates of behavioural aerobic dive limits in Cuvier's beaked whales. J. Exp. Biol. 225, jeb222109.

Rubber Duck, https://www.toyhalloffame.org/toys/rubber-duck, (accessed July 19, 2021), 2021.

Rush, S.A., Cooper, R.J., Woodrey, M.S., 2007. A nondestructive method for estimating the age of Clapper Rail eggs. J. Field Ornithol. 78 (4), 407–410.

Staff, T.O.I., 2018. Woman, 88, drowns in Dead Sea. The Times of Israel 27 May.

Tyack, P.L., Johnson, M., Soto, N.A., Sturlese, A., Madsen, P.T., 2006. Extreme diving of beaked whales. J. Exp. Biol. 209, 4238–4253.

van Aalderink, E., 2021. How did whales become the world's deepest-diving mammals? Whale Scientists https://whalescientists.com/whales-deepest-diving-mammals/.

Verschoof, R.A., van der Veen, R.C.A., Sun, C., Lohse, D., 2016. Bubble drag reduction required large bubbles. Phys. Rev. Lett. 117, 104502.

World Wildlife, https://www.worldwildlife.org/species/leatherback-turtle, (accessed October 1, 2021), 2021.

Chapter 9

Bernoulli flow

"Nature always tends to act in the simplest way."

Daniel Bernoulli

Chapter Objectives

- Differentiate streamline, streakline, and pathline.
- Apply Newton's second law to fluid flows.
- Appreciate the Bernoulli equation, its derivation, applications, and limitations.
- Invoke the Bernoulli equation to solve flow problems that can be assumed to be steady, inviscid, and incompressible along a streamline.
- Master the concepts and applications of static, stagnation, dynamic, and total pressures.
- Employ energy and hydraulic grade lines to solve flow problems.

Nomenclature

A	area
a	acceleration; $\mathbf{a}$ is the acceleration vector
C	constant
$C_{discharge}$	discharge coefficient
d	diameter
EGL, EL	energy (grade) line
F	force; $\mathbf{F}$ is the force vector
g	gravity
H	head; H_{in} is the head into the fluid, H_{loss} is the head loss, H_{out} is the head out of the fluid
HGL	hydraulic grade line
$\mathbf{i}$	unit vector in the x direction
$\mathbf{j}$	unit vector in the y direction
m	mass
P	pressure; P_0 is the stagnation pressure, P_S is the static pressure, P_{tot} is the total pressure
r	velocity along a streamline; $\mathbf{r}$ is the velocity vector
s	streamline or distance along a streamline
t	time
U	velocity or velocity in the x direction
u	magnitude of the velocity in the y direction

Thermofluids: From Nature to Engineering. DOI: https://doi.org/10.1016/B978-0-323-90626-5.00004-8

V	velocity; $\mathbf{V}$ is the velocity vector
v	magnitude of the velocity in the y direction
x	streamwise direction or distance
y	crosswise direction or distance
z	vertical direction or distance (elevation)

Greek and other symbols

ρ	density
$\forall$	volume; $\forall$' is the volume flow rate

9.1 Streamline, streakline, and pathline

Let us start this chapter by first distinguishing three fundamental flow field lines in fluid mechanics. They are the streamline, streakline, and pathline. They coincide, or are the same, when the flow is steady, that is, not changing with time.

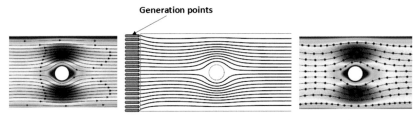

FIGURE 9.1 Streamlines (left figure), streaklines (middle figure), and pathlines (right figure) around a circular cylinder at Reynolds number based on the cylinder diameter of 500 (created by X. Wang). For steady flow, the streamline, streakline, and pathline are identical.

9.1.1 Streamline

A streamline is a line that is tangent to the local velocity vector at an instant in time. Fig. 9.1 depicts a snapshot of sample velocity vectors around a circular cylinder at a given instant. Being a line that is parallel to the direction of flow at a given moment in time, a streamline indicates the direction of the fluid motion at that point in time. It follows that no flow can cross a streamline.

9.1.2 Streakline

A streakline is a line connecting fluid particles[1] that have passed through a particular, that is, the same spatial point; see Fig. 9.1. This is most easily shown by continuously injecting dye at a fixed spatial point over a period of time. In

---

1. A fluid particle is a tiny volume of fluid that contains a large number of molecules, such that it is representative of the fluid in the vicinity that it occupies.

this way, the line formed by the dye (by connecting the dyed fluid particles) is a streakline.

9.1.3 Pathline

A pathline is the trajectory that a particular fluid particle traces out; see Fig. 9.1. It is the actual path traversed by a given fluid particle.

Imagine each one of us rides on one lifeboat of Dr. JAS' fleet of T&E lifeboats, each of which corresponds to a fluid particle. Let all the lifeboats start sequentially, following Dr. JAS' lead, from the same upstream location (the same loading dock) of a slow-moving river, that is, on the far left of Fig. 9.1. Because the current in the river is steady, each successive lifeboat, fluid particle, that passes through the same given point upstream will follow the same path going downstream. The line traced by a certain lifeboat is the pathline corresponding to that lifeboat or fluid particle. At any instant in time, the direction of each lifeboat is prescribed by the velocity vector, $\mathbf{V}$. A line that is tangent to the local velocity vector of a series of lifeboats or fluid particles is a streamline. When the flow is steady, the tangent of the velocity of the fluid particle is always in the direction of the flow path. This is most easily envisioned along the straight portion of the pathline, where the tangent is pointing directly forward or downstream. That being the case, this line acts a divider, separating the fluid on its two sides, that is, top and bottom sides, for the two-dimensional case shown in Fig. 9.1. On that account, no fluid can cross this streamline, as it is defined as a line that is tangent to the local velocity vector, that is, the normal component of the velocity is zero. A streamline indicates the direction of the fluid motion, and its non-fluid-crossing feature is true whether the flow is steady or not. While streamline, streakline, and pathline are the same for a steady flow, they are typically unique for unsteady flows.

Example 9.1 Dr. JAS' whereabouts from her velocity vector.

Given: In her effort to save the monarch butterflies, Dr. JAS sets her fleet of sailboats to sail, following their Atlantic Ocean migration route from Canada to Mexico during the southerly migrating season in autumn. With the help of artificial intelligence, she maps out the specifics of the voyage. The velocity vector can be described as $\mathbf{V} = (1.3 + 2.8x)\mathbf{i} + [1.5 + 2.8y]\mathbf{j}$, where $\mathbf{i}$ is the unit vector in the x or longitudinal direction and $\mathbf{j}$ is the unit vector in the y or latitudinal direction.

Find: The velocity of the sailboat when it is at coordinates x = 1 and y = 1. The expression for the streamlines. What can you say about the voyage, if she changes the velocity vector to

$$\mathbf{V} = (0.5 + 0.8x)\mathbf{i} + [1.5 + 2.5\ \sin(2\pi t) - 0.8y]\mathbf{J}?$$

Solution: At coordinates x = 1 and y 1, the velocity vector:

$$\mathbf{V} = (1.3 + 2.8x)\mathbf{i} + (1.5 + 2.8y)\mathbf{j} = 4.1\mathbf{i} + 4.3\mathbf{j}.$$

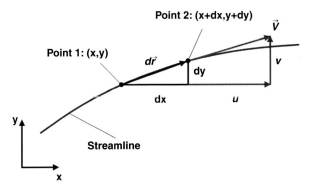

FIGURE 9.2   Describing a streamline in terms of the velocity components in a steady, two-dimensional flow (created by X. Wang). Over a short period of time, Dr. JAS moves from Point 1 with coordinates (x, y) to Point 2 with coordinates (x+dx, y+dy). This change in Dr. JAS' position is realized by the velocity vector V, whose x component is u and y component is v.

Consider Dr. JAS makes a slight turn with an infinitesimal curve stretch, $d\mathbf{r} = dx\mathbf{i} + dy\mathbf{j}$. Since this is along a streamline, $d\mathbf{r}$ is, thus, parallel to the local velocity and has a general mathematical expression $\mathbf{V} = u\mathbf{i} + v\mathbf{j}$. From Fig. 9.2, we can deduct from the triangles, one of which has sides dx, dy, and dr, and the other one with u, v, and V, that

$$dr/V = dx/u = dy/v.$$

The tangent of the streamline at Point 1 is:

$$dy/dx = v/u = (1.5 + 2.8\,y)/(1.3 + 2.8\,x).$$

This can be re-expressed as:

$$dx/(1.3 + 2.8\,x) = dy/(1.5 + 2.8\,y).$$

Integrating the equation leads to:

$$(1/2.8)\ln(1.3 + 2.8x) = (1/2.8)\ln(1.5 + 2.8y) + C.$$

Sample streamlines can be plotted by setting C to selected values.

Adopting $\mathbf{V} = (0.5 + 0.8x)\mathbf{i} + [1.5 + 2.5sin(2\pi t) - 0.8y]\mathbf{j}$ implies that Dr. JAS will be 'flowing' unsteadily. This is because the y or j component of the velocity vector varies with respect to time t.

9.2 Streamline, streamtube, and Bernoulli's Wig

How did Daniel Bernoulli come up with Bernoulli flow? From his utterance, "All birds need to fly are the right-shaped wings, the right pressure, and the right angle," it is clear that he was fascinated by natural fluid mechanics and aerodynamics. Legend has it that, on a windy day in 1735, the wind was so strong that it blew off Bernoulli's new wig, see Fig. 9.3. Apparently, the vision of his wig

FIGURE 9.3 Bernoulli's wig as a streamtube (created by K. Esmaeilifoomani, edited by D. Ting). Each strand of hair is a streamline, through which no fluid crosses. For steady flows, the streamlines are fixed lines in space. For unsteady flows, the streamlines change with respect to time. At any moment in time, the mass passing through any cross-section of a streamtube remains constant.

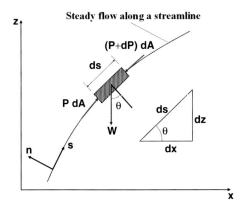

FIGURE 9.4   A straightforward derivation of the Bernoulli equation for a two-dimensional stream-line, based on Newton's second law of motion, F = ma (created by X. Wang). The forces acting on the infinitesimal streamline element are the pressure forces acting normal to the two ends and the weight of the fluid element. The sum of these forces is equal to the mass of the element times its acceleration.

in midair, or, more specifically, strands of hair streaming in the wind, stubbornly stayed in his mind and pestered him. It is not clear how long he pondered the vision until finally he realized that no flow could cross a strand of hair. The strand of hair can be replaced by a line in a flow field, which ultimately led to the dawning of the concept of a streamline. As the fluid velocity along the streamline is always tangent to it, the normal component of the velocity is zero and, thus, no fluid crosses a streamline. With the many strands of hair, Bernoulli's wig formed a tube in the blowing wind. A tube formed by a bundle of streamlines is called a *streamtube*. Since no fluid can cross any of the streamlines, it follows that no fluid can cross a streamtube, which is surrounded by streamlines. That being the case, the mass of fluid flowing through a cross-section of a streamtube must remain within the streamtube.

9.3   The Bernoulli equation

The expression for steady, inviscid, incompressible flow along a streamline is called the Bernoulli equation. Çengel and Cimbala (2021) possibly have the simplest and most straight-forward derivation. Let us follow their sage derivation. Consider the fluid element that is a small segment of a streamline, as shown in Fig. 9.4. The application of Newton's second law, force is equal to mass times acceleration, can be expressed as:

$$P\,dA - (P + dP)dA - mg \sin \theta = m\, UdU/ds, \tag{9.1}$$

where P is pressure, dP is infinitesimal pressure change, A is area, dA is the infinitesimal area of the streamline element of concern, m is mass, g is gravity, θ is the angle between the normal to the streamline and gravitational force direction, U is velocity, and s is distance along the streamline. The equation bespeaks that the force acting in the streamline direction, on an infinitesimal element of the streamline, plus that acting in the opposite direction (and, hence, negative) at other end of the segmental element at s+ds, plus that due to the weight of the fluid element in the negative z direction, is equal to mass times acceleration. From Fig. 9.4, we see from the right-angle triangle that sin θ = dz/ds. This, along with the fact that mass is equal to density times volume, that is, m = ρdAds, allows us to rearrange the equation into:

$$-dP \, dA - \rho dAds \, g \, dz/ds = \rho dAds \, UdU/ds. \tag{9.2}$$

Dividing all terms by the infinitesimal area, dA, results in:

$$-dP - \rho g \, dz = \rho UdU. \tag{9.3}$$

We note that $UdU = \frac{1}{2} d(U^2)$ and, therefore, we can recast the expression as:

$$dP + \frac{1}{2}\rho dU^2 + \rho gdz = 0. \tag{9.4}$$

Integrating this equation, with g as a constant, gives:

$$P + \frac{1}{2}\rho U^2 + \rho gz = \text{Constant}. \tag{9.5}$$

This relationship between pressure, velocity, and elevation, derived from Newton's second law, the conservation of linear momentum, is called the Bernoulli equation. We can divide every term by the density to obtain an alternate form described by:

$$P/\rho + \frac{1}{2}U^2 + gz = C. \tag{9.5a}$$

We note from the dimensions and/or units that the terms represent energy per unit mass of fluid. To put it another way, the flow energy plus the kinetic energy plus the potential energy of a unit mass of fluid remains constant for:

1) inviscid,
2) steady,
3) incompressible flow, and
4) along a streamline.

It is critical that we keep these assumptions in mind when applying the Bernoulli equation in practice. We should remind ourselves that the Bernoulli equation is not applicable for the flow or flow region that is unsteady and/or where viscosity is important. For example, we cannot apply the Bernoulli equation to the boundary layer or turbulent wake behind a bluff body.

Example 9.2 Archerfish licensed to jet and kill.
Given: Archerfish can accurately spit progressively faster droplets, by meticulously manipulating their mouth in a complex manner, at prey that is up to 2 m above water; see, for example, Dewenter et al. (2017)

Find: The required pressure, based on the Bernoulli equation, assuming the stream is a Bernoulli jet.

Solution: According to the Bernoulli equation, the total pressure:

$$P_{tot} = P + \frac{1}{2}\rho U^2 + \rho gz.$$

At the maximum jet height of 2 m, we have:

$P_{tot} = 101325 + 0 + 1000(9.81)(2) = 120,945$ Pa absolute or 19,620 Pa gage.

Instead of squandering such power and energy to sustain a continuous jet, Archerfish are equipped with a sophisticated mouth that controls droplets of particular sizes, throwing them along the trajectory at an increasing speed. In this way, the later droplets catch up with the earlier ones and coalesce into a powerful lump as it strikes the prey.

Archerfish require much more than accurate and timely control of the mouth when they jet to kill. According to Newton's third law, for every action there is a reaction. In this case, the shooting of the water jet leads to an opposing force that would sink the Archerfish some distance below the free surface, and possibly cause it to miss its game, if not for the simultaneous maneuvering of its fins to keep it in position to receive its reward with its mouth (Gerullis et al., 2021).

While not as sophisticated as archerfish, jumping water jets, especially under the illumination of colorful lights, are aesthetic displays of the Bernoulli principle.

9.4   Bernoulli's pressures

The constant associated with the Bernoulli equation expressed in Eq. (9.5) can be replaced by total pressure P_{tot}. Namely:

$$P_S + \frac{1}{2}\rho U^2 + \rho gz = P_{tot}. \tag{9.6}$$

This can be interpreted as static pressure, P_S, plus the dynamic pressure, $\frac{1}{2}\rho U^2$, plus the "hydrostatic pressure" on moving or dynamic fluid, ρgz, is equal to the total pressure, P_{tot}. The flowing fluid, as shown in Fig. 9.5, has some amount of static pressure, P_S. The dynamic pressure is the kinetic energy per unit volume of a moving fluid. If we slow down the flowing fluid until it stops, the static pressure and the dynamic pressure merge into stagnation pressure, in the ideal case without losses (no viscosity and, thus, no friction). The "hydrostatic pressure" on dynamic fluid, ρgz, denotes the "pressure" due to elevation and it increases with elevation z or decreases with fluid depth h in a body of water. This "hydrostatic pressure" on dynamic fluid is different from the hydrostatic pressure on static fluid, ρgh, which signifies the pressure exerted by the column of static fluid above it. The hydrostatic pressure on static fluid increases with fluid depth h, that is, it decreases with increasing elevation z. In other words, ρgz decreases with increasing depth, h, whereas ρgh increases with increasing depth, h.

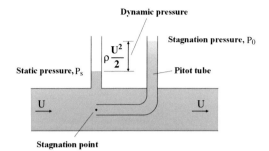

FIGURE 9.5 The Bernoulli equation can be viewed as pressures, where the total pressure is the sum of static pressure, dynamic pressure, and "hydrostatic pressure" on moving fluid (created by X. Wang). Static pressure, P_S, represents the pressure of the fluid when moving at the same velocity as the fluid. The dynamic pressure, $\frac{1}{2}\rho U^2$, is the kinetic energy per unit volume of the moving fluid. The "hydrostatic pressure" on the moving fluid, ρgz, accounts for changes in the elevation.

Before we put the two hydrostatic pressures to rest, let us envision them in a more applied manner. The "hydrostatic pressure" on dynamic fluid, ρgz, can be perceived as pressure due to its potential energy in helping to push the fluid along the flow. The higher up in elevation, larger z, the volume of moving fluid is, the larger the potential for it to move faster via a drop in elevation. In plain English, an increase in this pressure, the larger the available pressure to push the fluid. The hydrostatic pressure on static fluid, ρgh, can be viewed as the ability to squeeze. The deeper we dive into a body of fluid, the more we will be squashed.

Let us look at the static pressure and the dynamic pressure more closely, noting that the "hydrostatic pressure" on a dynamic fluid remains fixed for a horizontally flowing fluid. The pressure at the stagnation point, P_0, consists of the static pressure of the fluid, P_S, plus the dynamic pressure of the moving fluid, $\frac{1}{2}\rho U^2$, that is,

$$P_0 = P_S + \frac{1}{2}\rho U^2. \tag{9.7}$$

Because of this principle, the classic flow measurement method was devised. We can determine the flow velocity using a Pitot-static tube, as shown in Fig. 9.6. Hole 1 is parallel to the flow and, thus, the dynamic pressure associated with the moving fluid is not picked up by Hole 1. A tube with only a static hole is called a static tube, though the static pressure is most frequently measured by such a hole flushed to the wall of the conduit so that the flow is not disturbed at the measurement point. To rephrase it, Hole 1 corresponds to the static pressure, that is, $P_1 = P_S$, and this is the pressure if we move along with the flow. Hole 2 is confronting the moving fluid head on. To that end, Hole 2 is a stagnation point where the total pressure is sensed, that is, $P_2 = P_0 = P_S + \frac{1}{2}\rho U^2$. The dynamic pressure can be deduced simply by subtracting P_1 from P_2. Specifically:

$$P_2 - P_1 = P_0 - P_0 - \frac{1}{2}\rho U^2 \tag{9.8}$$

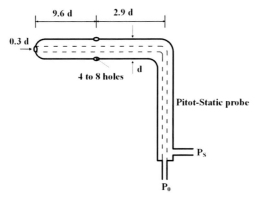

FIGURE 9.6   Pitot-static tube operation is based on the Bernoulli principle (created by X. Wang). The difference between the stagnation pressure, P_0, and the static pressure, P_S, is the dynamic pressure, $\frac{1}{2}\rho U^2$, from which the velocity can be determined. The specifics, such as stagnation port size, number and locations of the static ports, and the distance between the 90° bend, have been optimized to provide the most accurate measurements that are least sensitive to small misalignments of the probe.

For gaseous fluids, air being the one most commonly encountered, the density is relatively small and, thus, also the "hydrostatic pressure" over small changes in elevation such as the difference of a few centimeters of commercial Pitot-static tubes. The velocity of the flowing fluid can be determined by rearranging the equation into:

$$U = \sqrt{[2(P_2 - P_1)/\rho]} = \sqrt{[2(P_0 - P_S)/\rho]}. \qquad (9.9)$$

Example 9.3   Clogging of pitot-static tube and airspeed.
Given: Cadet T–E Kaya takes Airship T&E for an exploration trip of the Bermuda Triangle, under the supervision of Dr. JAS. Dr. JAS reminds cadet T–E Kaya that when Airship T&E is parked, the ram (stagnation) pressure picked up by the Pitot tube only includes the static pressure, that is, the dynamic pressure is zero. As such, the signal picked up by the Pitot tube equals that registered by the static tube and, thus, the Pitot-tube gives a zero reading.
Find: What happens when (a) the static port is clogged and (b) the Pitot tube is clogged.

Solution: When the static port is clogged, the ram or stagnation pressure is correctly sensed. The dynamic pressure is deduced based on this ram pressure with reference to the clogged, unchanging, static pressure. To that end, the airspeed indicator based on the dynamic pressure displays the right airspeed, as long as the atmospheric pressure stays the same. If cadet T–E Kaya climbs up in altitude, the static pressure will drop and, hence, the airspeed indicator will signal a higher speed even when Airship T&E is traveling at the same speed. The reverse is the case if cadet T–E Kaya descends.

If the Pitot port is clogged while the static port is open, the signal of the airspeed indicator based on the ram pressure minus the static pressure will remain zero irrespective of the Airship T&E traveling speed, at the altitude the clogging occurs. The false signal will be positive when ascending and negative when descending.

The message is that airplanes rely heavily on Pitot-Static or Pitot tubes for monitoring their flying speeds. A clogged Pitot-static or Pitot tube can give erroneous readings, leading to disasters. Icing and moisture are common causes of clogging; however, insects can also cause the same trouble. The solution is to check and clean the tubes frequently.

What is the significance of the static pressure? For an incompressible fluid flowing along a horizontal conduit of uniform cross-section, the total pressure, $P_{tot} = P_S + \frac{1}{2}\rho U^2 + \rho g z$, decreases in real life because of the omnipresent frictional losses. There is no change in the dynamic pressure, $\frac{1}{2}\rho U^2$, due to conservation of mass, or the continuity principle. The "hydrostatic pressure" of the dynamic fluid, $\rho g z$, also remains constant for a horizontal conduit. As such, the static pressure, P_S, is the component that decreases with losses along the conduit. On that account, static pressure may be viewed as the potential or capacity to keep the fluid moving. In other words, static pressure refers to the amount of pressure a fan or a pump must supply to overcome the losses to push or pull the fluid through.

The Bernoulli equation has assisted in explaining why the duration of urination does not change with body size, for a wide range of mammals (Yang et al., 2014). Yang et al. (2014) found that mammals above 3 kg universally empty their bladders for between 8 and 34 s. This is because the intelligently designed flow-enhancing device, the urethra, provides for a scaling of 3600 in volume while retaining its functionality. The discovery was made possible, in part, via the application of the Bernoulli equation, where the bladder pressure plays an important role. The other fluid mechanic principles in effect include the Reynolds number that denotes the relative significance of inertial force with respect to viscous force. Larger animals, with substantially larger bladders, are also equipped with bigger urethras. As such, large animals have large urination jets with large inertia while smaller animals drip with high viscosity.

Other highlights of the application of the Bernoulli equation in natural intelligent designs include prairie dogs who are thought to force fresh air through their underground burrows by erecting their fur and running through the passages like pistons (Vogel, 1978). This approach is expensive in terms of energy usage and, thus, not an intelligent method. It is found that, in reality, prairie dogs exploit the Bernoulli equation to vent fresh air through their burrows. The opening of the hole upwind, as far as the prevailing wind is concerned, is flush or level with the plane ground surface, whereas the downwind hole is flush or level with the top

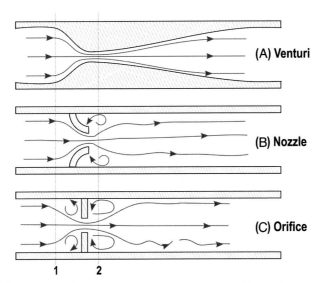

FIGURE 9.7 Common volumetric flow-rate measurement meters: (A) venturi, (B) nozzle, and (C) orifice flow meters (created by A. Raj). The smooth contraction followed by a gradual expansion of a venturi flow meter disturbs the flow marginally and, thus, results in minimal pressure drop. An orifice interferes with the flow abruptly, leading to a substantial pressure drop.

of a raised mound. The wind traveling up and then down the raised mound has to travel faster due to the extra distance created by the mound and, thus, a suction pressure is created to help pump the air out of the burrow or, equally correctly, to pull fresh air into the upwind hole[2].

Just when you think that no new engineering applications can come out of the Bernoulli principle, we learn that there are still many small breakthroughs in engineering practice when the engineer adeptly exploits the simple Bernoulli fundamentals to leapfrog technologies. For instance, Kupczyk et al. (2019) utilized the Bernoulli principle for modeling and advancing milking engineering.

9.5 Flow rate measurements

Some of the most-utilized volumetric flow rate measurement meters are the (1) venturi flow meter, (2) nozzle flow meter, and (3) orifice flow meter; see Fig. 9.7.

---

2. This is similar to the operation of an airfoil, where the "mound" on the upper surface forces air to travel a longer distance compared to the air traveling along the lower, flat surface. Continuity, or conservation of mass (air), requires the adjacent fluid particles that start at the leading edge at the same time to reunite at the trailing edge at the same time. This requires the fluid particles to move faster along the upper, longer surface. The increase in speed, U, implies a decrease in (static) pressure, P_S, according to $U = \sqrt{[2(P_0 - P_S)/\rho]}$; recall that a Pitot static tube measures the static pressure, P_S, via tiny holes that are flush with the surface over which the fluid is moving parallel to. This lower pressure sucks the airfoil upward, creating lift.

To minimize losses due to secondary and turbulent flow motions, the venturi flow meter is meticulously contoured so that the flow goes through it smoothly. The two pressure taps are appropriately located such that the respective pressures correspond to those of the largest and smallest cross-sections of the moving fluid. Note that Tap 1 is located as close to the converging section as possible, to minimize frictional losses between Tap 1 and Tap 2. Tap 2 coincides with the smallest cross-sectional area location, for sound static pressure sensing, as far as the moving fluid is concerned, and this is marginally downstream of the throat because of a slight delay in the moving fluid in responding to the narrowing throat. Expressing the Bernoulli equation for Points 1 and 2, we have:

$$P_1 + \frac{1}{2}\rho U_1^2 + \rho g z_1 = P_2 + \frac{1}{2}\rho U_2^2 + \rho g z_2. \tag{9.10}$$

For all practical purposes, z_1 is equal to z_2 and, hence, we can simplify the equation into:

$$P_1 - P_2 = \frac{1}{2}\rho\left(U_2^2 - U_1^2\right). \tag{9.11}$$

As we are dealing with incompressible fluid, the volumetric flow rate of fluid is conserved. Specifically, the volumetric flow rate at location 1 is equal to that at location 2, that is,

$$\forall' = A_1 U_1 = A_2 U_2. \tag{9.12}$$

From which we have $U_1 = A_2 U_2/A_1$, and this can be substituted into the pressure drop equation to obtain

$$P_1 - P_2 = \frac{1}{2}\rho\left[U_2^2 - (A_2/A_1)^2 U_2^2\right] = \frac{1}{2}\rho U_2^2\left[1 - (A_2/A_1)^2\right]. \tag{9.11a}$$

The velocity at the contraction can, thus, be expressed as:

$$U_2 = \sqrt{\left\{2(P_1 - P_2)/\left[\rho\left(1 - (A_2/A_1)^2\right)\right]\right\}}. \tag{9.11b}$$

Substituting this into the volumetric flow rate equation, we get:

$$\forall' = A_2\sqrt{\left\{2(P_1 - P_2)/\left[\rho\left(1 - (A_2/A_1)^2\right)\right]\right\}}. \tag{9.13}$$

The underlying physics is the same for nozzle flow meters. While the converging of the flow is reasonably smooth, this is not so for the diverging flow, which takes place suddenly. That being the case, there are notable losses, and these are accounted for by introducing a loss coefficient. Specifically, the volumetric flow rate:

$$\forall' = C_{discharge} A_2\sqrt{\left\{2(P_1 - P_2)/\left[\rho\left(1 - (A_2/A_1)^2\right)\right]\right\}}, \tag{9.13a}$$

where $C_{discharge}$ is a discharge coefficient that indicates the actual flow rate with respect to the theoretical one and it has a value of less than one. The other thing to note is that, even though the physical cross-section at the "flow throat" is the same as the pipe (at Tap 1), the cross-sectional area of the downstream-moving fluid is at its minimum. The flow area of this "flow throat" is actually slightly smaller than the smallest physical cross-section of the nozzle because of vena contracta. It is clear that nozzle flow meters are cheaper and require significantly less space than venturi flow meters. Cheaper still are the orifice flow meters, but they come at a substantially larger pressure loss. The discharge coefficient for orifice flow meters is around 0.6, compared to over 0.95 for typical nozzle flow meters. While the discharge coefficient can be accurately determined experimentally by the manufacturer, the pressure loss implies that additional pumping power is needed to move the fluid at the same flow rate as the unobstructed (without the orifice) pipe.

9.6 Energy line and hydraulic grade line

Another convenient form of the Bernoulli equation can be obtained by dividing Eq. (9.5) by the specific gravity of the involved fluid, ρg. Doing so gives:

$$P/(\rho g) + \frac{1}{2}U^2/g + z = H, \tag{9.14}$$

where H is the total head. We see that pressure head, $P/(\rho g)$, plus velocity head, $\frac{1}{2}U^2/g$, plus elevation head, z, is equal to the total head, H. The unit of head is length or, more explicitly, the height of the fluid, as illustrated in Fig. 9.8. The Bernoulli's head is also frequently referred to as the *Energy (Grade) Line*:

$$EGL = P/(\rho g) + \frac{1}{2}U^2/g + z = H. \tag{9.14a}$$

The *Hydraulic Grade Line*, on the other hand, is the sum of the pressure head and elevation head, that is,

$$HGL = P/(\rho g) + z. \tag{9.15}$$

Without the dynamic head, the sum of pressure head and elevation head is also referred to as the *Piezometric Head*. It is clear that subtracting the hydraulic grade line from the energy grade line gives the dynamic head, $\frac{1}{2}U^2/g$.

It is worth emphasizing that a Bernoulli flow does not lose any head because it is frictionless, that is, the total head is conserved. Fig. 9.8 depicts that the total head remains at the height of the free surface of water with respect to a common reference base. For a flow along the same elevation, the velocity head can increase due to a contraction in the flow cross-section. This increase is compensated by an equivalent decrease in the pressure head. An increase in local elevation will lead to a decrease in pressure head for a flow in a uniform cross-section pipe, in which case the velocity and, hence, the velocity head remain the same. The point is that the total head of an ideal Bernoulli flow is conserved.

In real life, the fluid has finite viscosity and there is at least frictional loss in a real flow. It follows that the total head or the energy grade line and, thus, also the hydraulic grade line, is always decreasing along the flow passage. This is illustrated in Fig. 9.9, which shows a continuous decrease in the total head. The decrease in the total head is faster in the smaller cross-section with faster-moving fluid because of the associated larger loss per unit pipe length. A pump

(A)

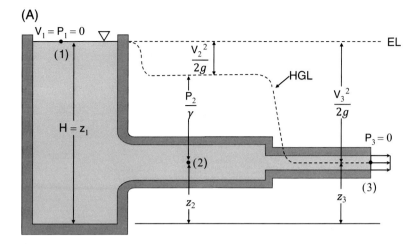

(B)

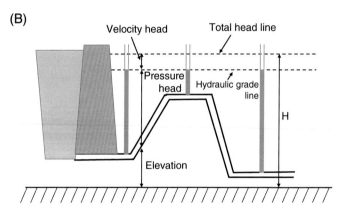

FIGURE 9.8 An illustration of static and flowing fluid heads, totaling the Bernoulli Head (created by O. Imafidon). In the absence of losses or gains the total (Bernoulli) head is conserved. (A) Demonstrates a flow along the same elevation, where the increase in velocity head with decreasing cross-sectional area of the pipe is equal to the decrease in the pressure head. For the fixed cross-section pipe in (B), the velocity head remains unchanged due to continuity. The pressure head decreases as the pipe is elevated but the hydraulic grade line remains fixed in the ideal frictionless flow.

can be used to increase the head, that is, energy is input into the flow via a pump and boosts the energy content (head) of the flow. On the other hand, a turbine draws energy out of a flowing fluid and, hence, drops the head of the flow. To put it succinctly, the total head at location 2, which is downstream of location 1:

$$H_2 = H_1 + H_{in} - H_{out} - H_{loss}, \tag{9.16}$$

where $H_2 = EGL_2 = P_2/(\rho g) + \frac{1}{2}U_2^2/g + z_2$ and $H_1 = EGL_1 = P_1/(\rho g) + \frac{1}{2}U_1^2/g + z_1$. Energy consuming devices such as pumps and fans input a portion

(A)

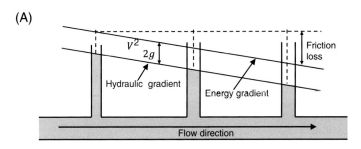

(B)

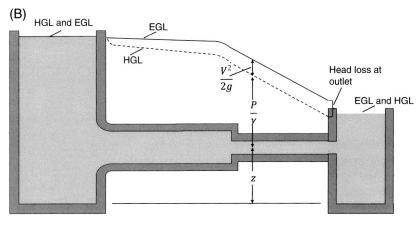

(C)

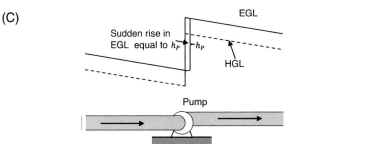

FIGURE 9.9 The variation of fluid head along the flow passage of a real fluid (created by O. Imafidon). With finite viscosity, we have finite losses; therefore, the total head decreases along the flow passage. For the uniform cross-section straight pipe, (A), the total head decreases linearly with pipe length. (B) shows that a reduction in pipe cross-section increases head loss, along with increasing flow velocity. (C) shows that the pump inputs energy into the fluid and hence, increases the fluid head.

of the energy they consume as head into the flow, whereas energy-producing devices such as water turbines harness part of head from the flow to produce energy. Frictional losses are the most common head loss; it is no wonder they are called major losses. Losses due to disturbance of flow through an inlet, bend, valve, exit, etc., are termed minor losses. Along a uniform cross-section segment,

$\frac{1}{2}U_1^2/g = \frac{1}{2}U_2^2/g$ and, thus, the hydraulic grade line at location 2 with respect to that at location 1 upstream is:

$$HGL_2 = HGL_1 + H_{in} - H_{out} - H_{loss}, \tag{9.17}$$

where $HGL_2 = P_2/(\rho g) + z_2$ and $HGL_1 = P_1/(\rho g) + z_1$.

Problems

9.1 Streamline.

A streamline is

A) the actual path traveled by an individual fluid particle over some time period

B) the locus of fluid particles that have passed sequentially through a prescribed point in the flow

C) a set of adjacent fluid particles that were marked at the same (earlier) instant in time

D) a line that is everywhere tangent to the instantaneous local velocity vector

9.2 Streakline.

A streakline is

A) the actual path traveled by an individual fluid particle over some time period

B) the locus of fluid particles that have passed sequentially through a prescribed point in the flow

C) a set of adjacent fluid particles that were marked at the same (earlier) instant in time

D) a line that is everywhere tangent to the instantaneous local velocity vector

9.3 Exhaust smoke.

The smoke (condensed water vapor) from an exhaust stack of a factory shows the

A) streamline

B) pathline

C) streakline

D) timeline

E) spaceline

F) sourceline

9.4 Bernoulli plunge.

Dr. JAS designed a large tank that can hold water up to a depth of 1 m and has a smooth 0.3-m-diameter exit at the bottom. During the annual Turbulence & Energy Lab research fund-raising event, Dr. JAS invites the president of her institution to sit below the tank, with his head at 2 m below the exit. Assuming Bernoulli flow, estimate the velocity of water pouring down on the president's head when the valve is knocked open by a ball tossed by a generous donor. Note that, as the tank is large, we can assume steady flow with the water height fixed at 1 m.

9.5 Applicability of the Bernoulli principle.

For a steady, incompressible, Newtonian fluid flowing across a bluff body, the Bernoulli principle is applicable

A) in the boundary layer

B) in the wake

C) in the free stream

D) All of the above

E) None of the above

The Bernoulli principle is not applicable:

A) in the boundary layer

B) in the wake

C) in the free stream

D) All of the above

E) None of the above

9.6 Bernoulli fire hose.

A fire hose with water at 500 kPa produces a horizontal jet exiting the 5-cm diameter nozzle at 25 m/s. Assuming no losses, what velocity is the water moving inside the 10-cm diameter hose?

9.7 Bernoulli vertical water jet.

A water pipe is broken, and a vertical jet of water reaches a height of 3 m. Assume that the opening is smooth and frictionless, and that the pipe is large and, thus, the water moves slowly in it. Estimate the pressure in the pipe based on the Bernoulli equation? What is this pressure if 20% of the head is lost due to non-ideal conditions and losses?

9.8 Bernoulli T-shaped standing-water-removal kit.

Some water is trapped in a crack with its bottom 7 cm below the floor. Inspired by the Bernoulli principle, you place a T-shaped pipe section with the base of the T touching the bottom of the crack where the water is standing. The idea is to blow air through the horizontal section of the T such that it sucks up the water. At what velocity must the air be blown through the horizontal passage?

9.9 Venturi flow meter.

For a venturi tube, the static pressure

A) is low where the cross-sectional area is small

B) is high where the cross-sectional area is small

C) does not change with a change in cross sectional area

D) is larger than the total pressure

9.10 Venturi flow measurement.

The volumetric flow rate of air, at standard temperature and pressure, in a 10-cm-diameter pipe is measured using a venturi flow meter with a throat

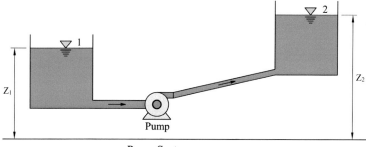

FIGURE 9.10 Energy grade line from a lower elevation tank to a higher one via a uniform pipe equipped with a pump (created by F. Kermanshahi).

of 7.5-cm diameter. What is the indicated pressure difference when the volumetric flow rate is 0.01 m^3/s?

9.11 Bernoulli wind on a 45° ramp.

Wind, which may be assumed to be of uniform profile, at 12 m/s blows along a flat roof of a building until it encounters a ramp at a 45° angle. The ramp is 10 m high and 10 m wide. Estimate the force exerted on the ramp. In which direction is this force acting?

9.12 Energy grade line and hydraulic grade line.

Water from Tank 1, with a free surface at z_1 above ground, is pumped steadily to Tank 2, with a free surface at a tank z_2, where z_2 is higher than z_1; see Fig. 9.10. Qualitatively draw the energy grade line and the hydraulic grade line between Tank 1 and Tank 2, with the pump between the two tanks connected with a uniform cross section pipe. Consider frictional losses only.

References

Çengel, Y., Cimbala, J., 2021. Fluid Mechanics: Fundamentals and Applications, Fourth Edition, McGraw-Hill, New York.

Dewenter, J., Gerullis, P., Hecker, A., Schuster, S., 2017. Archerfish us their shooting technique to produce adaptive underwater jets. J. Exp. Biol. 220 (6), 1019–1025.

Gerullis, P., Reinel, C.P., Schuster, S., 2021. Archerfish coordinate fin maneuvers with their shots. J. Exp. Biol. 224 (6), jeb233718.

Kupczyk, A., Gaworski, M., Szlachta, J., Tucki, K., Wojdalski, J., Luberański, A., Dróżdż, B., Krzywonos, M., 2019. A slug flow model in a long milk tube for designing a milking unit control system. J. Animal Plant Sci. 29 (5), 1238–1246.

Vogel, S., 1978. Organisms that capture currents. Sci. Am. 239 (2), 128–139.

Yang, P.J., Pham, J., Choo, J., Hu, D.L., 2014. Duration of urination does not change with body size. In: Proceedings of the National Academy of Sciences of the United States of America, 111, pp. 11932–11937.

Chapter 10

Dimensional analysis

"What is a society without a heroic dimension?"

Jean Baudrillard

Chapter Objectives

- Appreciate dimensions and dimensional homogeneity.
- Become familiar with dimensional analysis.
- Master the Buckingham Pi theorem.
- Recognize common dimensionless groups in fluid mechanics.

Nomenclature

A	area
Ar	Archimedes number
Bo	Bond number
C	quantity of light
C1, C2,	constants
C_D	drag coefficient
C_f	friction coefficient
C_L	lift coefficient
C_P	pressure coefficient
c	speed of sound
D	diameter
Eu	Euler number
F	force; F_D is drag force, F_L is lift force
f	friction or frequency
f_s	frequency or vortex shedding frequency
Fr	Froude number
g	gravity
Gr	Grashof number
I	current
L	length or characteristic length; [L] is the dimension of length, L_{leg} is leg length, L_{stride} is stride length
M	the dimension of mass
Ma	Mach number
N	amount of matter
n	number; n_d is number of dimension, n_v is number of variable

P	pressure; ΔP is pressure change
Re	Reynolds number
Ri	Richardson number
SI	International System of Units
St	Strouhal number
T	temperature dimension; T_H is the higher or hotter temperature, T_{ref} is reference temperature, T_s is temperature of the subject, T_∞ is ambient temperature
t	time; [t] is the dimension of time
U	velocity; velocity in the x direction
We	Weber number
z	vertical direction or distance

Greek and other symbols

Δ	difference
β	thermal expansion coefficient
γ	specific gravity
μ	dynamic viscosity
ρ	density; ρ_s is density of the submerged body
σ	surface tension.
τ	shear; τ_w is wall shear

10.1 Dimensional homogeneity

Parents (and teachers) are often reminded to not compare their children, as everyone is gifted differently. The everyday household maxim is to not compare apples with oranges. To put this in context with dimensional homogeneity, we cannot have an equation where not all the terms are of the same dimension. Let us consider Bernoulli's equation from the previous chapter,

$$P/\rho + \frac{1}{2}U^2 + gz = \text{constant}, \qquad (10.1)$$

where P is pressure, ρ is density, U is velocity, g is gravity, and z is elevation. This equation says that the sum of the pressure, kinetic, and potential energies is conserved along a streamline. For this to be valid, each term must have the same dimension. There are seven primary dimensions, as tabulated in Table 10.1 with their corresponding SI and English units. For Bernoulli's equation, we note that pressure is force per unit area, and force is mass times acceleration, therefore, the dimension of pressure is $[M][L/t^2]/[L^2] = [ML^{-1}t^{-2}]$. Density is mass per unit volume, that is, density is dimensionally $[M/L^3]$. Pressure divided by density is thus $[L^2t^{-2}]$. The second term is the square of velocity, it has the dimension of $[L^2t^{-2}]$, the same dimension as the pressure term. The dimension of the potential or elevation term is $[L/t^2][L] = [L^2t^{-2}]$, which is dimensionally homogeneous with the first two terms. That being the case, Bernoulli's equation is dimensionally sound.

10.2   Scaling and dimensional analysis

Have you ever wondered about the relationship between brain size and intelligence? Imagine if our intelligence scales with the size of our brain. If this were true, then, T&E Lab member Kaya, with a brain size equivalent to that of a 7-cm radius sphere, would be about 1.6 times smarter than T&E Lab member Rajin, with a brain that fits into a 6-cm radius sphere. There is no real backing to this hypothesis, nevertheless, interested readers can check out (Koch, 2016), among many other studies. What we are trying to achieve is "scaling" intelligence with brain size. Scaling, when it works, can manifest the underlying features and physical principles. A familiar use of scaling in biological systems that have been proven to be reasonable is the scaling of metabolic rate with body mass. Studies, including McNab (2008) and Ballesteros et al. (2018), have found that the logarithm of metabolic rate increases linearly with the logarithm of body mass; see Fig. 10.1. This log–log linear scaling can encompass all kinds of animals, from doves and rats at the lower end all the way up to cows and oxen (West and Brown, 2004).

In spite of the popularity of metabolic rate-body mass scaling, and the backing of this scaling with a wide range of data, there are some issues with this generalization from the dimensional perspective, see Butler et al. (1987). For instance, the "scaling" of metabolic rate with body mass is nonideal because of dimensional inhomogeneity between the two parameters we are correlating. In SI (International System of Units) units, metabolic rate can be expressed as watts and the body mass in kilograms. Watts is J/s or N·m/s or, in primary SI units, (kg·m/s/s)(m)/s or kg·m^2/s^3, which is dimensionally [ML^{2}t^{-3}]. The linear scaling of metabolic rate with body mass implies that kg·m^2/s^3 is logarithmically proportional to kg. This gives a slope that has units m^2/s^3 or dimensions of [L^{2}t^{-3}], which falls short of the dimensionless ideal. It is much better to correlate a nondimensional metabolic rate with a nondimensional body mass and, if a linear relation exists, the slope is also dimensionless. Such a general correlation

TABLE 10.1 The seven primary dimensions and the corresponding primary units in SI and English.

Dimension	SI unit	English unit
Mass [M]	kg (kilogram)	lbm (pound-mass)
Length [L]	m (meter)	ft (foot)
time [t]	s (second)	s (second)
Temperature [T]	K (kelvin)	R (rankine)
Amount of matter [N]	mol (mole)	mol (mole)
Electric current [I]	A (ampere)	A (ampere
Quantity of light [C]	cd (candela)	cd (candela)

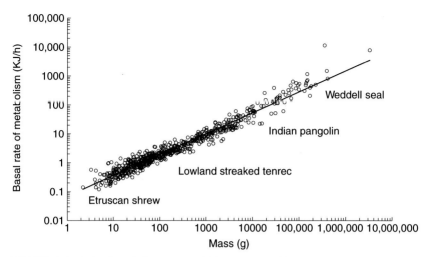

FIGURE 10.1 Animal metabolic rate versus body mass (created by Y. Yang). Data are taken from the appendix of McNab (2008). Within the scatter of the data, the log–log relationship between the metabolic rate and body mass of mammals extends from the tiny Etruscan shrews, where an average adult has a body mass of less than 2 g (Animalia, 2021), to the great Weddell seals that weigh between 400 and 500 kg when fully grown (Antarctica, 2021).

is much more versatile and tends to be applicable more generally. Dimensional analysis is the grouping of pertinent factors involved into a smaller number of nondimensional parameters. Let us look at the running speed of animals instead of the metabolic rate.

How fast can a dinosaur run? It is challenging for us to answer this question because dinosaurs are extinct. In spite of this, a great deal of work has been invested to answer this epochal question. Dimensional analysis has been successful in capturing the relation between nondimensional stride length and nondimensional speed; see Caton and Otts (1999), for example. Specifically,

$$L_{stride}/L_{leg} \propto \left[U/\left(L_{leg}g\right)\right]^{1/2}. \tag{10.2}$$

Here, L_{stride} is stride length, L_{leg} is leg length, U is speed, and g is the gravitational acceleration. Fig. 10.2 depicts this relationship. Dimensionally, we have

$$[L]/[L] = 1 \propto \left\{[L/t]/\left\{[L]\left[L/t^2\right]\right\}\right\}^{1/2} = 1, \tag{10.3}$$

where the symbol enclosed by square brackets signifies the dimension, L is the dimension of length, and t is the dimension of time. The derivation of the above normalized stride-speed relation was not a random exercise, but a logical endeavor based on physics and sound reasoning. In the said case, we are after stride efficiency as a function of speed. A larger stride length is likely advantageous in terms of running speed. To be fair, when we compare the running speed, of different animals, we normalize the stride length with the leg length. According to physics, our running speed is influenced by stride length

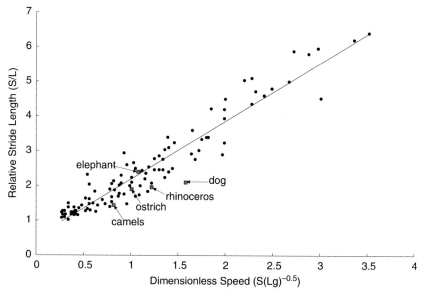

FIGURE 10.2 Nondimensional stride length versus running speed of various animals (created by Y. Yang). Data are taken from various sources; see, for example, (Alexander, 1989, 2004). As expected, the longer the normalized stride length, the higher the normalized running speed. Studies, including those conducted by Montanari (2017), Sellers et al. (2017), and Celine (2021), indicate that dinosaurs, such as Tyrannosaurus rex, cannot outrun humans. We can infer this from the relation, $L_{stride}/L_{leg} \propto [U/(L_{leg}\, g)]^{\frac{1}{2}}$, that this is because the much longer legs, L_{leg}, of the dinosaurs do not lead to a proportional increase in the stride length, L_{stride}. It follows that the speed, U, does not increase in proportion to L_{leg}.

and gravity. To put it another way, with everything else the same, the longer the stride length, the faster we run. Similarly, the less the gravitational pull, the faster we run. Yes, you and I can run much faster on the moon, if we do not have to wear the clumsy spacesuit. We cannot play with stride length and gravity alone and create a dimensionless product of any form. But another length will do the trick, and the choice of this length is obvious, that is, the same leg length used to normalize the stride length.

From the aforementioned example, we see that significant insight is needed before one can formulate a meaningful dimensionless parameter and, more so, for constructing a dimensionless relationship. Once contrived, a relation between dimensionless parameters provides the general behavior of similar systems. In the dinosaur running speed case, the $L_{stride}/L_{leg} = f\{[U/(L_{leg}\, g)]^{\frac{1}{2}}\}$ relation can, more or less[1], be applied universally to animals of all kinds and sizes. Based on

---

1. Obviously, factors such as whether the creature is running on two or four legs and other details of the biomechanics also come into play. Many studies have included these "secondary" factors. Nonetheless, the general relationship is very versatile and can be used as a first approximation until additional details are revealed.

this nondimensional relationship, we can deduce or predict an animal's running speed with known stride length (e.g., based on fossilized dinosaur footprints, leg length, and gravity). Because of this generalization, we can drastically reduce the effort in the experimental deduction of relations between pertinent variables. As an illustration, the general flow behavior of a fluid flowing in a pipe has been of engineering interest since the ancient Roman aqueduct system (Deming, 2020) in the first century, if not earlier. Thanks to the pioneering work of Osborne Reynolds, hydraulic engineers do not have to perform an experiment every time they are dealing with a different fluid (provided it is a Newtonian fluid such as air and water), flow rate, or a pipe of a different diameter. All that is needed is the Reynolds number, Re, which will be expounded shortly. Returning to the extinct dinosaurs, we can make scaled-down models of quetzalcoatlus, with up to 13 m wingspan, and study their aerodynamics in a wind tunnel, provided the actual Reynolds number is duplicated.

10.3 Buckingham Pi theorem

According to the Buckingham Pi theorem, a dimensionally homogeneous equation involving n_v variables can be reduced to a relationship among n_v minus n_d independent dimensionless products. Here, n_v signifies the number of relevant dimensional parameters and n_d denotes the number of primary dimensions involved. The dimensional analysis can be executed by carrying out the following steps.

Step 1: List all pertinent n_v dimensional variables. Include only variables that are independent of each other, that is, one cannot be expressed in terms of another, such as area and length, as the area is equal to the square of length.
Step 2: Express each variable in n_d primary dimensions.
Step 3: The number of nondimensional pi terms is equal to n_v minus n_d.
Step 4: Select the n_d repeating variables that have all the primary dimensions. No repeating variable should have the same net dimension differing only in the exponents. For example, do not select length and volume, as volume is dimensionally the cube of length.
Step 5: Multiply a nonrepeating variable with the product of repeating variables each raised to an exponent.
Step 6: Set the overall exponent of each dimension to zero to make the obtained pi term dimensionless.
Step 7: Go back and repeat Step 5 for each remaining nonrepeating variable.
Step 8: Check all the pi terms to make sure they are dimensionless and independent.
Step 9: Rearrange pi terms, if needed, to simplify the problem or to reflect known dimensionless parameters.

Step 10: Express the pi terms as an appropriate relation with appropriate physical meanings.

Example 10.1 Wind harps.
Given: Since her childhood, Dr. JAS always dreamed about duplicating shepherd boy David's wind harp. Recently, she realized that written records on wind harps dated all the way back to a handful of generations after Cain. Jubal was the pioneer who skillfully handled stringed and wind instruments [Genesis 4: 21]. Having some mastery of the violin, Dr. JAS makes up her mind to figure out how the musical notes that are based on frequencies work in a wind harp. Consider the strings as long circular cylinders. The factors that influence the frequency of the flow-induced (or vortex shedding-induced) vibrations are (1) diameter of the string, (2) wind speed, (3) density, and (4) viscosity of the moving fluid.
Find: The dimensionless parameters involved.

Solution:
Step 1: The pertinent variables are vortex shedding frequency, f, string diameter, D, wind speed, U, air density, ρ, and dynamic viscosity, μ. In other words, there are five pertinent variables, that is, $n_v = 5$.

Step 2: (Vortex shedding) frequency, f_s, has the dimension of $[t^{-1}]$.
  Diameter, D, has the dimension of [L].
  Speed, U, has the dimensions of $[Lt^{-1}]$.
  Density, ρ, has the dimensions of $[ML^3]$.
  Viscosity, μ, has the dimensions of $[ML^{-1}t^{-1}]$.

Step 3: We have 3 fundamental dimensions, M, L, and t, that is, $n_d = 3$. To this end, the number of pi terms $= 5 - 3 = 2$.

Step 4: Since $n_d = 3$, let us choose D, U, and μ as repeating variables.

Step 5:

$$\pi_1 = D^{C1} U^{C2} \mu^{C3} f_s$$

Step 6: For $\pi_1 = D^{C1} U^{C2} \mu^{C3} f_s$, we have $[L]^{C1} [Lt^{-1}]^{C2} [ML^{-1}t^{-1}]^{C3} [t^{-1}]$, that is,

$$[L] : C1 + C2 - C3 = 0$$

$$[t] : -C2 - C3 - 1 = 0$$

$$[M] : C3 = 0$$

We see from the [M] equation that C3 = 0. Substituting C3 = 0 to the [t] equation, we get C2 = −1. Substituting C3 = 0 and C2 = −1 into the [L] equation leads to C1 = 1. In short, $\pi_1 = D^1 U^{-1} \mu^0 f_s = f_s D/U$

Step 7: (repeat Step 5 and Step 6 for π_2):

$$\pi_2 = D^{C4} U^{C5} \mu^{C6} \rho$$

For $\pi_2 = D^{C4} U^{C5} \mu^{C6} \rho$, we have $[L]^{C4} [Lt^{-1}]^{C5} [ML^{-1}t^{-1}]^{C6} [ML^{-3}]$, that is,

$$[L] : C4 + C5 - C6 - 3 = 0$$

$$[t] : -C5 - C6 = 0$$

$$[M] : C6 + 1 = 0$$

We see from the [M] equation that C6 = −1. Substituting C6 = −1 into the [t] equation gives C5 = 1. Substituting C6 = −1 and C5 = 1 into the [L] equation results in C4 = 1. To sum up, $\pi_2 = D^1 U^1 \mu^{-1} \rho = DU\rho/\mu$

Step 8: $f_s D/U$ is independent of $(DU\rho/\mu)$

Step 9: If we had chosen ρ instead of μ as a repeating variable, we would have arrived at $\mu/(DU\rho)$ as π_2 instead of $DU\rho/\mu$. As $DU\rho/\mu$ is the Reynolds number, $Re = \rho UD/\mu$, this is the step we flip $Re^{-1} = \mu/(DU\rho)$ into $Re = DU\rho/\mu$.

Step 10: We have found $f_s D/U = f(DU\rho/\mu)$, that is, the dimensionless shedding frequency, the Strouhal number, St, is a function of the Reynolds number, Re.

Incidentally, St is approximately equal to 0.2 for Re between roughly 10^2 to 10^5, for a smooth circular cylinder; see Fig. 10.3. This gives Dr. JAS a wide enough range to design her Turbulence and Energy Lab Wind Harp, aiming to surpass the Lucia and Aristides Demetrios Wind Harp that stands at 28 m tall in San Francisco.

10.4 Prevailing nondimensional parameters in fluid mechanics

There are many pragmatic nondimensional parameters in fluid mechanics. By far the most well-known and utilized is the Reynolds number. Before we expound on the Reynolds number, let us be enlightened by something analogous called the Grashof number. Rising smoke, powered by buoyancy, starts up nice and smooth, as depicted in Fig. 10.4. As it rises, it tends to gain speed because of gravitational acceleration, though the diffusion of heat to the surrounding air decreases the buoyancy force. As importantly, in the rising smoke is the

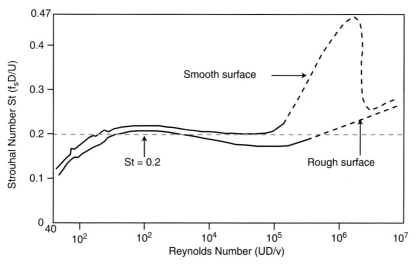

FIGURE 10.3 Nondimensional vortex shedding frequency, Strouhal number St = $f_s D/U$, as a function of Reynolds number for a circular cylinder (created by O. Imafidon). Note that for Reynolds number between 200 and 100,000 the Strouhal number remains approximately unaltered at 0.2.

amplification of naturally existing disturbances and, thus, the initially laminar rising smoke transitions into a turbulent plume. We note that temperature, or temperature difference between a volume (characterized by D^3) of hot smoke and the surrounding air, $T_s - T_\infty$, is a factor. Buoyancy does not come into effect unless there is gravity, g, and a density difference. The density difference is governed by the temperature difference, the coefficient of thermal expansion, β, and the density, ρ, of the involved fluid. By nature, viscosity, μ, is always there to oppose the rising smoke. A proper dimensional analysis would lead to the Grashof number:

$$Gr = g\beta(T_s - T_\infty)D^3/(\mu/\rho)^2. \tag{10.4}$$

We can perceive that Gr is the ratio of buoyancy force with respect to viscous force. When the Grashof number is small, there is not enough buoyancy force to overcome the restricting viscous force and, thus, a slightly lighter fluid in the midst of a marginally heavy fluid stays put. With increasing Grashof number, the lighter fluid starts rising in a laminar fashion. Transition into a turbulent plume takes place when the Grashof number is increased above the critical value. This usually occurs with an initially laminar rising plume because gravity, g, causes the rising plume to accelerate. The acceleration ultimately slows down with time as the temperature difference, $T_s - T_\infty$, diminishes due to diffusion heat transfer.

The phenomenon shown in Fig. 10.4 is a little complex in reality because of thermal and mass diffusion. An adiabatic (no heat transfer) buoyant fluid particle accelerates as it rises. Diffusion of thermal energy and mass, on the

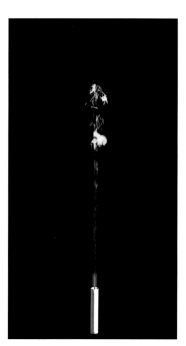

FIGURE 10.4   Laminar rising smoke transitions into turbulent smoke as it rises (created by X. Wang). The buoyant acceleration, which is of equal magnitude to the gravitational acceleration, causes the laminar smoke to become a turbulent plume before the acceleration eventually slows as the smoke loses heat to the surroundings. The smoke on the left is cooler than the one on the right. With a smaller temperature difference between the smoke and the ambient air, the left smoke is largely laminar. For the right smoke, the larger temperature difference powers it to accelerates upward rapidly and, hence, transitions into a turbulent smoke quickly.

other hand, weakens the buoyant force. Let us return to the pure fluid mechanic counterpart, the flow of an isothermal and homogeneous fluid that is governed by the Reynolds number. Professor Osborne Reynolds was the first to document a set of experiments with water flowing along a horizontal transparent pipe (Reynolds, 1883). With the help of dye, he noticed that the smooth and steady laminar flow at a low flow rate in a small pipe becomes unstable with increasing flow rate and/or pipe diameter. Thus, it is obvious that the velocity, U, and the pipe diameter, D, are participating factors. Dimensionally, we intuit that the fluid properties, specifically, density, ρ, and viscosity, μ, are also influencing factors. Performing a dimensional analysis ultimately leads to a nondimensional parameter of the form $\rho UD/\mu$, which is the Reynolds number. The physics underlying the Reynolds number has to do with the inertial force relative to the viscous force. Inertia is resistance to a change in motion. In context, a fluid particle of mass m moving at velocity U persists to move at velocity U. The product of this mass and velocity product can be expressed dimensionally as $mU = \rho \forall U = \rho D^3 U$, where D denotes the hydraulic diameter of the fluid container, flow passage, or volume of concern. The viscous force is prescribed by the fluid viscosity, μ. To compare apples to apples, we need to express the viscous effect in the same dimensions as the inertial one. The dimensions of μ are $[ML^{-1}t^{-1}]$ and that of mU are $[MLt^{-1}]$. That being the case, we can multiply μ with D^2 to make it an apple also, that is, having the same dimensions as mU. Doing so, we have

$$Re = (\rho D^3 U)/(\mu D^2) = \rho DU/\mu. \qquad (10.5)$$

At low Re, the viscous force dominates, and the flow is sticky, smooth, and steady, that is, laminar. Inertia increases with Re and, eventually, the flow becomes irregular or turbulent.

We have related the vortex shedding frequency in its normalized form, St, with Re, in Example 10.1. More common than vortex shedding in bluff body crossflow are the correlations between drag and lift with Reynolds number. In the wind harp example, the structure must be sturdy enough to withstand the drag imposed by the wind. Lift is caused by the alternate shedding of vortices off the cylinder the severity of which determines the vibration frequency and magnitude of each string and, thus, the melody. The nondimensional drag coefficient,

$$C_D = F_D/\left(\frac{1}{2}\rho U^2 A\right), \qquad (10.6)$$

where F_D is the drag force and A is the frontal or cross-sectional area. The denominator is the force produced by the dynamic pressure. In the same manner, the nondimensional lift coefficient,

$$C_L = F_L/\left(\frac{1}{2}\rho U^2 A\right), \qquad (10.7)$$

where F_L is the lift force.

Have you ever tried to sink a rubber ducky? The rubber ducky will accelerate up to the water surface because the buoyancy force is larger than the holding-back, viscous force. The nondimensional parameter involved is call the Archimedes number, Ar, credited to Greek mathematician Archimedes [287-212 BC], who discovered this during his eureka moment,

$$Ar = \rho_s gL^3(\rho_s - \rho)/\mu^2, \qquad (10.8)$$

where ρ_s is the density of the submerged body and L is the characteristic length (for example, the hydraulic diameter of the rubber ducky). We note that this is very similar to the Grashaf number, except that temperature does not come into play, in other words, we are dealing with isothermal flow. Instead of the rubber ducky, we can have a volume of lighter fluid and it rises increasingly faster with increasing Ar. Air bubbles rising in a column of water is possibly the most common illustration of significant Archimedes numbers in the workings.

To walk on water, like a green basilisk lizard, we have to make sure that the Bond number, which is also known as the Eötvös number, is small enough so that the surface tension can carry us through. The Bond[2] (or Eötvös) number is

$$\mathrm{Bo} = \mathrm{g}L^2 \Delta\rho/\sigma, \qquad (10.9)$$

where $\Delta\rho$ is the density difference between the two phases (typically gas and liquid) and σ is the surface tension. It describes the weight of the gravitational force with respect to the surface tension force. For a water lily pad, its weight, even with a not-too-heavy frog on top, is small compared to the surface tension force (note that this surface tension force increases with involved area) and, hence, it floats. Humans have indeed imitated and engineered lily pads that can support people afloat. The required surface area is large and, thus, not adroit as a suit for Dr. JAS to wear to walk on water when conducting water turbulence research. For Dr. JAS to stroll on water, she should mimic water striders instead of basilisks. This is because basilisks, unlike water striders and pygmy geckos, are significantly heavier and, thus, have to run to stay afloat. The first thing we notice about water striders are their long legs. A closer look will reveal that these legs are not only long and skinny but extremely hairy. These hydrofuge hairs trap many small air bubbles, making the legs hydrophobic. To have enough surface tension to hold their weight, notable portions of these hairy legs are placed parallel to the water surface to create an extra area on the water surface.

Somewhat similar to Bond or Eötvös number is the Weber number, named after German naval mechanics professor Moritz Weber [1871-1951]. The Weber number,

$$\mathrm{We} = \rho U^2 L/\sigma, \qquad (10.10)$$

can be perceived as the ratio of drag force to the cohesion force, the ratio of inertial force to interfacial (surface tension) force, or, more tangibly, the disruptive hydrodynamic forces with respect to the stabilizing surface tension force. It is the pertinent parameter for characterizing a droplet formation, both in terms of leaving a nozzle and when impacting a surface. An everyday example is a water fountain, where a small-We water jet is smooth and continuous while a large-We water jet is discontinuous, with blobs and/or droplets of water.

Table 10.2 summarizes the prevailing nondimensional parameters in fluid mechanics. To put it briefly, the Euler number, which can also be expressed as $\mathrm{Eu} = \Delta P/(\frac{1}{2}\rho U^2 A)$, indicates the pressure difference relative to the dynamic pressure. Among the many usages, Eu can be employed to characterize whether oil blobs will be trapped, pass clearly, or be dispersed as they pass through a restriction (Al-Shami et al., 2021). The map can be created by plotting Eu against Re or We. Eu has also been found to be useful in describing atomization

---

2. This is not attributed to 007 James Bond but English physicist Wilfrid Noel Bond [1897–1937]. The Eötvös number is for commemorating Hungarian physicist Loránd Eötvös [1848–1919].

TABLE 10.2 Prevailing nondimensional parameters in fluid mechanics.

Name	Definition	Underlying physics
Archimedes number	$Ar = \rho_s \, g \, L^3 \, (\rho_s - \rho)/\mu^2$	gravitational force/viscous force
Bond number, Eötvös number	$Bo = g \, L^2 \, \Delta\rho/\sigma$	gravitational force/surface tension force
Drag coefficient	$C_D = F_D /(\frac{1}{2}\rho U^2 A)$	drag force/dynamic force
Euler number	$Eu = \Delta P/(\rho U^2 A)$	pressure difference/dynamic pressure
Friction factor: Darcy friction factor Fanning friction factor	$C_f = 8\tau_w /(\rho U^2)$ $C_f = 2\tau_w /(\rho U^2)$	wall friction force/inertial force
Froude number	$Fr = U/\sqrt{(gL)}, \; U^2/(gL)$	inertial force/gravitational force
Lift coefficient	$C_L = F_L /(\frac{1}{2}\rho U^2 A)$	lift force/dynamic force
Mach number	$Ma = U/c$	flow speed/speed of sound
Pressure coefficient	$C_P = (P-P_\infty)/(\frac{1}{2}\rho U^2)$	static pressure difference/dynamic pressure
Reynolds number	$Re = \rho D L/\mu$	inertial force/viscous force
Richardson number	$Ri = g\beta(T_H - T_{ref})L/U^2 = Gr/Re^2$	buoyancy/flow shear
Strouhal number	$St = f_s L/U$	unsteadiness inertia/convective inertia
Weber number	$We = \rho U^2 L/\sigma,$	inertial force to interfacial force

regimes and liquid sheet thickness of a spray (Lachin et al., 2020). Friction factor describes losses in a flow due to wall friction with respect to the available inertial force. The more common Darcy friction factor is four times that of the Fanning friction factor. Froude number is the ratio of inertial force to gravitational force. It is important in free surface flow, such as water passing through a sluice gate. The flow speed with respect to the speed of sound is called the Mach number, Ma. The Mach number defines the compressibility of the flowing fluid. For values of Ma less than roughly 0.3, the flow can be assumed to be incompressible flow, the subject of this book. Next is the pressure coefficient, which signifies the difference between the local static pressure and the freestream, P-P$_\infty$, with respect to the dynamic pressure of the moving fluid. The Richardson number, Ri, is an everyday term in meteorology. Concerning atmospheric thermal convection, the Richardson number denotes the ratio of buoyancy to flow shear. It is equal to the Grashof number divided by the square of the Reynolds number.

10.5   Some remarks on dimensional analysis

To illustrate the need and utility of dimensional analysis, let us look at a worthwhile striving of figuring out the pressure drop per unit length of flow in a pipe. Considering the physics involved, we see that for flow in a smooth pipe, the pressure drop per unit length, $\Delta P/L = f(\mu, \rho, U, D)$. A conservative estimate of the number of required tests is $7 \times 7 \times 7 \times 7 = 2401$, in other words, over the range of interest to be covered, experiments for seven values of μ times seven values of ρ times seven values of U times seven values of D should be performed. The results can be plotted as $\Delta P/L$ versus μ for constant ρ, U and D; $\Delta P/L$ versus ρ while keeping μ, U, and D fixed; $\Delta P/L$ versus U for set μ, ρ, and D; and $\Delta P/L$ versus D for constant μ, ρ, and U. If we conduct a dimensional analysis, we will end up with

$$D\Delta P/L/(\rho U^2) = f(\rho UD/\mu). \qquad (10.11)$$

Dimensionally, the term on the left-hand side is a form of the Euler number, $\Delta P/(\rho U^2 A)$, and the right-hand side is Reynolds number. To rephrase it, dimensional analysis gives $Eu = f(Re)$. The specific relationship can be delineated by conducting the test at seven values of Re to give an explicit plot of Eu versus Re. All the above-mentioned 2401 points can be determined from this single plot of Eu versus Re!

  Thus, it is clear that dimensional analysis, when executed correctly, can save a lot of effort and, most of all, provides a generalized fundamental understanding of the underlying physics. To that end, significantly scaled-down (or scaled-up) models can be employed in a wind or water tunnel to study large-scale fluid mechanics such atmospheric flow over an urban area; see Cheng et al. (2021), for example. It must, however, be noted that some information may be lost in significant scaling, for instance, the range between the large and small turbulent eddies is substantially shrunk when going from atmospheric boundary layer flow over an actual urban area to the simulated flow over an appreciably scaled-down model. Like everything else out there, there are limitations and assumptions must be made. We should have a discerning mind, capitalizing on the usefulness while appreciating the limitations.

Problems

10.1 Pressure loss in a smooth pipe.
    The drag force in a smooth pipe is known to be a function of the density of the fluid, the velocity at which the fluid is moving, the diameter of the pipe, and the fluid viscosity, that is, $F_D = f(\rho, U, D, \mu)$. Conduct a dimensional analysis to obtain a relationship between the nondimensional drag force and the relevant dimensionless parameter.

10.2 Selection of the repeating terms in the Buckingham Pi theorem.

We have shown, through dimensional analysis, that the pressure drop in a smooth pipe can be expressed in terms of the Euler number as a function of the Reynolds number. Apply the Buckingham Pi theorem, by choosing different repeating variables (specifically, U, D, μ), and show that this can alternatively be expressed as $\Delta PD^2/(UL\mu) = f(Re)$. This tells us that a dimensional analysis can lead to a different set of pi terms.

10.3 Drag of a rectangular plate normal to the wind.

Rectangular signs subject to atmospheric wind load is a common occurrence especially on highways. Perform a dimensional analysis to determine the nondimensional parameters influencing the nondimensional drag. Hint: fluid density, viscosity, velocity, as well as the width and height of the sign are expected to be significant players.

10.4 Model car wind tunnel study.

The aerodynamic drag of a car moving at 60 km/h is to be predicted based on a one-third (1/3rd) scale model in a wind tunnel. To achieve similarity between the model and the car, the air in the wind tunnel should be moving at

A) 6 m/s

B) 17 m/s

C) 20 m/s

D) 33 m/s

E) 50 m/s

F) 60 m/s

G) 67 m/s

10.5 Study parachute in wind or water tunnel?

Parachutes are used to bring the parachuter's terminal velocity down to about 10% that of the free-falling value. To improve the design of an approximately 4-m diameter parachute, we can use scaled-down models in either a wind or water tunnel. If the target is for the parachute to help a 70-kg person to land with a terminal velocity of 20 km/h, what are the required conditions (wind speed) for a wind tunnel with a workable (keeping the blockage down to no more than 10%) cross-section of 1 m²? What about in a water tunnel with a small cross-section of 0.25 m²? For dynamic studies that include the deployment of the parachute, water has the advantage over air as it furnishes slower motion.

10.6 Visualizing knuckle ball.

A knuckle ball is presumably the hardest ball for a baseball batter to handle. This is because the ball is spun in such a way that the particular seam line jumbles the otherwise predictable ball path. For ease of visualization, a water tunnel is chosen for the study. An actual baseball spins in the order of 300 rpm while moving in the ballpark of 130 km/h. What should the water velocity be, and at what rpm should we spin the baseball, in the water tunnel?

10.7 Harnessing vortex shedding energy.

Spherical buoys are commonly encountered in many ocean engineering applications. As such, there is a need for a better understanding of vortex-induced vibrations of a sphere (Govardhan and Williamson, 2005; Sareen et al., 2018). Determine the vibration frequency, f, as a nondimensional function of the key parameters such as the diameter of the sphere, D, the mass of the sphere, m, the specific weight of the sphere γ, the velocity of the water current, U, and the dynamic viscosity of the water, μ. How does the vibration frequency vary with respect to a change in the mass of the sphere?

10.8 Appropriate rupturing of air balloons underwater.

Air can be stored at the hydrostatic pressure under water in an accumulator. Rupturing of this accumulator, similar to puncturing a balloon with a needle, can result in a beautiful buoyant vortex ring. Dr. JAS is commissioned to design an appropriate scale model for recreational water surfing purposes. The parameters of interest are the upward speed of the vortex ring, densities of air and water, viscosities of air and water, surface tension between the air–water interface, radius of the vortex ring radius of the accumulator, and gravity.

Part I)

Perform the dimensional analysis to deduce the nondimensional groups.

Part II)

Assign the originator's names to the common fluid mechanics dimensionless parameters involved.

Part III)

If the expansion and acceleration of the buoyant vortex ring are also of interest, are there other essential nondimensional parameters that needed to be included?

References

Alexander, R.M., 1989. Optimization and gaits in the locomotion of vertebrates. Physiological Rev. 69 (4), 199–227.

Alexander, R.M., 2004. Bipedal animals, and their differences from humans. J. Anat. 204, 321–330.

Al-Shami, T.M., Juffar, S.R., Negash, B.M., Abdullahi, M.B., 2021. Impact of external excitation on flow behavior of trapped oil blob. J. Pet. Sci. Eng. 196, 108002.

Animalia, https://animalia.bio/etruscan-shrew, (accessed October 6, 2021), 2021.

Antarctica, https://www.antarctica.gov.au/about-antarctica/animals/seals/weddell-seals/, (accessed October 6, 2021), 2021.

Ballesteros, F.J., Martinez, V.J., Luque, B., Lacasa, L., Valor, E., Moya, A., 2018. On the thermodynamic origin of metabolic scaling. Sci. Rep. 8 (1448), 1–10.

Butler, J.P., Feldman, H.A., Fredberg, J.J., 1987. Dimensional analysis does not determine a mass exponent for metabolic scaling. Am. J. Physiol.—Regulatory, Integrative Comparative Physiol. 253 (1), 195–199.

Caton, R., Otts, C., 1999. Fossil footprints: How fast was that dinosaur moving? Am. Biol. Teacher 61 (7), 528–531.

Celine, T., 2021. Tyrannosaurus rex speed found to be surprisingly slow that even humans can 'outwalk' them. The Science Times April 22.

Cheng, W-C., Liu, C-H., Ho, Y-K., Mo, Z., Wu, Z., Li, W., Chan, L.Y.L., Kwan, W.K., Yau, H.T., 2021. Turbulent flows over real heterogeneous urban surface: win tunnel experiments and Reynolds-averaged Navier-Stokes simulations. Build. Simul. 14 (5), 1345–1358.

Deming, D., 2020. The aqueducts and water supply of ancient Rome. Ground Water 58 (1), 152–161.

Govardhan, R., Williamson, C.H.K., 2005. Vortex-induced vibrations of a sphere. J. Fluid Mech. 531, 11–47.

Koch, C., 2016. Does brain size matter? Scientific Am. Mind 27 (1), 22–25.

Lachin, K., Turchiuli, C., Pistre, V., Cuvelier, G., Mezdour, S., Ducept, F., 2020. Dimensional analysis modeling of spraying operation—impact of fluid properties and pressure nozzle geometric parameters on the pressure-flow rate relationship. Chem. Eng. Res. Des. 163, 36–46.

McNab, B.K., 2008. An analysis of the factors that influence the level and scaling of mammalian BMR. Comparative Biochem. Physiol.—Part A: Molecular Integrative Physiol. 151 (1), 5–28.

Montanari, S., 2017. Actually, You Could Have Outrun a T. rex. National Geographic. https://www.nationalgeographic.com/science/article/tyrannosaur-trex-running-speed, (accessed February 2, 2022).

Reynolds, O., 1883. An experimental investigation of the circumstances which determine whether the motion of water shall be direct or sinuous, and of the law of resistance in parallel channels, Phil. Trans. Royal Soc. London 174: 935–982.

Sareen, A., Zhao, J., Lo Jacono, D., Sheridan, J., Hourigan, K., Thompson, M.C., 2018. Vortex-induced vibration of a rotating sphere. J. Fluid Mech. 837, 258–292.

Sellers, W.I., Pond, S.B., Brassey, C.A., Manning, P.L., Bates, K.T., 2017. Investigating the running abilities of Tyrannosaurus rex using stress-constrained multibody dynamic analysis. J Life Environ. 5, e3420.

West, G.B., Brown, J.H., 2004. Life's universal scaling laws. Phys. Today 57 (9), 36–42.

Chapter 11

Internal flow

"The quality of the imagination is to flow and not to freeze."

Ralph Waldo Emerson

Chapter Objectives

- Cherish internal flow of Newtonian fluids.
- Correlate flow type (laminar, transition, turbulent) with Reynolds number.
- Appreciate flow development at pipe entrances.
- Fathom the relation between pressure drop and wall shear.
- Become familiar with the velocity profile of fully developed laminar pipe flow.
- Note the role of fluid weight in non-horizontal pipe flow.
- Understand energy conservation and head loss in pipe flow.
- Differentiate minor losses from major losses.
- Comprehend friction factor and master the Moody chart (diagram).

Nomenclature

A	area; A_{cross} is cross-sectional area, A_{surf} is surface area
a	acceleration
C	constant; C_1, C_2 ... are constants
D	diameter; D_e is hydraulic or equivalent diameter
d	difference or differential
F	force; F_i is inertial force, F_τ is viscous force
f	friction factor
g	gravity
h	head; h_L is head loss, $h_{L,major}$ is major head loss, $h_{L,minor}$ is minor head loss
î	unit vector in the x direction
K_L	loss coefficient
L	characteristic length; L_e is entrance length, ΔL is section length
m	mass; m' is mass flow rate
MERV	Minimum Efficiency Reporting Value
P	pressure; ΔP is pressure change
p_e	perimeter
R	radius
r	radial distance
Re	Reynolds number
t	time

Thermofluids: From Nature to Engineering. DOI: https://doi.org/10.1016/B978-0-323-90626-5.00014-8

U   velocity; velocity in the x direction, U_{avg} is average velocity, U_c is centerline velocity, $\vec{U}$ is velocity vector

x   distance or streamwise direction

y   distance or direction

z   vertical direction or distance

Greek and other symbols

α kinetic energy coefficient

γ specific weight

Δ difference

θ angle of inclination

μ dynamic viscosity

ρ density

τ shear; τ_w is wall shear

ε roughness

$\forall$ volume; $\forall'$ is volume flow rate

11.1   Flow in a channel

Flow in nature and in engineering can be categorized into (1) internal flow, (2) external flow, and (3) open-channel flow. We introduce internal flow in this chapter. An internal flow is a flow that is walled by a solid surface such as a tube. Blood flowing in a blood vessel is an essential internal flow for sustaining life. In engineering, water flowing in a pipe is the most obvious internal flow example. It is critical for quenching thirst, maintaining hygiene, producing produce, and so forth. In this book, we will only give an account of Newtonian fluids, where the shear is proportional to the velocity gradient, that is,

$$\tau = \mu dU/dy \qquad (11.1)$$

Here, μ is the dynamic viscosity, U is the velocity in the streamwise direction, and y is the distance in the cross-stream direction. The point here is that for Newtonian fluids, the viscosity of the fluid does not vary with the flow or velocity gradient. Furthermore, we are limiting our discussion to incompressible fluid, that is, the density of the fluid remains unchanged. That is to say that we are only concerned with subsonic and isothermal flows. With these in mind, we proceed to dig into the underlying physics of the internal flow.

11.2   The Reynolds number and the type of pipe flow

What Reynolds number denotes is clearly perceptible for external flows such as pouring honey out of a container and water falling down a big waterfall. For the flow of honey, the stickiness is very tangible, and it dominates the flow. The rumbling sound produced by water tumbling down a waterfall depicts inertia[1] of

1. Recall that inertia is resistance to change in motion and, within context, to slow down. To put it another way, the many chunks of moving fluid continue to move, while fluid viscosity tries to slow them down. For the free-falling fluid chunks flowing down a waterfall, gravity furthers their motion, causing them to accelerate.

the hastening fluid. Irish mathematician and physicist Osborne Reynolds [1842–1912] is credited with the first explicit demonstration of the relative importance of inertial versus viscous forces in fluid flows. The acclaimed Reynolds apparatus, illustrating flow in a horizontal pipe, is still in use today for demonstrating the transition from the laminar flow with increasing velocity (and hence inertial force) in a transparent pipe (Reynolds, 1883). This is because the fundamental driver dictating the key flow characteristics is the inertial-viscous force ratio. Having said that, much more details have since been revealed, mostly via detailed numerical simulations, see Wu et al. (2015), for example.

There are different ways to derive the Reynolds number. Here is one straight-forward approach. The Reynolds number,

$$\text{Re} = \text{inertial force/viscous force.} \tag{11.2}$$

Dimensionally, the inertial force associated with an element of moving fluid, according to Newton's second law of motion, is

$$F_i = ma = m\prime U = \rho \forall\prime U = \rho(L^2 U)U = \rho L^2 U^2, \tag{11.3}$$

where m is mass, a is acceleration, m' is mass flow rate, ρ is density, $\forall$ is volume, $\forall$' is volume flow rate, and L is the characteristic length, that is, the hydraulic diameter of the conduit. We may envision the moving fluid element as a sphere of diameter L and, thus, the volume flow rate, $\forall$', is $L^2 U$, dimensionally. For the viscous or shear force, we can write,

$$F_\tau = \tau A = (\mu dU/dy)A = (\mu U/L)(L^2) = \mu UL, \tag{11.4}$$

where A is the surface area of concern. Dividing the inertial force, Eq. (11.3), by the viscous force, Eq. (11.4), gives the Reynolds number,

$$\text{Re} = (\rho L^2 U^2)/(\mu UL) = \rho LU/\mu. \tag{11.5}$$

For a fluid of fixed density and viscosity, such as water at room temperature, the Reynolds number can be varied by changing the characteristic length and/or the flow velocity. For a constant cross-sectional circular pipe, the characteristic length is the diameter of the pipe. The hydraulic diameter is the characteristic length for noncircular pipes.

To minimize flow disturbance, Osborne Reynolds employed a trumpet entrance and, for widening the range of Reynolds numbers, 6-foot-long glass tubes with three different diameters were utilized (Reynolds, 1883). To ease visualization, he injected color at the entrance of the tube. The injected color streak remained steady at low velocity and/or in a small tube, as illustrated in Fig. 11.1A. Small disturbances are taken care of by viscosity at this low Reynolds number. This can be envisioned as a large group of peacekeepers (viscosity) keeping a small group of mellow demonstrators (small disturbances) in check, resulting in orderly laminar flow furnishing steady streaks along the flow passage. With increasing flow rate, by elevating the velocity and/or increasing the

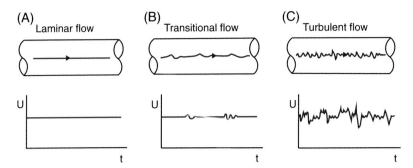

FIGURE 11.1   Laminar, transitional, and turbulent flow regimes for pipe flow (created by Y. Yang). Typically, the flow is steady and laminar for Re < 2100, it is transitional for 2100 < Re < 4000, and it becomes fully turbulent for Re > 4000.

diameter of the tube, the flow showed an intermittent behavior; see Fig. 11.1B. Occasional oscillations in the otherwise smooth and steady flow are exhibited by the color streaks. This is the Reynolds number range where the inertial force and the viscous are quite even, where the flow has entered the transitional regime. The intermittent, mildly unsteady streaks signify the instances when the inertia (demonstrators) has a slight upper hand. Further increases in Reynolds number, via an increase in flow velocity and/or pipe diameter, give rise to turbulent flow, as depicted in Fig. 11.1C. With inertia dominating, the flow is highly disturbed, with eddying motions of various sizes everywhere in the flow except the near-wall region. The parallel for this in the peacekeeping analogy is that when the inertia of the mob surpasses the "viscous" capacity of the peace-keepers, things go out of control and the group of people goes randomly in all directions.

The critical Reynolds number above which flow in a pipe becomes unsteady is approximately 2000. That said, it is possible to delay this occurrence to a much larger Reynolds number, if extreme care is taken to smoothen the pipe, reduce vibrations, and minimize upstream fluctuations and flow agitations. Pfenninger (1961) was able to delay the critical Reynolds number to a value of the order of 10^5. There are many imperfections and disturbances in typical engineering applications and, hence, the flow in a pipe is assumed to be turbulent for Reynolds numbers larger than 2300.

Example 11.1   Water channel for studying natural gas flow.
Given: TransCanada Pipelines approached Dr. JAS to further their natural gas delivery system. As a first step, she wishes to have a preliminary understanding of natural gas flow in a 1.2 m diameter pipe using her 0.2-m diameter water channel in the Turbulence and Energy Laboratory. In the real pipeline, natural gas ($\rho \approx 28$ kg/m^3, $\mu \approx 1.2 \times 10^{-6}$ kg/(s·m^2)) travels at about 30 m/s.

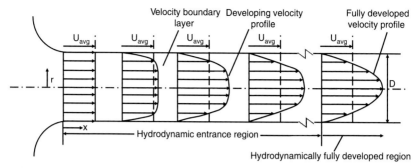

FIGURE 11.2 The development of pipe flow with a uniform velocity at the entrance (created by Y. Yang). The boundary layer is defined by the layer from the wall where the velocity is less than 99% the freestream velocity. The flow becomes fully developed when the boundary layer reaches the center of the pipe. The parabolic profile signifies that the flow is laminar; a turbulent profile is fuller or flatter.

Find: If Dr. JAS' 0.2-m diameter water channel can run at the same Reynolds number.

Solution: The Reynolds number of the natural gas pipe flow,

$$\text{Re} = \rho LU/\mu = 28(1.2)(30)/1.2 \times 10^{-6} = 8.4 \times 10^{8}$$

To duplicate this in Dr. JAS' 0.2 m diameter water channel, we have

$$\text{Re} = \rho LU/\mu = (1000)(0.2)U/0.001 = 8.4 \times 10^{8}$$

This requires the water to be moving at a velocity, U, of 4200 m/s. This is an impossible task! In engineering practice, the upper limit for water velocity is in the neighborhood of 6 m/s, even though this maximum velocity can be pushed up to 9 m/s for larger pipes (Kim, 2021). Cavitation erosion, where water vapor cavities are formed at high flow velocities, tends to dictate the upper limit of water velocity (Kim, 2021). It is also worth noting that the speed of sound in water is approximately 1500 m/s; see, for example, Bilaniuk and Wong (1993) and Lubbers and Graaff (1998).

11.3 Developing pipe flow

To better appreciate pipe flow, let us start from the entrance of a horizontal pipe with a uniform velocity profile, as shown in Fig. 11.2. A gradually converging funnel is employed to guide the flow smoothly into the pipe, where x = 0 is the beginning of the pipe with a perfectly uniform velocity profile. Imagine Dr. JAS temporarily shrinks all her students into infinitely small beings each riding on a fluid element without interfering with the flow. The students line up across, along the diameter, of the entrance section. Just inside the pipe, the scholars along the wall get stuck to the wall due to viscosity. This is called the "no-slip" condition.

The remaining pupils, at varying distances from the wall, continue to cruise along the pipe. A little into the pipe, the trapped-to-the-wall students drag their nearby classmates and slow them down. The layer by the wall, where viscosity drags the moving fluid particles (and thus the individuals riding on them) down, is called the boundary layer. That being the case, the boundary layer is the layer within which viscous effects are important. Since the effect of viscosity fades gradually from the near-wall boundary layer to the freestream, it is convenient to define the boundary layer as the distance from the wall where velocity reaches 99% of the freestream value. Outside this boundary layer is the core region, where viscous effects are negligible, and, thus, it is referred to as the inviscid core.

For the incompressible fluid moving in a constant cross-section pipe, as depicted in Fig. 11.2, continuity requires the volume flow rate at any cross-section to remain the same. That being the case, the significant near-wall slowdown has to be compensated by an increase in the flow at the remaining cross-section of the pipe. The boundary layer development and the corresponding speeding up of flow away from the wall, progress until the thickening boundary layer merges along the center of the pipe. Beyond this distance from the entrance, called the hydrodynamic *entrance length*, the velocity profile remains unchanged and, hence, the flow is *fully developed.*

Wall shear is of interest because it directly relates to friction and, hence, the power required to push the fluid along the pipe. For Newtonian fluids, the shear is directly related to the velocity gradient, dU/dy, where y is the distance from the wall. We note that in the entrance region where the boundary layer develops, this velocity gradient decreases with respect to distance from the entrance, x. under the circumstances, according to Eq. (11.1), the shear decreases with increasing x. Due to the asymptotic nature of the flow development, the entrance length, where the flow is considered fully developed, corresponds to the location where the wall shear stress, the friction factor, reaches within 2% of the fully developed value.

11.3.1   Laminar pipe flow entrance length

For values of Reynolds number less than about 2300, the flow in a pipe is laminar. For laminar flow, the nondimensional hydrodynamic entrance length is found to be

$$L_e/D \approx 0.06\rho LU/\mu = 0.06\,Re. \qquad (11.6)$$

We see that L_e increases linearly with the Reynolds number and it can be up to 138D at Re = 2300. This indicates that laminar pipe flow takes a long time to become fully developed. To put it another way, typical laminar pipe flows encountered in real life do not have enough length to become fully developed. This is more so after a bend downstream of which large secondary flow structures, such as Dean's vortices, are formed. These organized flow structures can skew the velocity profile over an extended distance. In short, we should be aware that the flow may not be in the ideal fully developed condition in real life. Thankfully, invoking the fully developed flow assumption does not usually have any significant detrimental effects in engineering practice.

11.3.2 Turbulent pipe flow entrance length

When the flow becomes turbulent, there is a lot of inertia in the flow and the inertia is random, that is, fluctuating in all directions. Because of this, the boundary layer develops significantly faster than its laminar counterpart. To put it another way, the inertia in the cross-stream direction hastens the thickening of the boundary layer, such that the boundary layer from the wall converges timely. It is found that the entrance length for turbulent pipe flow is:

$$L_e/D = 1.359\,Re^{1/4}. \tag{11.7}$$

In contrast to its laminar counterpart, the entrance length for turbulent pipe flow is only weakly dependent on the Reynolds number; specifically, it only increases with the fourth root of the Reynolds number. As a consequence of this weak dependence, the entrance length for turbulent pipe flow can be approximated as

$$L_e \approx 10\,D. \tag{11.7a}$$

In short, it only takes about 10 pipe diameters for turbulent pipe flow to become fully developed. It is worth mentioning that flow turbulence also tends to disintegrate organized flow structures such as secondary flows generated in an elbow due to centrifugal force or by a bluff body due to flow shear. Because of that, the fully developed turbulent velocity profile is re-established shortly. In other words, invoking the fully developed flow assumption for internal turbulent flow is much more readily valid compared to when the flow is laminar. Most recently, however, the study by Bopp and Weiss (2022) indicates that the entrance length for turbulent pipe flow can be much longer than forty diameters.

Example 11.2 Entrance length of a microchannel.
Given: Kucukal et al. (2021) studied blood flow coagulation in a microchannel. They employed a 4-mm wide, 0.05-mm high, and 25-mm long channel. The velocity range of interest was 200 to 600 μm/s.
Find: The entrance length according to the classical correlation expressed as Eq. (11.6), assuming blood behaves like water.

Solution: According to the classical correlation, the hydraulic entrance length for laminar pipe flow is a linear function of the Reynolds number, namely,

$$L_e/D \approx 0.06\rho LU/\mu = 0.06\,Re.$$

The hydraulic diameter of the rectangular channel is

$$D_e = 4A/p_e = 4(0.004)(0.00005)/[2(0.004 + 0.00005)] = 9.88 \times 10^{-5}\text{m}.$$

Here, A is the cross-sectional area of the rectangular channel, and p_e is the perimeter. With this hydraulic diameter, along with $\rho = 1000$ kg/m³ and $\mu = 0.001$ Ns/m² for water at 20°C, the corresponding Reynolds number is,

$$\text{Re} = \rho D_e U / \mu = (1000)(9.88 \times 10^{-5})U/(0.001).$$

For $U = 200$ μm/s,

$$\text{Re} = \rho D_e U / \mu = (1000)(9.88 \times 10^{-5})(200 \times 10^{-6})/(0.001) = 0.02.$$

For $U = 600$ μm/s,

$$\text{Re} = \rho D_e U / \mu = (1000)(9.88 \times 10^{-5})(600 \times 10^{-6})/(0.001) = 0.06.$$

The entrance length,

$$L_e/D \approx 0.06\,\text{Re} = 0.0012 \text{ to } 0.0036.$$

This indicates that the entrance length is extremely short and can be neglected.

Note that microchannels of larger sizes, with fluid flowing at higher velocities, are also common. The corresponding Reynolds numbers are in the tens or hundreds; see, Chen et al. (2020), for example. If the Reynolds number, $\text{Re} = 500$. Then, we have

$$L_e/D \approx 0.06\text{Re} = 0.06(500) = 30.$$

This suggests that care should be exercised for some microchannel applications where the total length is less than 10^2. Having said that, there is still significant debate concerning the validity of the classical expression, $L_e/D \approx 0.06\text{Re}$, at this low Re end; see Li et al. (2009) and Galvis et al. (2012), for example.

It is also worth mentioning that blood is more viscous than water (Klabunde, 2021) and its viscosity changes with temperature, vessel size (Secomb and Pries, 2013), etc. Interested readers may also wish to read the work of Yamamoto et al. (2020), where a falling needle rheometer is employed to measure the viscosity of human blood.

11.3.3 Pressure and shear stress

To circulate the blood inside our body, with a total length of over 160 km (Franklin Institute, 2021), a powerful natural pump, our heart, has to work hard day and night. With that in mind, a significant amount of work is required to force the flow of blood. Similarly, a mechanical pump is needed to deliver water through a network of pipes. The simplest pipe flow involves a horizontal pipe, where gravity does not come into play. A pump is still needed to overcome the pressure drop associated with friction because every fluid has a finite viscosity. Losses due to secondary flows, such as recirculating eddies in a bend, demand additional pumping power.

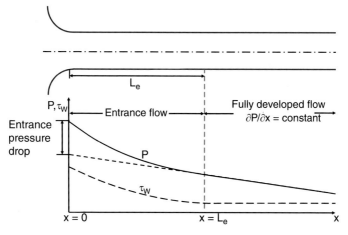

FIGURE 11.3 The variation of pressure drop and wall shear for a developing horizontal pipe flow (created by Y. Yang). The rate of pressure drop with distance decreases in the entrance region. It reaches a constant value once the flow is fully developed. The corresponding wall shear remains constant beyond the entrance length.

According to Newton's second law, the net force acting on a fluid element moving at a constant velocity in a straight horizontal pipe is zero because there is no acceleration (due to gravity). This is the case when the flow is fully developed, and the velocity profile remains constant. For that reason, the opposing viscous force is equal to the pumping force, for a flowing volume of fluid with a fixed velocity profile moving at a fixed velocity. Specifically, the pressure drop in a segment of a straight horizontal pipe section of moving fluid,

$$\Delta P A_{cross} = \tau A_{surf} = \tau 2\pi r \Delta L, \tag{11.8}$$

where A_{cross} is the inner cross-sectional area of the pipe and A_{surf} is the inner surface area of the ΔL segment of the pipe of interest. Remember that multiplying pressure by area gives us force, that is, the expression is about the change in the force of the flowing fluid. Dividing it by the cross-sectional area, A_{cross}, gives the pressure drop,

$$\Delta P = 2\tau \Delta L/r. \tag{11.8a}$$

Specifically, the pressure change (drop) per unit pipe length is directly related to the wall shear, that is, dividing it by ΔL gives

$$\Delta P/\Delta L = 2\tau_w/r \tag{11.8b}$$

The pressure gradient and the wall shear are plotted as a function of the distance from the entrance of a horizontal pipe, in Fig. 11.3. Recall that the near-wall velocity gradient, dU/dy, is the highest at the entrance and, thus, also the wall shear τ_w. As the flow develops, due to the thickening boundary layer and the consequential smoothing out of the velocity profile, the velocity gradient decreases to an asymptotic value at the entrance length. The direct fallout of this is the decreasing wall shear to the corresponding asymptotic value. It follows that the pressure gradient decreases asymptotically to its fully developed value. This is displayed by the decreasing pressure drop rate in Fig. 11.3 until a constant pressure drop rate beyond the entrance length.

11.4 Fully developed horizontal pipe flow

Earlier, we hinted that an entrance, bends, and elbows can have significant impacts on the resulting flow. In practice, there are many different fittings and, thus, their respective characteristics are experimentally quantified and supplied by the manufacturer. Straight pipe sections that typically make up the largest share of a fluid distribution network are quite generic and can be understood via first principles. Let us scrutinize a section of straight pipe.

Consider a steady, fully developed flow in a horizontal pipe section, shown in Fig. 11.4. Since the flow is fully developed, the velocity at any radial location is fixed and, thus, there is no local acceleration. Namely,

$$\partial \vec{U}/\partial t = 0, \tag{11.9}$$

where $\vec{U}$ is the velocity vector and t is time. Noting that $\partial U/\partial t = (\partial x/\partial t)$ $(\partial U/\partial x)$, we can express the streamwise component as

$$U\partial U/\partial x\,\hat{\imath} = 0, \tag{11.9a}$$

where $\hat{\imath}$ is the unit vector in the x direction. Physically, this, as illustrated in Fig. 11.4, says that fluid at any part of the cross-section flows along its streamline, parallel to the pipe, at a constant local velocity, notwithstanding that the local velocity decreases to zero from the core to the wall.

11.4.1 Pressure drop

The importance of pressure drop is worth reemphasizing and hereby, let we go through the derivation step-by-step with the help of an illustrative figure. We can apply Newton's second law, $F = ma$, to the cylindrical element of fluid of length L in the pipe, as depicted in Fig. 11.5. As the pipe is horizontal, we can neglect the force of gravity. Furthermore, the pressure at any cross-section is uniform, that is, $\partial P/\partial r = 0$. In other words, the pressure only decreases in the streamwise direction, $P = P(x)$, due to losses as the fluid makes its way along the flow passage. Without acceleration, the fluid is moving at a constant velocity, Newton's second law says that the net force in the streamwise direction is zero. Namely, the force balance of the free body shown in Fig. 11.5 is

$$P\pi r^2 - (P - \Delta P)\pi r^2 - \tau 2\pi rL = 0, \tag{11.10}$$

where P is the pressure at the beginning of the cylindrical fluid element, ΔP is the pressure drop over L distance downstream, and τ is the shear acting on a

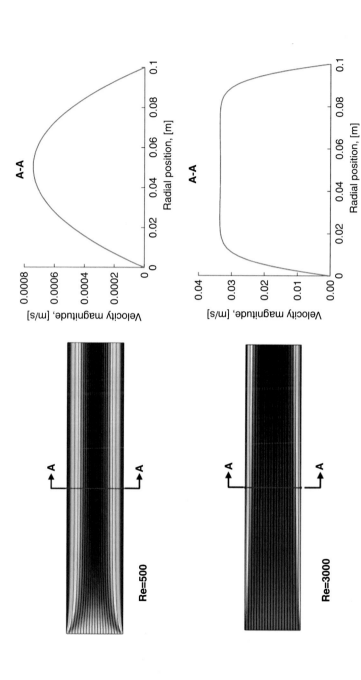

FIGURE 11.4 Fully developed flow in a horizontal pipe (created by X. Wang). Note that the velocity profile remains unchanged along the pipe. The smooth, parabolic profile corresponds to laminar flow at Re = 500, while the flatter one is turbulent at Re = 3000.

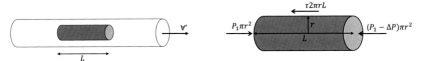

FIGURE 11.5   Forces acting on a cylindrical fluid element of length L (created by O. Imafidon). The net force is zero because the fluid is moving at a constant velocity, that is, the acceleration is zero, according to Newton's second law of motion.

cylindrical area defined by $2\pi rL$. The equation can be reduced into

$$\Delta P/L = 2\tau/r. \tag{11.10a}$$

We note that, since the flow is fully developed, the left-hand side of Eq. (11.10a) is a constant, and the pressure drop rate does not change with respect to the streamwise distance, x. It follows that the right-hand side, $2\tau/r$, is also a constant, that is,

$$\tau = Cr, \tag{11.11}$$

where C is a constant which can be determined from the boundary conditions. At the core, $r = 0$, the velocity gradient, $\partial U/\partial r$, is zero due to symmetry in the velocity profile and, thus, the shear, τ, is zero. The other known boundary condition is at the wall, where the maximum shear stress occurs. To put it another way, at $r = D/2$, $\tau = \tau_w$, where D is the inner pipe diameter and τ_w is wall shear. Substituting this boundary condition into Eq. (11.11) gives $\tau_w = C\,(D/2)$ or $C = 2\tau_w/D$. That being the case, we have

$$\tau = 2\tau_w r/D. \tag{11.11a}$$

This conveys that the shear increases linearly with radial distance from the center, where the shear is zero, to τ_w at the wall. Substituting Eq. (11.11a) into Eq. (11.10a) gives us the pressure drop across the segment of pipe, that is,

$$\Delta P = 4L\tau_w/D. \tag{11.12}$$

This asserts that the pressure drop increases linearly with pipe length and wall shear, and inversely with the diameter of the pipe.

11.4.2   Velocity profile

Let us further the analysis to deduce an expression for the fully developed velocity profile in terms of wall shear. For Newtonian fluids, $\tau = \mu dU/dy$, where y is the distance from the wall into the core of the pipe and, thus, the radial distance from the center, $r = -y$. We will only expound on the laminar case here, noting that the added, and more consequential, turbulent eddy viscosity significantly flattens the velocity profile when the flow is turbulent (Ting, 2016).

For laminar flow of a Newtonian fluid, the shear is related to the velocity gradient in the form

$$\tau = -\mu dU/dr. \tag{11.13}$$

This can be rearranged into

$$dU/dr = -(1/\mu)\tau. \tag{11.13a}$$

From $\Delta P/L = 2\tau/r$, we have $\tau = \frac{1}{2} r \, \Delta P/L$, which can be substituted into Eq. (11.13a) to get

$$dU/dr = -(\Delta P/(2\mu L)) r. \tag{11.14}$$

This can be integrated to give the streamwise velocity variation along the radial distance from the center as

$$U(r) = -(\Delta P/(4\mu L)) r^2 + C_1, \tag{11.15}$$

where C_1 is a constant. At the wall, $r = D/2$, the velocity, $U = 0$, due to no-slip at the wall. It follows that $C_1 = (\Delta P/(16\mu L))D^2$ and the streamwise velocity can be written as

$$U(r) = \left(\Delta P D^2/(16\mu L)\right)\left[1-(2r/D)^2\right] = U_c\left[1-(2r/D)^2\right]. \tag{11.15a}$$

In normalized form, this fully developed laminar velocity profile in a circular pipe is simply

$$U(r)/U_c = 1 - (r/R)^2, \tag{11.15b}$$

where R is the pipe radius. For fully developed turbulent pipe flow, the generic form is

$$U(r)/U_c = (1 - r/R)^C, \tag{11.15c}$$

where the exponent C is in the neighborhood of 0.2.

There is one more expression for fully developed laminar pipe flow. We note that the centerline velocity, $U_c = \Delta P D^2/(16\mu L)$. Since $\Delta P = 4L\tau_w/D$, the streamwise velocity variation with respect to radial distance from the center can alternatively be written as

$$U(r) = (\tau_w D/(4\mu)) \left[1-(r/R)^2\right], \tag{11.16}$$

where R is the inner radius of the pipe.

11.4.3   Volumetric flow rate and average velocity

The volumetric flow rate can be deduced by integrating the velocity with respect to the cross-sectional area. Specifically,

$$\forall' = \int U(r)dA = \int U_c\left[1 - (r/R)^2\right]2\pi r dr = \frac{1}{2}\pi R^2 U_c = \pi \Delta P R^4/(8\mu L)$$

$$= \pi \Delta P D^4/(128\mu L). \tag{11.17}$$

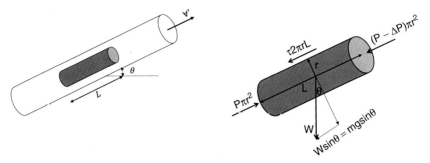

FIGURE 11.6 Forces acting on a cylindrical fluid element of length L at an inclination angle, θ (created by O. Imafidon). Compared to the horizontal case, the specific weight of the fluid, $\gamma = \rho g$, is the additional term.

We see that the volumetric flow rate is proportional to the diameter to the fourth power. It is proportional to the pressure change and inversely proportional to viscosity and pipe length. The average velocity is simply the volumetric flow rate divided by the cross-sectional area, that is,

$$U_{avg} = ∀'/A = \Delta P D^2/(32\mu L). \tag{11.18}$$

11.5 Fully developed inclined pipe flow

With inclination, gravity comes into play. If the fluid is a gas such as air, the gravitational effect is typically small and can be neglected. Nonetheless, it is good practice to keep the gravitational term in and let the numbers do their work. Consider the inclined circular pipe, as shown in Fig. 11.6. Everything is the same as the horizontal pipe (Fig. 11.5) except there is the gravitational term associated with the weight of the cylindrical element of the fluid. The force balance of the free body shown in Fig. 11.6 is

$$P\pi r^2-(P - \Delta P)\pi r^2-\tau 2\pi rL - mg \sin \theta = 0, \tag{11.19}$$

where the inclination angle, θ, is positive if the fluid is flowing uphill and negative when it is moving downhill. If the pipe is vertical and the flow is vertically upward, then this weight will act directly downward; that is, $\theta = 90°$ and $\sin 90° = 1$. The mass of the element is equal to the fluid density times the volume, $m = \rho ∀ = \rho(\pi r^2 L)$. With that in mind, the force balance equation can be simplified into

$$\Delta P\pi r^2-2\tau\pi rL - \rho(\pi r^2 L)g \sin \theta = 0. \tag{11.19a}$$

This expression can be rearranged into

$$[\Delta P - \rho g L \sin \theta]/L = 2\tau/r. \tag{11.19b}$$

We notice that the second term on the left-hand side of Eq. (11.19b) drops out when the pipe is horizontal. Given that the entire left hand is a constant for a particular inclined pipe flow, we can, as we did for the horizontal pipe, write

$$\tau = C_2 r, \tag{11.20}$$

where C_2 is a constant. At the center of the pipe, $r = 0$, the velocity gradient, $\partial U/\partial r$, is zero and thus also the local flow shear, τ. At the wall, $r = D/2$, the flow shear corresponds to the wall shear, that is, $\tau = \tau_w$. Substituting this into Eq. (11.20), we get $C_2 = 2\tau_w/D$. Thence, $\tau = 2\tau_w r/D$, which can be substituted into Eq. (11.19b) to get

$$\Delta P - \rho g L \sin\theta = 4L\tau_w/D. \tag{11.21}$$

Recall that the velocity gradient, $dU/dr = -(1/\mu)\,\tau$. From $[\Delta P - \rho g L \sin\theta]/L = 2\tau/r$, the shear can thus be expressed as $\tau = \frac{1}{2}r\,(\Delta P - \rho g L \sin\theta)/L$. With this, the velocity gradient becomes

$$dU/dr = -\big[(\Delta P - \rho g L \sin\theta)/(2\mu L)\big]\,r. \tag{11.22}$$

Integrating this leads to

$$U(r) = -\big[(\Delta P - \rho g L \sin\theta)/(4\mu L)\big]\,r^2 + C_3, \tag{11.23}$$

where C_3 is a constant. With no-slip at the wall, we have $U = 0$ at $r = D/2$, from which we get $C_3 = [(\Delta P - \rho g L \sin\theta)/(16\mu L)]D^2$. That being so, the streamwise velocity becomes

$$U(r) = \big[(\Delta P - \rho g L \sin\theta)D^2/(16\mu L)\big]\big[1-(2r/D)^2\big] = U_c\big[1-(2r/D)^2\big]. \tag{11.23a}$$

Noting that the centerline velocity, $U_c - (\Delta P - \rho g L \sin\theta)\,D^2/(16\mu L)$. Since $\Delta P = 4L\tau_w/D + \rho g L \sin\theta$, the streamwise velocity variation with radial distance from the center can also be expressed as Eq. (11.16), the same expression for the horizontal pipe. That is to say that the inclination angle does not directly come into play when it comes to the velocity profile. Indirectly, however, the fluid weight effect associated with inclination influences the wall shear, as can be seen from Eq. (11.21).

Integrating the velocity with respect to the cross-sectional area gives the volumetric flow rate,

$$\forall' = \int U(r)dA = \pi(\Delta P - \rho g L \sin\theta)R^4/(8\mu L)$$
$$= \pi(\Delta P - \rho g L \sin\theta)D^4/(128\mu L). \tag{11.24}$$

Dividing this by the cross-sectional area gives the average velocity,

$$U_{avg} = \forall'/A = (\Delta P - \rho g L \sin\theta)D^2/(32\mu L). \tag{11.25}$$

Example 11.3 Pumping blood up or down a Giraffe's long neck.
Given: With its head up to 3 m away from its heart, a giraffe is bestowed with a powerful pump and unique circulatory mechanisms; see, for example, Zhang (2006). An adult giraffe's carotid artery is around 1.2 cm in diameter (Goetz, 1960).

Find: The pressure change, ΔP, when the giraffe stands tall, while grazing the premium greens at the top of a tree. How does that compare with when the giraffe has its neck horizontally level, such as during galloping (Dagg, 1962)? What about when its head is vertically down when the giraffe is quenching thirst? Assume the blood to be flowing at 4×10^{-5} m³/s, its viscosity is 0.0012 kg/(m·s), its density is 1000 kg/m³, and the vessel is smooth and straight.

Solution: The volumetric flow rate is related to the pressure drop and fluid weight, that is,

$$\forall' = \pi(\Delta P - \rho g L \sin\theta)D^4/(128\mu L).$$

For the horizontal case, with $\theta = 0$, we have

$$4 \times 10^{-5} = \pi(\Delta P - 0)(0.012)^4/[128(0.0012)(3)].$$

This leads to $\Delta P = 0.28$ kPa, which is not a challenge, even for a human heart to supply the pressure to overcome this pressure drop.

When the giraffe holds its neck vertically up, $\theta = 90°$, we have

$$4 \times 10^{-5} = \pi[\Delta P - (1000)(9.81)(3)\sin(90°)](0.012)^4/[128(0.0012)(3)].$$

This gives a pressure drop, $\Delta P = 29.71$ kPa. To overcome this amount of pressure drop by pumping blood up this 3-m rise in elevation is a challenge even for the large 30-cm heart of a giraffe. For this reason, giraffes are equipped with distinctive ingenious features. Specifically, they are blessed with valves in their jugular veins to prevent the backflow of blood along their 3-m long vertical necks.

When its neck is vertically down, $\theta = -90°$, we have

$$4 \times 10^{-5} = \pi[\Delta P - (1000)(9.81)(3)\sin(-90°)](0.012)^4/[128(0.0012)(3)].$$

This results in a pressure drop, $\Delta P = -29.15$ kPa. The negative ΔP denotes a pressure rise instead of a pressure drop, that is, blood tends to plummet freely down from the giraffe's heart to its head, which is 3-m below. To avoid blood rushing to the head, the blood vessels are equipped with sphincters that work as speed bumps to slow down the free fall of the giraffes' blood.

11.6 Energy conservation and head loss in pipe flow

Analyzing pipe flow from the energy perspective has proven to be revealing. According to the first law of thermodynamics, energy cannot be created nor destroyed. To put it in context, the energy associated with a volume of moving fluid is conserved. Consider Points 1 and 2 on an inclined straight pipe, as shown in Fig. 11.7. In the ideal case, with uniform profile velocity and no losses along

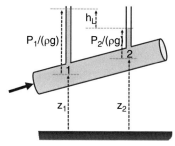

FIGURE 11.7 Energy per unit mass of a moving fluid in a pipe (created by D. Ting). The three components are pressure energy, kinetic energy, and potential energy.

the pipe, conservation of energy gives

$$P_1 \forall + \frac{1}{2}\rho\forall U_1^2 + \rho g\forall z_1 = P_2 \forall + \frac{1}{2}\rho\forall U_2^2 + \rho g\forall z_2, \qquad (11.26)$$

where z is the elevation. The first term is the pressure energy, the second term is the kinetic energy, and the third term is the potential energy. In reality, the velocity profile is parabolic for fully developed laminar flow and flatter and closer to uniform for fully developed turbulent flow. To account for this nonuniform velocity profile, we introduce the kinetic energy coefficient,

$$\alpha = \int \frac{1}{2}\rho U^3 dA \Big/ \left(\frac{1}{2}\rho\forall U_{avg}^2\right). \qquad (11.27)$$

For laminar flow, $\alpha = 2$, and for turbulent flow, α is close to one. Including this kinetic energy correction coefficient, we have

$$P_1 \forall + \frac{1}{2}\alpha_1\rho\forall U_1^2 + \rho g\forall z_1 = P_2 \forall + \frac{1}{2}\alpha_2\rho\forall U_2^2 + \rho g\forall z_2. \qquad (11.28)$$

For fully developed incompressible flows, $\alpha_1 = \alpha_2$, and $\frac{1}{2}\alpha_1\rho\forall U_1^2 = \frac{1}{2}\alpha_2\rho\forall U_2^2$. On that account, the energy conservation equation is reduced to

$$P_1 \forall + \rho g\forall z_1 = P_2 \forall + \rho g\forall z_2. \qquad (11.29)$$

Dividing every term by the fluid weight, $\rho\, g\, \forall$, leads to

$$P_1/(\rho g) + z_1 = P_2/(\rho g) + z_2. \qquad (11.30)$$

11.6.1 Head loss

Losses are inevitable in real life and, thus, must be accounted for. Combining all the losses into a loss term, h_L, we have

$$P_1/(\rho g) + z_1 = P_2/(\rho g) + z_2 + h_L, \qquad (11.31)$$

where it has been implicitly applied that Point 2 is downstream of Point 1. The head loss, h_L, signifies energy dissipated from Point 1 to Point 2 via viscosity;

see Fig 11.7. We also note from Fig 11.7 that $P_2 = P_1 - \Delta P$ and $z_2 - z_1 = L \sin \theta$. The head loss of key concern in practice can thus be expressed as

$$h_L = \left[P_1/(\rho g) + z_1\right] - \left[(P_1 - \Delta P)/(\rho g) + z_2\right] = \Delta P/(\rho g) - L \sin \theta. \quad (11.32)$$

Earlier, from force balance analysis, we found $[\Delta P - \rho g L \sin \theta]/L = 2\tau/r$. With this in mind, head loss can be described as

$$h_L = \Delta P/(\rho g) - L \sin \theta = (\Delta P - \rho g L \sin \theta)/(\rho g) = 2\tau L/(\rho g r). \quad (11.33)$$

As deduced earlier, the shear, $\tau = 2\tau_w r/D$ and, thereupon,

$$h_L = 2\tau L/(\rho g r) = 4\tau_w L/(\rho g D). \quad (11.34)$$

We note that the head loss increases with wall shear, length of pipe, and inversely with pipe diameter.

With the head loss expression, we can generalize the energy conservation of pipe flow as

$$P_1 \forall + \frac{1}{2}\alpha_1 \rho \forall U_1{}^2 + \rho g \forall z_1 = P_2 \forall + \frac{1}{2}\alpha_2 \rho \forall U_2{}^2 + \rho g \forall z_2 + \rho g \forall h_L. \quad (11.35)$$

This can be expressed in terms of per unit volume, that is,

$$P_1 + \frac{1}{2}\alpha_1 \rho U_1{}^2 + \rho g z_1 = P_2 + \frac{1}{2}\alpha_2 \rho U_2{}^2 + \rho g z_2 + \rho g h_L. \quad (11.35a)$$

In fluid mechanics, it is cool to use jargon such as head, when talking about the flow energy. Dividing the energy per unit volume equation by the specific weight, $\gamma = \rho g$, gives

$$P_1/(\rho g) + \frac{1}{2}\alpha_1 U_1{}^2/g + z_1 = P_2/(\rho g) + \frac{1}{2}\alpha_2 U_2{}^2/g + z_2 + h_L. \quad (11.35b)$$

The heads are pressure head, velocity head, elevation head, and head loss. Let us examine the head loss in more detail.

11.7 Major and minor losses in pipe flow

In engineering practice, a good estimation of the pressure drop is key. For instance, the appropriate pump, blower, or fan can only be sized when this piece of information is available. The pressure change due to elevation, such as that due to a giraffe raising or lowering its head, is simply the product of fluid density, gravity, and elevation difference. With known conditions, this elevation pressure drop can be straightforwardly determined. The pressure change due to variation in kinetic energy can also be readily deduced from the flow rate and pipe diameter. What is left is the pressure drop caused by head loss. The total head loss is the sum of two head losses, that is, the major and minor losses.

Major head loss describes head loss due to viscous effects. It is generally, but not necessarily, the larger loss in a typical piping network with extended straight flow passages. An obvious example is fossil fuel pipelines, many of

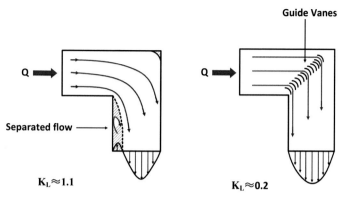

FIGURE 11.8   Turning or guide vanes for reducing minor head loss (loss coefficient, K_L) in a 90° bend (created by X. Wang). The guide vanes eliminate or lessen the formation of recirculating flow and, hence, the associated losses.

which span thousands of kilometers. The major head loss associated with fossil fuel transportation is so appreciable that numerous pumping stations have to be implemented to overcome the loss between consecutive stations.

Minor head loss is head loss caused by components such as fittings, valves, elbows, tees, the entrance, and exit. The loss is largely associated with the flow disturbance induced by these parts. As the flow disturbance is complex, a theoretical deduction of minor head loss is beyond challenging. Instead, standard testing procedures are employed to quantify the minor head loss associated with each of these components, under the intended nominal working conditions. Namely, the minor head loss,

$$h_{L,minor} = K_L \frac{1}{2} U_2{}^2, \qquad (11.36)$$

where K_L is the experimentally deduced loss coefficient. As expected, this loss coefficient is a function of geometry, including workmanship, surface finish and/or roughness, and Reynolds number. For illustration purposes, the loss coefficient, K_L, of a 90°-bend air duct, such as the one shown in Fig. 11.8, can be reduced from roughly 1.1 to 0.2 with a set of turning vanes.

Example 11.4 High-MERV, low-loss air filter.

Given: MERV is the abbreviation for minimum efficiency reporting value. A MERV rating indicates the effectiveness of a filter in trapping small particles. On a scale of 1 to 16, the higher the MERV rating, the smaller the size of particles the filter can trap. The price for increasing MERV is increasing pressure loss. After spending a lot of time helping to mitigate the spread of virus in a pandemic, Dr. JAS is dawned with a novel idea about air filters that can more effectively filtering out small air particles while minimizing the pressure drop across the filter. To test her idea, she designed an experimental air filter performance evaluation facility,

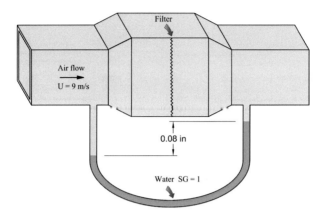

FIGURE 11.9 The setup for measuring the loss coefficient of an air filter (created by F. Kerman-shahi). There are well established standards to specify exactly how the measurements should be conducted, including the definite locations to flush mount the pressure taps.

as shown in Fig. 11.9, in the Turbulence and Energy Lab wind tunnel. For the first set of tests, the air velocity is maintained at the nominal air circulation velocity of 9 m/s and the pressure drop is measured to be 0.08 inches of water.

Find: The loss coefficient of Dr. JAS' new air filter at an airspeed of 9 m/s.

Solution: To deduce the loss coefficient, we call upon the flow head (conservation) equation,

$$P_1/(\rho g) + \frac{1}{2}\alpha_1 U_1{}^2/g + z_1 = P_2/(\rho g) + \frac{1}{2}\alpha_2 U_2{}^2/g + z_2 + h_L.$$

Due to the negligible section involved between the pressure taps upstream and downstream of the filter, the major head loss, that is, frictional losses along the wind tunnel walls, is negligible. With that in mind, the minor head loss can be expressed as $h_{L,minor} = K_L \frac{1}{2} U_2{}^2$. Furthermore, the velocity head remains the same before and after the filter. This is because wind tunnels are designed such that a constant and uniform velocity profile across a constant cross-section is maintained. In addition, for a horizontal wind tunnel test section, the elevation head is zero. Under those circumstances, the head equation can be simplified into

$$\Delta P/(\rho g) = K_L \frac{1}{2} U_2{}^2.$$

Substituting the values, we have

$$19.91/[(1000)(9.81)] = K_L \frac{1}{2}(9)^2.$$

This gives $K_L = 0.33$, which is significantly lower than that associated with typical air filters. Because of the needed mechanical tightness to filter particles, there is not much more we can do to further reduce the pumping power required to push the air through these filters. Instead, a newer approach is to utilize an electrostatic precipitator, a filter-less device for removing particles via the force of an induced electrostatic charge; see Tian et al. (2018), for example.

As inferred at the beginning of this example, face masks are another type of air filter. Face masks are rated according to filtration effectiveness for blocking bacteria and particles. While there is significant variation, typical bacteria are of the order of 1 μm and viruses are one or two orders of magnitude smaller. Accordingly, face masks are evaluated based on how effective they are in blocking 3 μm, 0.3 μm, and/or 0.1 μm (the size of COVID-19 virus) particles. Typical standards require at least 95% of all particles equal to or greater than the targeted size to be blocked.

11.7.1   The Moody chart (diagram)

Dimensional analysis, covered in the previous chapter, can be invoked to identify the parameters contributing to the viscous losses (major head loss) in a straight pipe section. By examining the parameters that take part in pipe flow, we can come to the conclusion that the pressure drop is likely a function of fluid viscosity, pipe length, fluid velocity, pipe diameter, and fluid density. In other words,

$$\Delta P = f(\mu, L, U, D, \rho, \varepsilon), \tag{11.37}$$

where ε is the roughness of the conduit. Experiments have confirmed the completeness of the list of parameters above for the pressure drop in a duct. It has also been found that pipe roughness does not come into play when the flow is laminar. Dimensional analysis can be performed for *laminar* pipe flow to give

$$D\Delta P/(\mu U) = f(L/D). \tag{11.38}$$

This can be expressed explicitly as

$$\Delta P/L = C_4 \mu U/D^2, \tag{11.38a}$$

where C_4 is an empirical constant.

The volumetric flow rate can be obtained from the velocity times the cross-sectional area, that is,

$$\forall' = UA = \left[\Delta PD^2/(LC_4\mu)\right]\left(\pi D^2/4\right) = \pi \Delta PD^4/(4LC_4\mu). \tag{11.39}$$

Recall that for fully developed laminar flow in a circular pipe, the integration of the velocity across the cross-section, $\forall' = \int U(r)dA = \pi \Delta PD^4/(128\mu L)$. Comparing this with Eq. (11.39) gives $C_4 = 32$. On that account, the pressure drop per unit length can be expressed as

$$\Delta P/L = 32\mu U/D^2. \tag{11.38b}$$

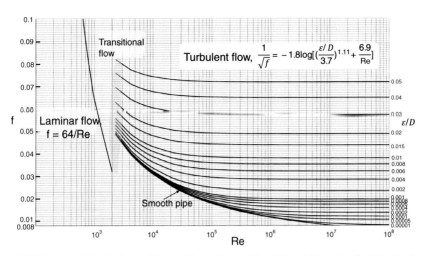

FIGURE 11.10   Moody chart or Moody diagram (created by Y. Yang). Note that for fully developed laminar pipe flow, the friction factor, f = 64/Re.

Dividing this by $\frac{1}{2}\rho U^2/L$ gives

$$\Delta P / \left(\frac{1}{2}\rho U^2\right) = 64(\mu/\rho UD)(L/D). \tag{11.38c}$$

This pressure drop is often expressed as

$$\Delta P = 64/Re(L/D)\left(\frac{1}{2}\rho U^2\right) = f(L/D)\left(\frac{1}{2}\rho U^2\right), \tag{11.40}$$

where the dimensionless friction factor, $f = \Delta P \, (D/L) / (\frac{1}{2}\rho U^2)$. For *fully developed laminar pipe flow*, this friction factor is also referred to as the Darcy friction factor,

$$f = 64/Re. \tag{11.41}$$

This relation forms the laminar portion of the Moody Chart or Moody diagram, as shown in Fig. 11.10. We can also relate this friction factor with wall shear. From Eq. (11.12), the pressure drop, $\Delta P = 4L\tau_w/D$. Comparing this with Eq. (11.40), we see that

$$f = 8\tau_w/(\rho U^2). \tag{11.42}$$

We note that the friction factor is linearly related to the wall shear, τ_w, and inversely correlated with the dynamic pressure, $\frac{1}{2}\rho U^2$.

Proper execution of dimensional analysis for the turbulent case can lead to

$$\Delta P / \left(\frac{1}{2}\rho U^2\right) = f(\rho UD/\mu, L/D, \varepsilon/D). \tag{11.43}$$

We note that the pressure drop is normalized by the dynamic pressure, $\frac{1}{2}\rho U^2$, associated with the high-Reynolds-number moving fluid. The dynamic pressure is the appropriate denominator, instead of the characteristic viscous shear stress, $\mu U/D$, when the flow is turbulent. This is because the turbulent shear depends more on ρ (associated with flow inertia) than on μ (fluid viscosity). We can bring L/D from the right-hand side to the left-hand side, that is,

$$\Delta P(D/L)/\left(\frac{1}{2}\rho U^2\right)= f(\rho UD/\mu, \varepsilon/D). \qquad (11.43a)$$

From the previous paragraph, the dimensionless friction factor, f = ΔP (D/L)/ $(\frac{1}{2}\rho U^2)$, which is the left-hand side. In other words, the friction factor for *turbulent pipe flow*,

$$f = \Delta P(D/L)/\left(\frac{1}{2}\rho U^2\right)= f(\rho UD/\mu, \varepsilon/D). \qquad (11.43b)$$

This conveys that the friction factor for turbulent pipe flow depends on both Reynolds number, Re = $\rho UD/\mu$, and pipe roughness, ε/D. As you may have expected, it was an enervating undertaking to compile the values of friction factor for turbulent pipe flow. The fruit of the labor of those trailblazers, nicely packaged into the Moody Chart or Moody Diagram, such as that shown in Fig. 11.10, is for us to enjoy; it seriously eases internal flow analysis. Readers who are interested in the history of the Moody Diagram and the latest alternative methods in place of the iterative approach based on Moody Diagram can refer to LaViolette (2017). More importantly, it is important to note that the Moody Diagram is still essential in both engineering practice and research; see Madeira (2020), for example.

With the friction factor explicitly defined, we can express the head equation in terms of the non-dimensional friction factor, f. For a section of a pipeline, the head can be described as

$$P_1/(\rho g) + \frac{1}{2}\alpha_1 U_1^2/g + z_1 = P_2/(\rho g) + \frac{1}{2}\alpha_2 U_2^2/g + z_2 + h_{L,major} + h_{L,minor}.$$
$$(11.44)$$

The total head loss is the sum of major and minor head losses, $h_L = h_{L,major} + h_{L,minor} = \Delta P_L/(\rho g)$, where ΔP_L is pressure drop associated with the losses. We can rearrange the friction factor equation, f = ΔP (D/L) / $(\frac{1}{2}\rho U^2)$ into $\Delta P/(\rho g) = f$ (L/D) $(\frac{1}{2}\rho U^2)$ / $(\rho g) = f$ (L/D)$(\frac{1}{2}U^2/g)$. This last term is the major head loss and can be substituted into Eq. (11.44) to give

$$P_1/(\rho g) + \frac{1}{2}\alpha_1 U_1^2/g + z_1$$
$$= P_2/(\rho g) + \frac{1}{2}\alpha_2 U_2^2/g + z_2 + f(L/D)\left(\frac{1}{2}U^2/g\right) + h_{L,minor}. \quad (11.45)$$

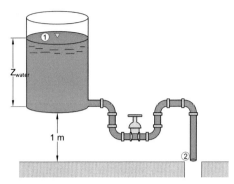

FIGURE 11.11 Water delivery from a tank (created by F. Kermanshahi). The entrance into the pipe, the four 90° elbows, the valve, and the exit into the atmosphere contribute to minor losses.

Example 11.5 Water delivery system.

Given: A 9.5-m deep tank with a 16-m² cross-section is filled with water at room temperature. The tank is connected to a 7-m long piping system at the bottom of the wall, with an inner diameter of 2 cm, as depicted in Fig. 11.11. The loss coefficient for the entrance, each elbow, the valve, and the exit are 0.5, 1.5, 10, and 1, respectively.

Find: The flow rate, the rate at which the water depth decreases in the tank, and if the pipe flow is laminar or turbulent. Assume the friction factor of the pipe is 0.03.

Solution: The appropriate equation to apply is the head equation between the free surface of the water in the tank, Point 1, and the exit of the piping system, Point 2, that is,

$$P_1/(\rho g) + \frac{1}{2}\alpha_1 U_1^2/g + z_1 = P_2/(\rho g) + \frac{1}{2}\alpha_2 U_2^2/g$$
$$+z_2 + f(L/D)\left(\frac{1}{2}U^2/g\right) + h_{L,minor}.$$

We note that $P_1 = P_2 =$ atmospheric pressure and, thus, zero gage pressure, $z_1 = z_2 + 1 + z_{water}$. Let us assume $U_1 = 0$, since the tank cross-section is significantly larger than that of the pipe and, also, $\alpha_1 = \alpha_2 = 1$. With these assumptions, we have

$$1 + z_{water}\frac{1}{2}U_2^2/g + (0.03)(7/0.02)\left(\frac{1}{2}U_2^2/g\right) + [0.5 + 5(1.5) + 10 + 1]\frac{1}{2}U_2^2.$$

Substituting for $z_{water} = 9.5$ m, we get $U_2 = 1$ m/s.

For continuity or conservation of mass, we have $U_1 A_1 = U_2 A_2$. Accordingly,

$$U_1 = U_2 A_2/A_1 = 1(\pi)(0.01)^2/16 = 0.000002 \, \text{m/s}.$$

In other words, the speed at which the free surface decreases, U_1, is many orders of magnitude smaller than the velocity at the exit of the piping system, U_2. For this reason, we can neglect U_1 in the head calculation.

The Reynolds number for the involved pipe flow,

$$\text{Re} = \rho DU/\mu = 1000(0.02)(1)/0.001 = 20{,}000,$$

that is, the pipe flow is turbulent. As the flow is highly turbulent, it validates the assumption that the kinetic energy coefficient at the exit of the piping system, α_2, is roughly one. The kinetic energy coefficient in the tank, α_1, is approximately two, as the flow is laminar. Fortunately, the assumption that α_1 is equal to one does not come into play because it only appears in the term $\frac{1}{2}\alpha_1 U_1^2/g$, where U_1 is zero and, thus, the term disappears from the equation.

If the pipe has a roughness of 0.001 m, the friction coefficient from the Moody chart at $\text{Re} = 20{,}000$ is $f = 0.026$.

Problems

11.1 Keep the flow laminar.

In a heating, ventilation, and air conditioning system, conditioned air at a volumetric flow rate of 0.03 m³/s is required. What is the appropriate diameter for the air duct, if you wish to minimize noise by keeping the flow laminar? Is this a maximum or minimum diameter? What is the corresponding velocity?

11.2 Entrance length.

Hot water at 45°C is delivered via a 0.635-cm diameter pipe at 1×10^{-4} m³/s. If the water from the hot water tank enters the 0.635-cm diameter pipe smoothly, what is the entrance length?

11.3 Laminar flow velocity at 10% radius from the wall.

You are asked to select a velocity sensor that is to be fixed at 0.1R from the wall of a pipe to monitor the flow of water at 10°C. The pipe radius, R, is 0.05 m. The flow is fully developed at the location where the sensor is to be installed. What is the velocity range that the sensor will pick up, if the Reynolds number associated with the pipe flow varies between 1000 and 2000?

11.4 Friction factor with increasing flow rate.

A long circular pipe of diameter D is placed horizontally, and the flow is increased from zero to a very large value. If the pipe roughness, ε, is about 1% of D, describe the change in the friction factor with respect to the increasing flow rate.

11.5 Friction increase when doubling the flow velocity.

Part I) In a fully developed laminar pipe flow, doubling the flow velocity would result in

A) no change in major loss

B) halving of major loss

C) doubling of major loss

D) quadrupling (4 × increase) of major loss

Part II) In a fully developed turbulent pipe flow, doubling the flow velocity would result in

A) no change in major loss

B) halving of major loss

C) doubling of major loss

D) quadrupling (4 × increase) of major loss

11.6 Applying Moody chart for flow in a commercial pipe.

A Newtonian fluid with $\rho = 1000$ kg/m^3 and $\mu = 1.3 \times 10^{-3}$ kg/m·s flows at 1.05 m/s in a 0.375-m diameter galvanized iron (roughness, $\varepsilon = 0.15$ mm) pipe, the corresponding friction factor, f, is approximately

A) 0.011

B) 0.014

C) 0.016

D) 0.019

E) 0.021

F) 0.024

G) 0.03

H) 0.034

I) 0.037

J) 0.041

K) 0.051

L) 0.062

11.7 Fully rough zone on moody chart.

For pipe flow in the fully rough zone, the head loss is independent of

A) the friction factor

B) the length of pipe

C) the Reynolds number

D) the flow rate

E) the (degree of) roughness

F) All of the above

G) None of the above

11.8 Blood donation.

Dr. JAS is blessed as a blood donor, that is, her blood is type O negative, the universal blood type that can be received by everyone. How much longer does it take her to donate 450 mL of blood when the pressure in her vein is 50 mm Hg gage versus 40 mm Hg gage? Assume that a plastic tube of 1-m length, with an inner diameter of 0.5 cm, connects the needle (0.2 cm inner diameter and 4 cm long) that is pricked into a vein with a 0.3-cm inner diameter. To assist the flow of blood, the receiving bag is placed 0.4 m below the vein.

11.9 Required head for the intended flow delivery.

Water ($\rho = 1000$ kg/m^3, $\mu = 1.3 \times 10^{-3}$ kg/(m·s)) from the bottom (at 120 m above sea level) of a reservoir is delivered at 0.001 m^3/s through a 10-cm cast iron (roughness, $\varepsilon = 0.26$ mm) pipe to a resort at sea level. The pipeline consists of an entrance (K = 0.4), 90° elbow (K = 0.3), 120-m vertical pipe, 90° elbow (K = 0.3), 1.1-km horizontal pipe, gate valve (K = 0.2 when open), and exit (K = 0.25). What is the minimum reservoir water height to ensure that there is enough head to overcome all losses along the flow passage and to deliver the required flow rate? What is this minimum reservoir water height, if there is enough flow energy to drive a 1.2-kW turbine along the pipeline?

References

Bilaniuk, N., Wong, G.S.K., 1993. Speed of sound in pure water as a function of temperature. J. Acoust. Soc. Am. 93 (3), 1609.

Bopp, J., Weiss, D.A., 2022. On the origin of the centreline velocity overshoot in the entrance region of a turbulent pipe flow. Int. J. Therm. Sci. 172, 107256.

Chen, Y., Chen, X., Liu, S., 2020. Numerical and experimental investigations of novel passive micromixers with fractal-like tree structures. Chem. Phys. Lett. 747, 137330.

Dagg, A.I., 1962. The role of the neck in the movements of the giraffe. J. Mammal. 43 (1), 88–97.

Franklin Institute, 2021. https://www.fi.edu/heart/blood-vessels, (accessed June 22, 2021).

Galvis, E., Yarusevych, S., Culham, J.R., 2012. Incompressible laminar developing flow in microchennels. J. Fluids Eng. 134 (1), 014503.

Goetz, R.H., Warren, J.V., Gauer, O.H., Patterson Jr., J.L., Doyle, J.T., Keen, E.N., McGregor, M., 1960. Circulation of the giraffe. Circ. Res. 8, 1049–1058.

Kim, S., 2021. Maximum allowable fluid velocity and concern on piping stability of ITER Totamak cooling water system. Fusion Eng. Des. 162, 112049.

Klabunde R.E., "Viscosity of blood" in Cardiovascular physiology concepts, 2021. https://www.cvphysiology.com/Hemodynamics/H011, (accessed October 8, 2021).

Kucukal, E., Man, Y., Gurkan, U.A., Schmidt, B.E., 2021. Blood flow velocimetry in a microchannel during coagulation using particle image velocimetry and wavelet-based optical flow velocimetry. J. Biomech. Eng. 143, 091004.

LaViolette, M., 2017. On the history, science, and technology included in the Moody Diagram. J. Fluids Eng. 139 (3), 030801.

Li, C., Jia, L., Zhang, T., 2009. The entrance effect on gases flow characteristics in micro-tube. J. Therm. Sci. 18 (4), 353–357.

Lubbers, J., Graaff, R., 1998. A simple abd accurate formula for the sound velocity in water. Ultrasound Med. Biol. 24 (7), 1065–1068.

Madeira, A.A., 2020. Major and minor head losses in a hydraulic flow circuit: experimental measurements and a Moody's diagram application. Eclética Química J. 45 (3), 47–56.

Pfenninger, W., 1961. Boundary layer suction experiments with laminar flow at high Reynolds numbers in the inlet length of tubes of various suction methods. In: Lachmann, G.V. (Ed.), Boundary Layer and Flow Control Pergamon Press, Oxford, pp. 961–980.

Reynolds, O., 1883. An experimental investigation of the circumstances which determine whether the motion of water shall be direct or sinuous, and of the law of resistance in parallel channels. Philosophical Trans. Royal Soc., A 174, 935–982.

Secomb, T.W., Pries, A.R., 2013. Blood viscosity in microvessels: experiment and theory. C.R. Phys. 14 (6), 470–478.

Tian, E., Mo, J., Li, X., 2018. Electrostatically assisted metal foam coarse filter with small pressure drop for efficient removal of fine particles: effect of filter medium. Build. Environ. 144, 419–426.

Ting, D.S-K., 2016. Basics of Engineering Turbulence. Academic Press, New York, NY.

Wu, X., Moin, P., Adrian, R.J., Baltzer, J.R., 2015. Osborne Reynolds pipe flow: direct simulation from laminar through gradual transition to fully developed turbulence. Proc. Nat. Acad. Sci. USA 112 (26), 7920–7924.

Yamamoto, H., Yabuta, T., Negi, Y., Horikawa, D., Kawamura, K., Tamura, E., Tanaka, K., Ishida, F., 2020. Measurement of human blood viscosity using falling needle rheometer and the correlation to the modified Herschel-Bulkley model equation. Heliyon 6, e04792.

Zhang, Q.G., 2006. Hypertension and counter-hypertension mechanisms in giraffes. Cardiovascular Hematological Disorders—Drug Targets 6 (1), 63–67.

Chapter 12

External flow

"Discovery consists of seeing what everybody has seen and thinking what nobody has thought."

Albert Szent-Gyorgyi

Chapter Objectives

- Become aware of everyday external flows.
- Appreciate lift and drag.
- Comprehend fluid viscosity and boundary layers.
- Master laminar, transition, and turbulent boundary layer development along a flat plate.
- Be familiar with bluff body aerodynamics, including vortex shedding and streamlining.
- Fathom the flow around a circular cylinder.

Nomenclature

A	area
Bo	Bond number
C	coefficient; C_D is the drag coefficient, $C_{D,x}$ is the local drag coefficient at x, C_f is the average friction coefficient, $C_{f,x}$ is the local friction coefficient at x, C_L is the lift coefficient
D	diameter
F	force; F_D is the drag force, F_L is the lift force
f	frequency
g	gravity
L	length
P	pressure
Re	Reynolds number; Re_L is the Reynolds number based on characteristic length L, Re_x is the Reynolds number based on distance x
St	Strouhal number
U	velocity; velocity in the x direction, U_∞ is the freestream velocity
u	local velocity; local velocity in the x direction, u_∞ is the local freestream velocity
V	velocity; velocity in the y direction
W	velocity; velocity in the z direction
We	Weber number
w	width

x   direction, distance, distance in the x direction
y   y direction, perpendicular to the x direction
z   z direction, perpendicular to both x and y directions

Greek and other symbols

Δ difference
δ thickness of the boundary layer
θ angle
μ dynamic viscosity
ρ density
σ surface tension
τ shear; τ_w is wall shear
ω vorticity

12.1   Everyday external flow

External flows are flows of principally unbounded, as far as the object of interest is concerned, fluids over an object. The boundary layer within which viscosity is of significance presents only around the object or surface under study. In this sense, the flow is free, that is, unlike internal flows where the boundary layer caused by the constraining walls containing the flow fluid is just about every-where. For example, atmospheric wind flowing across Burj Khalifa of Dubai is an external flow. For tall man-made structures such as this, an obvious reason that we are interested in fluid mechanics is to understand flow-induced vibrations. In a relatively large river, the water current around a resting fish is also an external flow example. Why would a fish want to master fluid mechanics in this case? Without that mastery, it would drift with the current and easily subjected to prey. External flows such as these pertain to a fluid flowing across a stationary body or surface. The predominantly unbounded flow stream away from the body or surface is called the free stream. The alternative to a flowing fluid over a stationary object is a moving object in an otherwise-motionless fluid. A fish swimming in a large standing-water lake is one such example. The third case involves both the body and the fluid in motion but moving at different velocities such that there is a relative velocity between the moving body and the flowing fluid. A salmon swimming upstream against the forceful down-pouring current of a great rapids is truly an inspirational external flow. Anthony Liccione says it well, "Some people will just go with the flow of things and sway in life, while others will fight against the currents and go upstream to reach their destiny."

Birds flying in the sky is presumably the most captivating everyday external flow exhibition that has enlightened uncountable inquisitive minds. It was not until 1903 that the Wright brothers were able to imitate birds and flew the first airplane (Gray, 2002). Okay, airplanes are not new, what about wind turbines? Centuries before the Wright brothers, our ancestors started harnessing a prevailing external flow, the wind, using windmills, where records showed windmills existed as early as 200 BC. According to the U.S. Energy Information

Administration (EIA, 2021), our ancestors have exploited wind energy to propel boats along the Nile River since 5000 BC. Moreover, wind was harnessed to power water pumps in China, and windmills were employed for grinding grain in Persia and the Middle East by 200 BC (EIA, 2021). Ecclesiastes, Chapter 1, says it well, "There is nothing new under the sun." and "It was already here long ago, in the ages long before us." This is also the case concerning exploiting natural external flow to sail far and wide. The ancient Egyptians are thought to have made use of sailing vessels before 3000 BC and, around the same time, the Austronesian[1] peoples had crossed oceans in the first sea-going sailing ships. Yet, there is much concerning the fascinating fluid mechanics that we are still trying to unravel.

Some common external flows that are exceedingly more important than the aforementioned man-made inventions are those involved in pollination. You and I would not be talking about external flows if there were no food to sustain us. To name but a few, wind-pollinated plants include wheat, rice, corn, rye, and barley. Young et al. (2018) found, among other interesting results, that the number of blackberry fruitlets positively correlated with the wind speed. Most fruits are pollinated by bees and birds, and neither bees nor birds can pollinate apart from the enabling power of an external flow. In this case, wind may or may not be needed but these flying creatures could not pollinate without atmospheric air. As mentioned above, a bee flying around in largely still air is a classic example of external flow. For bat, mango, or durian lovers, there are over five hundred plant species that rely on bats to pollinate their flowers (Fleming et al., 2009). This serviceability of bats is often overshadowed by their unsightly side; no wonder Aristotle uttered, "For as the eyes of bats are to the blaze of day, so is the reason in our soul to the things which are by nature most evident of all."

12.2 Lift and drag

In its simplest form, the wing of a bird, airplane, or wind turbine can be regarded as an airfoil, as shown in Fig. 12.1. Whether you are a bird, an airplane, a wind turbine, or a human being, you want lift and not drag. For the two-dimensional airfoil of a unit span out of the page, lift is the upward force induced by the pressure and shear associated with the flow. For a small, differential area, dA, the uplifting force can be expressed as

$$dF_L = -P \, dA \sin \theta - \tau_w dA \cos \theta. \tag{12.1}$$

Here, the negative sign signifies that the pressure is acting in the opposite direction with respect to the surface normal and θ is the angle between the surface

---

1. Being born and growing up in the rainforest of Borneo, the author is astonished to learn from publications such as (Kumar, 2011) that the earlier people from his birthplace have presumably exploited natural wind and water and voyaged to far-away places such as Madagascar. Interested readers are encouraged to read the book by Doran, Jr (1981).

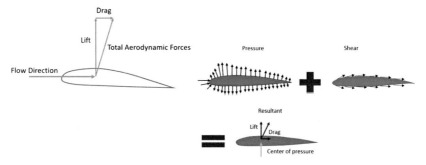

FIGURE 12.1 A two-dimensional airfoil where lift and drag are described (created by S. Khademi). The flow (friction) always tries to drag the object along. Lift is produced by the pressure difference; pressure is negative on the upper surface and positive on the lower surface and, thus, pushes the airfoil upward, creating lift.

normal and flow direction. The total lift force for the airfoil is simply the sum of the elemental lift forces, dF_L. The summation is described mathematically as an integral, that is,

$$F_L = \int_A dF_L = -\int_A (P \sin\theta + \tau_w \cos\theta)dA. \tag{12.2}$$

It is more convenient to express lift in the nondimensional form. Dividing the lift force by the product of dynamic pressure of the flow and the participating surface area gives the non-dimensional lift coefficient, that is,

$$C_L = F_L / \left(\frac{1}{2}\rho U^2 A\right), \tag{12.3}$$

where ρ is the density of the medium the airfoil is in, U is the streamwise velocity, and A is the frontal area, that is, the area projected on a plane normal to the flow direction.

Similarly, the usually unwanted drag for a differential area, dA, is

$$dF_D = -P\,dA\cos\theta + \tau_w dA \sin\theta. \tag{12.4}$$

The corresponding drag force for the airfoil is

$$F_D = \int_A dF_D = \int_A (-P\cos\theta + \tau_w \sin\theta)dA. \tag{12.5}$$

The non-dimensional drag coefficient,

$$C_D = F_D / \left(\frac{1}{2}\rho U^2 A\right), \tag{12.6}$$

where the same force caused by the dynamic pressure has been used for normalizing.

Flying birds and airplanes have to live with a certain amount of drag. In engineering applications, much effort has been invested into reducing the drag and augmenting the lift. For that reason, interest in biomimicry has heightened in recent years. After all, there is much for us to learn from the abundance of intelligent designs in nature. Both birds and airplanes, however, call upon additional drag when they need to slow down. It is a relief to see the flaps of airplane wings fully deployed when the airplane first touches the ground. Research on learning from flying creatures to improve airplane landings remains relatively scarce compared to that aiming at furthering lift or the lift- over-drag coefficient ratio.

12.3 Boundary layer

The vast unperturbed fluid, moving or stationary, is of little concern when it comes to external flows. What is of interest is the fluid around the body under study. Viscosity is the underlying parameter that leads to boundary layer formation and development. This gives rise to drag, which slows down a moving object. For a bluff body, boundary layer development can give rise to a wake, when the inertial force is sufficiently large. In general, drag increases drastically in the presence of a wake.

In the 1800s AD, a number of mathematical geniuses managed to elegantly solve some inviscid fluid flow problems, based on Euler's groundbreaking theoretical hydrodynamics (Euler, 1755), in an exact manner. As viscosity was completely ignored in solving these problems, the solutions were of little use in practice. On the other hand, there were hardcore application-oriented hydraulic engineers who had to painfully conduct detailed experiments before every undertaking, due to a lack of general solutions. Naturally, instead of working together to converge toward a solution, these two groups disdained each other. It was not until 1904 that these two diverging developments were brought together by Prandtl (1904). Prandtl recognized that viscosity is only significant within the very thin layer now known as boundary layer. Outside of this thin layer, the flow is essentially inviscid. To that end, Prandtl elegantly connected the inviscid freestream with the no-slip condition at the wall via the boundary layer.

Fig. 12.2 illustrates the formation and development of a boundary layer along a flat surface. This can be realized by passing a uniform flow over a very thin plate, or, by passing a very thin plate across still fluid. At a low velocity and/or high viscosity, the fluid starts to slow down ahead of the leading edge, Fig. 12.2A, and a laminar boundary layer promptly forms. Within the laminar boundary layer, the velocity gradient in the direction normal to the plate is more or less even. As the appropriate characteristic length of the Reynolds number is the distance from the leading edge, the boundary layer ultimately becomes turbulent farther downstream. At higher velocities, the freestream does not slow down until it reaches the plate, where a boundary layer develops, Fig. 12.2B. If the velocity is high, the stretch of the laminar boundary layer is short, that is, the boundary layer quickly becomes a turbulent one. To put it another way, the dominating inertial force for this high-Reynolds-number flow pushes the fluid faster and

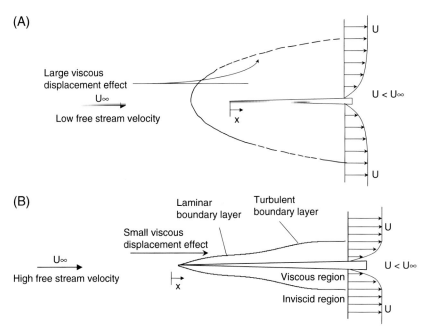

FIGURE 12.2 Boundary layer over a thin plate at (A) a very low freestream velocity and/or high viscosity and (B) a very high freestream velocity (created by O. Imafidon). When you ride on a slow-moving fluid particle along the streamline that will meet the leading edge head on, you will slow down ahead of the leading edge, as many fluid particles ahead of you also slow down to a stop at the stagnation point at the pointed edge. On the other hand, if you ride on a fast-moving fluid particle ahead of the pointed edge, you, like all other fast-moving particles ahead of you, are moving so fast that you do not slow down until viscosity within the forming boundary layer on the top or lower surface drags you down.

closer to the plate, resulting in a boundary layer with a large velocity gradient next to the plate. For typical engineering practice, the critical Reynolds number at which the boundary layer undergoes transition from laminar to turbulent is taken to be 5×10^5. It should be noted that this critical value depends heavily on factors such as surface roughness and free-stream disturbances. Minimizing disturbances and roughness can delay this critical Reynolds number to much farther downstream, to more than an order of magnitude higher in value.

12.3.1 Disturbance boundary layer

While there are a few types of boundary layers, we are only concerned with the most common one called the disturbance boundary layer in this book. Disturbance boundary layer is the layer where free-stream fluid is being slowed down, via viscosity, to zero at the no-slip surface. This is illustrated in Fig. 12.3, where the non-moving fluid particles that are "stuck" to the wall are slowing down the adjacent moving fluid particles. The boundary layer thickness, δ, is the

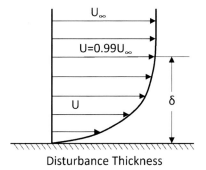

Disturbance Thickness

FIGURE 12.3   Disturbance boundary layer (created by N. Bhoopal). The disturbance boundary layer is defined by $U = 0.99U_\infty$. Interested readers would enjoy watching Lesics (2021), especially from 4:20 min. into the video clip.

perpendicular distance from the wall where the local velocity U reaches 99% of the freestream value. Thin as it is, it continues to perplex experts in the field, after more than a century of extensive scrutinization.

12.4   Flat plate boundary layer development

Let us look at the ideal incompressible steady fluid of uniform velocity flowing parallel to a smooth, flat plate introduced in the previous section. If the viscosity is not too high or the freestream velocity is not too low, a laminar boundary layer is initiated at the leading edge, as shown in Fig. 12.4. With distance from the leading edge, the laminar boundary layer thickens until it becomes unstable at a sufficiently large Reynolds number. The location where the flow inside the boundary layer becomes irregular is the transition point. Naturally-present instabilities begin to amplify until the boundary layer becomes fully turbulent at the critical Reynolds number, $\text{Re}_c \approx 5 \times 10^5$, beyond which the turbulent boundary layer continues to thicken.

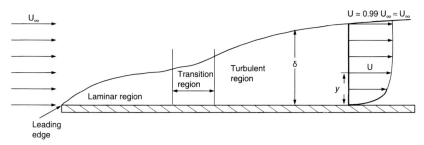

FIGURE 12.4   Boundary layer development, from laminar to transition to turbulent, over a flat plate (created by O. Imafidon). Due to inherent imperfections, such as disturbances in the flow and surface roughness, the critical Reynolds number, $\text{Re}_r \approx 5 \times 10^5$ in practice.

While the effect of viscosity dominates throughout the entire laminar boundary layer, this is not the case when the boundary layer becomes turbulent. In general, three sublayers can be identified from a turbulent boundary layer (Ting, 2016). Next to the freestream is the outer boundary layer; see Fig. 12.4. The velocity is relatively uniform, and the flow is characterized by random, fluctuating motion. Inertia dominates and viscosity plays an inconsequential role in this outer boundary layer. Next to the wall, on the other hand, is the viscous sublayer. We can infer from the name of this sublayer that fluid viscosity plays a dominant role here. In between the inertia-controlling outer layer and the viscosity-dominating viscous sublayer is a buffer zone. As expected, both inertia and viscosity are important in the buffer layer.

For steady, two-dimensional, incompressible flow over a flat plate in the absence of gravity and other forces, we can express the conservation of mass in terms of the continuity equation,

$$\frac{\partial U}{\partial x} + \frac{\partial V}{\partial y} = 0, \tag{12.7}$$

where U is the streamwise velocity in the x direction, and V is the normal-to-the-wall velocity in the y direction. The momentum in the streamwise direction can be expressed as

$$\rho \left(U \frac{\partial U}{\partial x} + V \frac{\partial U}{\partial y} \right) = -\frac{\partial P}{\partial x} + \mu \left(\frac{\partial^2 U}{\partial x^2} + \frac{\partial^2 U}{\partial y^2} \right), \tag{12.8}$$

where μ is the dynamic viscosity. Similarly, the y-momentum expression is

$$\rho \left(U \frac{\partial V}{\partial x} + V \frac{\partial V}{\partial y} \right) = -\frac{\partial P}{\partial y} + \mu \left(\frac{\partial^2 V}{\partial x^2} + \frac{\partial^2 V}{\partial y^2} \right). \tag{12.9}$$

For high-Reynolds-number flows, the shear layer is very thin. Therefore, the cross-stream velocity is substantially smaller than the stream-wise counterpart, that is,

$$V \ll U. \tag{12.10}$$

We also note that the rate of change in velocity in the stream-wise direction is much smaller than that in the cross-stream direction. Namely,

$$\frac{\partial U}{\partial x} \ll \frac{\partial U}{\partial y} \tag{12.11}$$

$$\frac{\partial V}{\partial x} \ll \frac{\partial V}{\partial y}. \tag{12.12}$$

From these, we can compare the magnitudes of the five terms in the y-momentum equation, that is,

$$\text{small} + \text{small} = -\partial P/\partial y + \text{very small} + \text{small}. \tag{12.9a}$$

We further observe that there is a negligible pressure gradient in the y direction, that is, $\partial P/\partial y \approx 0$. As such, we are left with the continuity equation and the x-momentum equation. Concerning the x-momentum equation, we can apply Bernoulli's equation to the outer inviscid flow to get

$$\partial P/\partial x = dP/dx = -\rho U_\infty dU_\infty/dx, \tag{12.13}$$

where U_∞ is the freestream velocity, as compared to the local velocity U. Substituting this into the x-momentum equation, we get

$$U\frac{\partial U}{\partial x} + V\frac{\partial V}{\partial y} \approx U_\infty\frac{dU_\infty}{dx} + \frac{1}{\rho}\frac{\partial \tau}{\partial y}, \tag{12.14}$$

where, for laminar flow, the shear,

$$\tau = \mu\frac{\partial U}{\partial y}, \tag{12.15}$$

and, for turbulent flow,

$$\tau = \mu\frac{\partial U}{\partial y} - \overline{\rho uv}. \tag{12.16}$$

At the wall, $U = 0$ because of no-slip, and $V = 0$ because no fluid can move into or out of an impermeable surface. The other boundary condition is at the boundary layer where $U = 0.99U_\infty \approx U_\infty$. With these boundary conditions, we can solve for the two-dimensional velocity field, U(x, y) and V(x, y). The lift and drag can then be deduced from the velocity field.

12.4.1 Laminar boundary layer

The Chinese saying, "Great teachers produce brilliant students," came true in the case of Prandtl and Blasius. Blasius, a brilliant student of Prandtl, was the first to elegantly solve the flat plate laminar boundary layer problem (Blasius, 1908). For constant U_∞ ($dU_\infty/dx = 0$) over a flat surface, Blasius unraveled that the laminar boundary layer thickness varies as

$$\delta/x \approx 5.0/Re_x^{1/2}, \tag{12.17}$$

where x is the distance from the leading edge. This approximation has been found to be sound for $10^3 < Re_x < 10^6$. The corresponding local skin friction coefficient, see White (2011), for example,

$$C_{f,x} = 2\tau_w/(\rho U_\infty^2) = 0.664/Re_x^{1/2}. \tag{12.18}$$

The corresponding drag induced by the friction can also be expressed in terms of the drag coefficient,

$$C_{D,A} - 2C_{f,n} = 1.328/Re_Y^{1/2}, \tag{12.19}$$

FIGURE 12.5   Dr. JAS' flat-bottom canoe for demonstrating friction drag (created by A. Raj). The pressure drag is minimized via streamlining. Specifically, the incoming flow is guided along a gradually declining smooth plate and the flow leaves the end of the canoe smoothly over a plate with a small incline.

The wall shear as a function of distance from the leading edge, x, is

$$\tau_w(x) = 0.332\rho^{1/2}\mu^{1/2}U_\infty^{1.5}/x^{1/2}. \tag{12.20}$$

For a flat plate of width w, the total drag force,

$$F_D = w\int^L \tau_w dx = 0.664w\rho^{1/2}\mu^{1/2}U_\infty^{1.5}L^{1/2}, \tag{12.21}$$

where L is the length of the plate. For practical purposes, it may be more convenient to use the average friction coefficient (Çengel and Cimbala, 2006),

$$C_f = (1/L)\int^L C_{f,x}dx = 1.33/Re_L^{1/2}, \tag{12.22}$$

for $Re_L > 5 \times 10^5$.

Example 12.1   Sustaining a flat surface against a current.

Given: Dr. JAS has a flat-bottom canoe, as illustrated in Fig 12.5, that she uses to demonstrate friction drag. During one of her Thermofluids classes, Dr. JAS places the 1-m wide and 2.5-m long canoe in the Turbulence and Energy Laboratory water flume and T&E Lab member Kaya volunteers to paddle the canoe. Dr. JAS turns on the water flume to create a steady 0.2 m/s current, where the flowing water is at 20°C.

Find: The drag force T&E Lab member Kaya has to overcome to keep the canoe in position.

Solution: Check the boundary layer/flow regime based on the Reynolds number. The Reynolds number at the end of the canoe,

$$Re_L = UL/\nu = 0.2(2.5)/1 \times 10^{-6} = 500,000.$$

This value corresponds to the critical Reynolds number for flow over a flat plate. In other words, the boundary layer over the entire bottom of the canoe is laminar. This being the case, the drag force equation for a laminar boundary layer over a flat plate is the appropriate formula to use. Specifically,

$$F_D = 0.664 w \rho^{1/2} \mu^{1/2} U_\infty^{1.5} L^{1/2}.$$

Substituting for the values, we get the total drag force,

$$F_D = 0.664(1)(1000)^{1/2} \left(1 \times 10^{-3}\right)^{1/2} (0.2)^{1.5} (2.5)^{1/2} = 0.09 \, N.$$

We see that, for this laminar case, only minimum effort is needed to keep the canoe in position.

12.4.2 Transitional boundary layer

Under typical real-life conditions, the boundary layer becomes unstable at a local Reynolds number, Re_x of roughly 5×10^5. Schubauer and Skramstad (1943) determined that the free-stream disturbance, in terms of turbulence intensity, has to be greater than 0.2% to notably decrease the critical Reynolds number from a value of a few million. This critical Reynolds number, Re_c decreases further, from 3×10^6 to roughly 10^6, when the turbulence intensity is augmented to 0.5%, and it drops to the order of 10^5 at a turbulence intensity of 3% (Dryden, 1947). In engineering practice, Re_c is generally assumed to be 5×10^5, and the boundary layer is expected to be fully turbulent at Re of 1×10^6. As the range of transitional boundary layer is short, there is no special treatment for this section. To put it another way, calculations are only developed for laminar flow stretch and beyond that extent, turbulent flow analysis is recommended.

12.4.3 Turbulent boundary layer

With at least three sublayers at play, the analysis of the turbulent boundary layer is more involved and, thus, is deferred to more advanced monographs such as Schlichting and Gersten (2000). It can be shown, see, for example, White (2011), that the boundary layer thickness, δ, with respect to the distance from the leading edge, x, can be expressed in terms of the local Reynolds number, Re_x, as

$$\delta/x \approx 0.16/Re_x^{1/7}. \tag{12.23}$$

From this, we can deduce the local friction coefficient to be

$$C_{f,x} \approx 0.027/Re_x^{1/7}. \tag{12.24}$$

The induced drag in terms of the drag coefficient has been found to be

$$C_{D,x} = 7C_{1,x}/6 = 0.031/Re_x^{1/7} \tag{12.25}$$

The wall shear as a function of distance from the transitional point can be expressed as

$$\tau_w(x) = 0.0135\rho^{6/7}\mu^{1/7}U_\infty^{13/7}/x^{1/7}. \qquad (12.26)$$

For a flat plate of width w, the total drag force,

$$F_D = w\int^L \tau_w dx = 0.0158w\rho^{6/7}\mu^{1/7}U_\infty^{13/7}L^{6/7}, \qquad (12.27)$$

where L is the length over which the boundary layer is turbulent. It is important to note that turbulence increases the friction and, thus, frictional drag. For that reason, engineers continue to work on different ways to keep the boundary layer laminar or, delay the transition to turbulent boundary layer, to minimize frictional losses.

It is worth mentioning that quite different expressions have been used to describe the primarily empirical turbulent boundary layer. For example, Çengel and Cimbala (2006) describe the local turbulent boundary thickness as

$$\delta/x \approx 0.38/Re_x^{1/5}. \qquad (12.28)$$

The corresponding local friction coefficient is

$$C_{f,x} \approx 0.059/Re_x^{1/5}. \qquad (12.29)$$

The numerical values obtained, nevertheless, are quite close to those deduced based on the expressions given in White (2011). Following Çengel and Cimbala (2006), the average friction coefficient over a stretch of turbulent boundary layer of length L can be expressed as

$$C_f \approx 0.074/Re_L^{1/5}. \qquad (12.30)$$

For turbulent flows parallel to a flat plate, where the laminar portion is significant, the average friction coefficient can be estimated using

$$C_f \approx 0.074/Re_L^{1/5} - 1742/Re_L. \qquad (12.31)$$

Example 12.2 Sustaining a flat surface with a turbulent boundary layer against a current.

Given: After reading about the use of ribs and small-scale roughness for reducing drag on a cylinder by Skeide et al. (2020), T&E Lab member Rajin went and applied similar roughness to the bottom surface of Dr. JAS' 1-m wide by 2.5-m long, flat-bottom canoe. Not aware of this endeavor, Dr. JAS placed the canoe in the Turbulence and Energy Laboratory water flume and set the 20°C water to flow at 0.2 m/s. Once again, T&E Lab member Kaya jumped into the canoe and volunteered to demonstrate his paddling skill.

Find: The drag force T&E Lab member Kaya has to overcome to keep the canoe in position.

Solution: Assuming the bottom surface of the canoe is roughened sufficiently that the 0.2 m/s current over it is fully turbulent. (If not, we can imagine T&E

Lab member Rajin placed some obstacles upstream to disturb the water, creating a turbulent current in the water flume.) As such, Eq. (12.27) is the appropriate equation to determine the total drag force acting on the canoe. Specifically, the total drag,

$$F_D = 0.0158 w \rho^{6/7} \mu^{1/7} U_\infty^{13/7} L^{6/7}$$
$$= 0.0158(1)(1000)^{6/7} \left(1 \times 10^{-3}\right)^{1/7} (0.2)^{13/7} (2.5)^{6/7} = 0.24 \, \text{N}.$$

This is more than double the effort needed to overcome laminar frictional drag.

It is important to note that an approach that can lead to a reduction in the drag experienced by a bluff body such as a cylinder does not necessarily apply equally to a flat plate. Here, we see that the particular roughening of the surface has resulted in a contrasting outcome.

We see that typical roughening, such as not taking efforts to smoothen the bottom of a canoe including polishing and coating it with a smooth layer of paint, tends to perturb the boundary layer into a turbulent one. Intelligent designs which abound in nature enlighten us that there are many unique and organized roughening patterns that mysteriously keep the boundary layer laminar. A conspicuous example is the beautiful array of fish scales; see Muthuramalingam et al. (2019, 2020). Not so obvious may be the denticles on shark skin (Oeffner and Lauder, 2012). Both of these uneven surfaces abstrusely reduce the drag appreciably. There is immeasurably more for us to learn from the uncountable amazing designs in nature. Tian et al. (2021) are but one of the most recent papers disclosing the state-of-the-art research and engineering on drag-reduction surfaces. A word of caution here. Most natural systems have multifaceted features that are coupled together to effectively function under the designed settings and within the imposed limitations. Choosing selective features to imitate different environments and working conditions may not always give the best results.

12.5 Bluff body aerodynamics

Thus far, for flow parallel to a flat plate, we have only considered the drag caused by fluid viscosity, that is, the frictional drag. For Dr. JAS' canoe, see Fig. 12.5, the moving water will induce a force on the frontal area, in the absence of the tilted shield just ahead of the canoe. Moreover, the rear surface of the canoe will encounter a suction, without the inclined shield at the back. The front and back shields in Fig. 12.5 have taken on both pressure and frictional drag of the approaching and departing flow, respectively, leaving only the flat surface frictional drag for the canoeist to overcome.

Example 12.3 The load on a square piling.

Given: Dr. JAS signed a contract with an engineering firm to design 2-m by 2-m square pilings for supporting a bridge across a river. The water is 10-m deep and the nominal current is 0.2, but for safety purposes, a 0.3 m/s maximum current is considered

Find: The maximum and minimum drag forces exerted on each piling by the 0.3 m/s currents.

Solution: The Reynolds number based on the hydraulic diameter,

$$Re_D = \rho UD/\mu = 1000(0.3)(2)/\left(1 \times 10^{-3}\right) = 6 \times 10^5.$$

For Reynolds numbers between approximately 1×10^4 and 1×10^6 (Liu et al., 2007; White, 2011), with one of the surfaces normal to the incoming flow, the drag coefficient, $C_D = 2.1$. This is the most bluff or blunt condition and, thus, the maximum normal load on the piling. The corresponding drag force,

$$F_D = C_D\left(\frac{1}{2}\rho U^2 A\right) = 2.1\left[\frac{1}{2}(1000)(0.3)^2(2)(10)\right] = 1890\,\text{N}.$$

If one of the edges is facing the incoming flow, then this forms the most streamlined condition and the corresponding drag coefficient, C_D, decreases to 1.6. Accordingly, the minimum drag force,

$$F_D = C_D\left(\frac{1}{2}\rho U^2 A\right) = 1.6\left[\frac{1}{2}(1000)(0.3)^2(2)(10)\right] = 1440\,\text{N}.$$

Therefore, if feasible, we should align the piling in the most streamline orientation with respect to the flowing stream. This will not only lead to about a 24% reduction in the steady load, but it will also significantly minimize the dynamic load due to vortex shedding. Vortex shedding and streamlining will be expounded shortly.

12.5.1 Steady flow across a smooth circular cylinder

The flow characteristics of a steady flow around a smooth circular cylinder are quite well established. A key difference between flow around a bluff body, the most common being a circular cylinder, and that parallel to a flat surface is the pressure gradient. Specifically, the flow around the front half of the cylinder experiences a favorable pressure gradient, keeping the boundary layer thin and attached to the cylinder; see Fig. 12.6. This favorable pressure gradient deteriorates as the flow moves around the cylinder, downstream from the front stagnation point. Beyond 90° from the front stagnation point, the pressure gradient becomes unfavorable, and the boundary layer thickens rapidly until the flow eventually separates. Recirculating or back flow occurs following the flow separation at the separation point. The flow separation and the subsequent recirculating flow are closely linked to the formation of a vortex. The flow separation and vortex formation from each side of the cylinder ultimately lead to alternating shedding of vortices.

Let us follow Coutanceau and Defaye (1991) and delineate the flow and pressure around a circular cylinder in detail, with the help of Fig. 12.6. The free stream fluid is brought to rest at the forward stagnation point, with an accompanying rise in pressure. From the forward stagnation point, the pressure

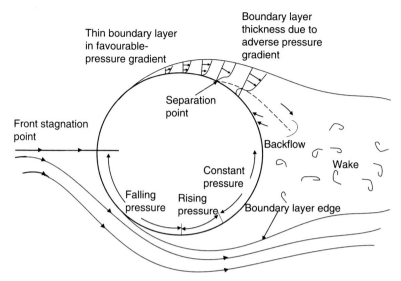

FIGURE 12.6 Boundary layer around a circular cylinder in steady flow (created by O. Imafidon). The flow separates from the surface of the cylinder due to an adverse pressure gradient. This is somewhat like someone riding on a fluid particle roller coaster along the curved streamline around the side of the cylinder and "flying off" the track at the separation point.

decreases with increasing streamline coordinate, x, and the boundary layer develops under the influence of a favorable pressure gradient (dP/dx < 0). The pressure eventually reaches a minimum and, hence, further boundary layer development occurs toward the rear of the cylinder, in the presence of an adverse pressure gradient (dP/dx > 0). Unlike conditions for the flat plate in parallel flow, the free stream velocity around the cylinder, $u_\infty(x)$, varies with x and is different from the upstream velocity U_∞. From Euler's equation for an inviscid flow, $u_\infty(x)$ must exhibit behavior opposite to that of P(x). From $u_\infty(x) = 0$ at the stagnation point (x = 0), the fluid accelerates because of the favorable pressure gradient, i.e., $du_\infty/dx > 0$ when dP/dx < 0. The fluid reaches a maximum velocity when dP/dx = 0 and decelerates farther downstream because of the adverse pressure gradient, that is, $du_\infty/dx < 0$ when dP/dx > 0. As the fluid decelerates after reaching its maximum velocity, the velocity gradient at the surface, $du/dy|_{x=0}$, eventually becomes zero. This occurs at the separation point, where the fluid near the surface lacks sufficient momentum to overcome the pressure gradient and, hence, continued downstream movement becomes impossible. Separation point is the location where $du/dy|_{x=0} = 0$, a condition where the boundary layer detaches from the surface, and a wake is formed in the downstream region. Flow in the wake is characterized by vortex formation and is highly irregular. The Reynolds number dictates the occurrence of boundary layer transition and hence, influences the position of the separation point. The occurrence of separation is delayed when the boundary layer becomes

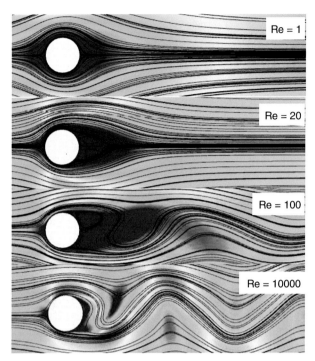

FIGURE 12.7 Flow regimes of a two-dimensional, circular cylinder in steady flow for Re = 1 to 1×10^4 (created by Y. Yang). Regime 1: creeping flow (Re $\lesssim$ 5); Regime 2: closed near-wake, standing vortex pair (5 $\lesssim$ Re $\lesssim$ 45); Regime 3: periodic laminar vortex street (45 $\lesssim$ Re $\lesssim$ 190); Regime 4: transition in shear layers (190 $\lesssim$ Re $\lesssim$ 3×10^5); Regime 5: critical regime (3×10^5 $\lesssim$ Re $\lesssim$ 3×10^6); Regime 6: postcritical regime (Re $\gtrsim$ 3×10^6).

turbulent, for the momentum of the fluid in the larger Reynolds number turbulent flow is larger than that in the laminar boundary layer. The boundary layer remains laminar if the Reynolds number is less than approximately 2×10^5, and separation occurs at $\theta \approx 80°$. If the Reynolds number is larger than 2×10^5, boundary layer transition occurs, and separation is delayed to $\theta \approx 140°$.

Let us expound on the flow characteristics around and downstream of a circular cylinder as a function of Reynolds number, with the help of Fig. 12.7. In general, there are five flow regimes, or more correctly, wake regimes, as described below.

Regime 1: Creeping flow (Re $\lesssim$ 5).
When the approaching flow is very slow and/or the fluid is highly viscous, there is not enough inertia to alter the "layered" flow stream around the cylinder. This is the case when the Reynolds number is less than about five. The streamlines remain attached around the cylinder with no visible wake on the leeward side. This creeping flow regime is also referred to as the unseparated flow regime.

Regime 2: Closed near-wake, standing vortex pair (5 $\lesssim$ Re $\lesssim$ 45).

Increasing the Reynolds number above five results in enough flow inertia to form a pair of counter-rotating vortices on the leeward side. The inertia is still weak and, thus, these vortices, known as Föppl vortices, stay put in the near wake but elongate with increasing Reynolds number.

Regime 3: Periodic laminar vortex street ($45 \lesssim Re \lesssim 190$).
As the inertial force increases with increasing Reynolds number, the standing recirculating wake becomes unstable, starts to oscillate, and laminar vortices begin to shed alternately. The periodic vortex shedding results in the formation of the celebrated von Kármán vortex street, which strictly should be called von Kármán–Bénard vortex street.

Regime 4: Transition in shear layers ($190 \lesssim Re \lesssim 3 \times 10^5$).
The laminar vortices become wrinkled and irregular when the Reynolds number is increased above 190. The free shear layers become turbulent, spawning forth turbulent eddies. The near wake becomes three-dimensional as the Reynolds number approaches 3×10^5, the pre-critical regime.

Regime 5: Critical regime ($3 \times 10^5 \lesssim Re \lesssim 3 \times 10^6$).
Regime 5 is called the critical regime because of a drastic drop in the drag coefficient due to a sudden windward shift of the separation points. This regime may be subdivided into two subregions, that is, subcritical and super-critical. A subtle bias occurs with the formation of a separation bubble on one side of the cylinder, resulting in a preferential lift. A return to an unbiased symmetry takes place with a further increase in the Reynolds number, with the formation of the other separation bubble. The abrupt drastic drop in drag is a fallout of the laminar boundary layer transforming into a turbulent one. The turbulent boundary layer randomizes the inertia of the flowing fluid, delaying the flow separation farther back. To that end, the wake narrows and no regular vortex shedding is discernable.

Regime 6: Post-critical regime ($Re \gtrsim 3 \times 10^6$).
Not too many laboratories around the world are capable of pushing the Reynolds number to over 3×10^6. Back in the later part of the 20th century, the curiosity-driven Canadian-born physicist and engineer, Anatol Roshko, capitalized a wind tunnel with such capacity just before it was scrapped. Roshko (1961) discovered the reestablishment of the turbulent vortex street when the Reynolds number exceeds three million.

12.5.2 Vortex shedding

We have been enlightened that vortex shedding prevails when a bluff body is subjected to a flow. Here is an experiment that can be conducted by everyone. Walk through a knee-high body of water or draw one of your arms across a bathtub filled with water from right to left or vice versa. When you execute either action at a moderately high speed, you will sense two forces in operation. First, there is the opposing drag that is always there to slow you down, giving you a good workout in a short lapse of time. More interestingly, there is an oscillating force moving your leg or arm from side to side as you traverse through the water. It is worth noting that both these forces, drag and oscillating lift, are also working as we move in the air. We do not realize these everyday forces as we go about our daily business because air is a much lighter fluid compared to water. Only a sufficiently strong wind brings to mind wind-induced drag and the passing of a big truck at high speed compels us to appreciate the "suction" caused by the oscillating lift behind the speedy truck. This oscillating lift is augmented when we move through the much heavier fluid, water. The regular alternate shedding of vortices from the two sides of the bluff body, our leg or arm, induce this oscillating lift. While there are variations at the detailed level concerning these rhythmical vortices, the general name associated with a street of such vortices is von Kármán–Bénard vortex street. A classical von Kármán–Bénard vortex street behind a long or two-dimensional circular cylinder is depicted in Fig. 12.7, for Reynolds numbers between approximately 45 and 1×10^5.

But what is a vortex? Within context, a vortex is simply the rotating motion of fluid particles around a center or line. For the primarily two-dimensional vortices depicted in Fig. 12.7, the vortex center associated with each vortex is, in fact, a vortex line coming out of (and/or into) the page. The right side of Fig. 12.8 shows the propagation of a laminar and a turbulent vortex ring generated from compressed air underwater (Yan et al., 2018). For vortex rings such as these, the vortex line is the closed circular line around which the fluid rotates, propagating the vortex ring forward, or upward in the buoyant vortex ring case. The left side of Fig. 12.8 illustrates the different regions of a buoyant vortex ring. Whether an organized vortex ring can be formed, or the formed vortex ring is laminar, slightly unstable (transition), turbulent, or disintegrate, is a function of Reynolds number, Re, Bond number, Bo, and Weber number, We. Bond number is also referred to as Eötvös number it describes the relative significance of the gravitational force with respect to the capillary force. Namely,

$$\text{Bo} = \Delta\rho g L^2 / \sigma, \tag{12.32}$$

where g is gravity, L is the characteristic length, and σ is the surface tension. Weber number signifies the magnitude of drag force with respect to cohesion force. Mathematically, we can write,

$$\text{We} = \rho V^2 / \sigma. \tag{12.33}$$

An applied mind may wonder if there is any use of these beautiful vortex rings in nature. We know that intelligent designs are not just for esthetics but also for utility. Curiosity-oriented research continues to reveal the beauty of life, and the subtle but potent functionality of each beauty. Take, for example, the ocean's most efficient swimmer, the moon jellyfish, which craftily exploits vortex rings to claim this throne; see Yong (2013) and Gemmell et al. (2013). It should be stressed that the jellyfish is not the only sea creature taking advantage of vortex rings, squids, salps and many fishes also excel in this; see Linden and Turner (2004). We talked about sustaining life on earth via effective

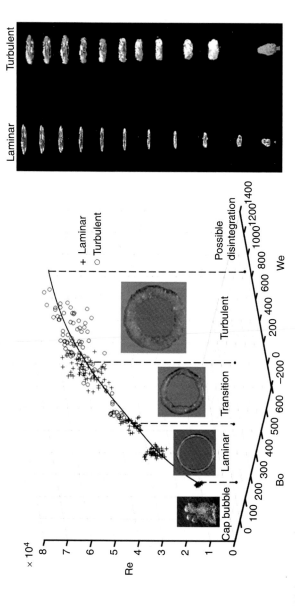

FIGURE 12.8 A map of a buoyant vortex ring in terms of Re, Bo, and We (created by X. Yan). At the low end, it indicates that some minimum inertial and/or gravitational force is required for the creation of a buoyant vortex ring. At the high end, it suggests that too much drag and/or inertia will lead to disintegration of the turbulent vortex ring.

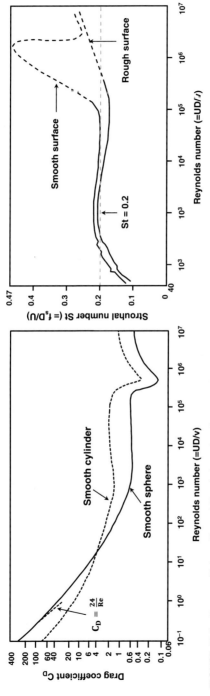

FIGURE 12.9 Drag coefficient, C_D, and Strouhal number, St, versus the Reynolds, number, Re, for common bluff bodies such as a smooth sphere and a smooth circular cylinder (created by O. Imafidon). Note that over a wide range of Reynolds numbers between 10^3 and 10^5, both C_D and St remain largely unchanged.

pollination earlier. Even a dandelion bristle has been bestowed with a vortex ring to extend the pollination and, hence, the procreation of the good[2] dandelion (Cummins et al., 2018).

A key characteristic of a vortex is vorticity, that is, a vortex contains vorticity. The stronger the vortex, the higher the concentration of vorticity. Mathematically, vorticity is the curl of a velocity vector, that is,

$$\tilde{\omega} = \text{curl}\tilde{V} = \nabla \times \tilde{V}. \tag{12.34}$$

In Cartesian coordinates, vorticity can be expressed as

$$\tilde{\omega} = \left(\frac{\partial W}{\partial y} - \frac{\partial V}{\partial z}\right)\hat{i} + \left(\frac{\partial U}{\partial z} - \frac{\partial W}{\partial x}\right)\hat{j} + \left(\frac{\partial V}{\partial x} - \frac{\partial U}{\partial y}\right)\hat{k}, \tag{12.35}$$

where U, V, and W correspond to the velocity component in the x, y, and z direction, respectively, and i, j, and k with hats signify unit vectors in the x, y, and z direction, respectively.

Coming back to regular vortex shedding, the regular shedding frequency can be conveniently expressed nondimensionally in terms of the Strouhal number,

$$\text{St} = fL/U, \tag{12.36}$$

where f is the vortex shedding frequency in Hertz, L is the characteristic length, and U is the velocity. Fig. 12.9 shows the variation of the drag coefficient and the Strouhal number of some common bluff bodies with respect to the Reynolds number. We see that over a wide range of Reynolds number, both C_D and St remain approximately constant for bluff bodies such as a circular cylinder and a sphere.

12.5.3 Streamlining

You do not have to be a fish to appreciate streamlining. The most familiar human inventions that lean heavily on streamlining are in transportation: cars, ships, and airplanes. There remains much more intelligence in nature for engineers to leverage. For instance, unique bird-like morphing wing structures can be imitated to substantially improve the aerodynamics of an airplane (Hui et al., 2020). Also, rainbow trout adjust, embrace, and exploit regular vortices with their whole bodies when they are downstream of a bluff body that spawns a von Kármán-Bénard vortex street (Liao et al., 2003). Other times, an intelligently designed beak can take you speedily through both air and water (Crandell et al., 2019). As an illustration, the beak of the kingfisher has seriously advanced high-speed bullet trains, mitigating high-speed-induced noise, vibration, and structural damage.

---

2. There are studies that indicate many health benefits of dandelion. Its root extract can potentially suppress cancer; see, for example, Ovadje et al. (2016) and Zhu et al. (2017).

Problems

12.1   Orientation of advertisement sign on car roof.

You are commissioned by Dr. JAS to promote an event to engage the general public with engineering research at the Turbulence and Energy Laboratory. The promotion is nicely displayed on both sides of a thin, 2-m long and 1-m tall rectangular board to be attached to the roof of your car. There are advantages, in terms of visibility, in orienting the sign parallel and perpendicular to the driving direction. How much drag force does the rectangular board have to withstand when you drive your car at 60 km/h? The drag coefficient of the board perpendicular to the wind can be assumed to be 1.2.

12.2   Ocean liner frictional drag.

Dr. JAS is on Symphony of the Seas, the world's largest cruise liner, with her family and external flow dawned on her. She wishes to carry out a detailed computational fluid dynamics simulation. Please provide Dr. JAS with an estimate of the haul. The cruise liner is 360 m long and roughly 66 m wide. The draught of the ship's hull is 9 m. Assume that the cross-section under the water is an inverted triangle of 66 m wide and 9 m high. In other words, estimate the frictional drag of two 34.2 m wide flat surfaces that are 360 m long at 20 knots (37 km/h). It is interesting to learn that small air bubbles are generated from the keel to reduce drag and noise. Interested readers can refer to Elbing et al. (2008), Ceccio (2010), Verschoof et al. (2016), Qin et al. (2017), Park and Lee (2018), for example, concerning some of the latest research on this topic.

12.3   Wind load on a giraffe's neck.

A giraffe's neck can be over 2 m long. Assume the neck of a giraffe as a 2-m long circular cylinder with a diameter of 0.3 m. Estimate the load induced by winds at 7 m/s and 14 m/s. What are the corresponding vortex shedding frequencies? Note that a giraffe's neck is not a constant-diameter smooth cylinder. The tapering feature, along with other details, most likely alleviate vortex-induced vibration. This was probably more so the case for the much bigger Behemoths with a tail like a cedar trunk, especially when they were in surging floodwaters (Job, Chapter 40).

12.4   Mitigating chimney wind loads.

Flue-gas stacks are chimneys used for exhausting flue gases farther up from the ground, away from where human beings resides, into the air, to enhance dispersion and dilution of pollutants to acceptable levels. Calculate the load induced by a 12-m/s wind on a 5-m diameter and 200-m tall flue-gas stack. This drag force and, more so, the oscillating lift due to vortex shedding demand sound reinforcement of flue-gas stacks. A pragmatic means of lessening the wind-induced forces is to clothe the chimney with a spiral. The spiral diverts the approaching flow upward on one side and downward on the other side, disrupting the otherwise well-organized vortex roll-up along the chimney and, thus, vortex-induced vibration. There is also a lot to be learned from a cactus, which is often subjected to severe desert winds. Recent studies on cactus-shaped cylinders, such as Wang et al. (2014), Wang et al. (2021), Zhdanov et al. (2021), reveal that the intelligently designed cactus shape is excellent for reducing drag and, equally so, for mitigating flow-induced vibrations.

12.5 Pufferfish hydrodynamics.

Pufferfish can be up to about 1 m in size. Consider a pufferfish as a 0.3-m diameter sphere. Estimate the force it has to exert against a water current of 1 m/s and 2 m/s. When puffed up, presumably in an effort to deter potential predators, a pufferfish is far from being a smooth and cozy bouncy ball. Most puffer fish have many spines on their skin and are poisonous. The spiky surface probably has a significant effect in lowering the drag. Talking about unique fish, boxfish stand out with their box-like body with many unique maneuvering features; see, for example, van Wassenbergh et al. (2015), also of interest is Godoy-Diana and Thiria (2018).

12.6 Golf ball drag reduction.

Ever wonder why a golf ball has dimples? The dimples on the golf ball are fashioned to advance the drag crisis, the sharp drop in the drag coefficient with the leeward shift of the separation point (or line) with increasing Reynolds number. For a smooth sphere, this drag crisis takes place at a critical Reynolds number of around 3×10^5, where C_D drops from about 0.5 to 0.1. For the dimpled golf ball, this critical Reynolds number is advanced to roughly 5×10^4, though C_D drops to only about 0.25. Estimate the increase in velocity after drag crisis for a 4.27-cm golf ball otherwise traveling at 50 m/s.

12.7 Vortex-induced stay cable vibration.

Stay cables of cable-stayed bridges are frequently subjected to wind-induced vibrations. When designing such stay-cable systems, meticulous care is exercised to make sure the vortex shedding frequencies and the natural frequencies of the cable system are far apart. Consider a 0.3-m diameter steel ($\rho = 7900$ kg/m^3, Young's modulus = 205 GPa) stay cable with a length of 180 m as a simply supported beam. What range of wind speed will result in vortex shedding that is in sync with the first, second, or third mode of vibration?

References

Blasius, H., 1908. Grenrschichten in Flussgkeiten mit Kleiner Reibung, Zeitschrift für Angewandte Mathematik und Physik, 56, pp. 1–37.

Ceccio, S.L., 2010. Friction drag reduction of external flows with bubble and gas injection. Annu Rev. Fluid. Mech. 42, 183–203.

Çengel, Y.A., Cimbala, J.M., 2006. Fluid Mechanics: Fundamentals and Applications. McGraw-Hill, New York, NY.

Coutanceau, M., Defaye, J-R., 1991. Circular cylinder wake configuration: a flow visualization survey. Appl. Mech. Rev. 44 (6), 255–305.

Crandell, K.E., Howe, R.O., Falkingham, P.L., 2019. Repeated evolution of drag reduction at the air-water interface in diving kingfishers. J. Royal Soc., Interface 16 (154), 20190125.

Cummins, C., Seale, M., Macentc, A., Certini, D., Mastropaolo, F., Viola, I.M., Nakayama, N., 2018. A separated vortex ring underlies the flight of the dandelion. Nature 562, 414–418.

Doran Jr., E., 1981. Wangka: Austronesian Canoe Origins. Texas A&M University Press, Texas.

Dryden, H.L., 1947. Some recent contributions to the study of transition and turbulent boundary layers. NACA Technical Notes 1168.

EIA, 2021. https://www.eia.gov/energyexplained/wind/history-of-wind-power.php, U.S. Energy Information Administration, (accessed October 9, 2021).

Elbing, B.R., Winkel, E.S., Lay, K.A., Ceccio, S.L., Dowling, D.R., Perlin, M., 2008. Bubble-induced skin-friction drag reduction and the abrupt transition to air-layer drag reduction. J. Fluid Mech. 612, 201–236.

Euler, L., 1755. Institutiones calculi differentialis cum eius usu in analysi finitorum ac doctrina serierum. Volume 1, Academiae Imperialis Scientiarum Petropolitanae (1755) 1–880.

Fleming, T.H., Geiselman, C., Kress, W.J., 2009. The evolution of bat pollination: a phylogenetic perspective. Ann. Bot. 104 (6), 1017–1043.

Gemmell, B.J., Costello, J.H., Colin, S.P., Stewart, C.J., Dabiri, J.O., Tafti, D., Priya, S., 2013. Passive energy recapture in jellyfish contributes to propulsive advantage over other metazoans. Proc. Nat. Acad. Sci. USA 110 (44), 17904–17909.

Godoy-Diana, R., Thiria, B., 2018. On the diverse roles of fluid dynamic drag in animal swimming and flying. J. Royal Soc., Interface 15 (139), 20170715.

Gray, C.F., 2002. The five first flights: the slope and winds of Big Kill Devil Hill - the first flight reconsidered. J. Early Aeroplane 177, 26–39.

Hui, Z., Zhang, Y., Chen, G., 2020. Tip-vortex flow characteristics investigation of a novel bird-like morphing discrete wing structure. Phys. Fluids 32, 035112.

Kumar A., "The single most astonishing fact of human geography: Indonesia's far west colony," Indonesia, 92: 59-95, 2011.

Lesics, 2021. https://www.youtube.com/watch?v=AfCyzIbpLN4, Tesla turbine – the interesting physics behind it. (Accessed on October 9, 2021).

Liao, J.C., Beal, D.N., Lauder, G.V., Triantafyllou, M.S., 2003. The Kármán gait: novel body kinematics of rainbow trout swimming in a vortex street. J. Exp. Biol. 206 (6), 1059–1073.

Linden, P.F., Turner, J.S., 2004. 'Optimal' vortex rings and aquatic propulsion mechanisms. Proceedings of the Royal Society London B: Biological Science 271, 647–653.

Liu, T.C., Ge, Y.J., Cao, F.C., Zhou, Z.Y., Zhang, W., 2007. Reynolds number effects on the flow around sqaure cylinder based on lattice Noltzmann method. In: New Trends in Fluid Mechanics Research, Proceedings of the Fifth International Conference on Fluid Mechanics, 170. Tsinghua University Press & Springer, Shanghai.

Muthuramalingam, M., Puckert, D.K., Villemin, L.S., Bruecker, C., 2020. Transition delay using biomimetic fish scale arrays. Sci. Rep. 10, 14534.

Muthuramalingam, M., Villemin, L.S., Bruecker, C., 2019. Streak formation in flow over biomimetic fish scale arrays. J. Exp. Biol. 222 (16), jeb205963.

Oeffner, J., Lauder, G.V., 2012. The hydrodynamic function of shark skin and two biomimetic applications. J. Exp. Biol. 215 (5), 785–795.

Ovadje, P., Ammar, S., Guerrero, J-A., Arnason, J.T., Pandey, S., 2016. Dandelion root extract affects colorectal cancer proliferation and survival through the activation of multiple death signalling pathways. Oncotarget 7 (45), 73080–73100.

Park, S.H., Lee, I., 2018. Optimization of drag reduction effect of air lubrication for tanker model. Int. J. Naval Architecture Ocean Eng. 10 (4), 427–438.

Prandtl, L., 1904. Motion of fluids with very little viscosity. German History Intersections. Verhandlung des III Internationalen MathematikerKongresses, (Heidelberg, 1904), pp. 484–491.

Qin, S., Chu, N., Yao, Y., Liu, J., Huang, B., Wu, D., 2017. Stream-wise distribution of skin-friction drag reduction on a flat plate with bubble injection. Phys. Fluids 29, 037103.

Roshko, A., 1961. Experiments on the flow past a circular cylinder at very high Reynolds number. J. Fluid Mech. 10, 345–356.

Schlichting, H., Gersten, K, K., 2000. Boundary Layer Theory, 8th ed. Springer, Berlin.

Schubauer, G.B., Skramstad, H.K., 1943. Laminar-boundary-layer oscillations and transition on a flat plate. NASA Report ATI 9595. (Also in Journal of the Aeronautical Sciences 14 (2), 69–78, 1947).

Skeide, A.K., Bardal, L.M., Oggiano, L., Hearst, R.J., 2020. The significant impact of ribs and small-scale roughness on cylinder drag crisis. J. Wind Eng. Ind. Aerodyn. 202, 104192.

Tian, G., Fan, D., Feng, X., Zhou, H., 2021. Thriving artificial underwater drag-reduction materials inspired from aquatic animals: progresses and challenges. Royal Soc. Chemistry Advances 11, 3399–3428.

Ting, D.S-K., 2016. Basics of Engineering Turbulence. Academic Press, New York, NY.

van Wassenbergh, S., Van Manen, K., Marcroft, T.A., Alfaro, M.E., Stamhuis, E.J., 2015. Boxfish swimming paradox resolved: forces by the flow of water around the body promote manoeuvrability. J. Royal Soc., Interface 12 (103), 20141146.

Verschoof, R.A., van der Veen, R.C.A., Sun, C., Lohse, D., 2016. Bubble drag reduction requires large bubbles. Phys. Rev. Lett. 117, 104502.

Wang, S.F., Liu, Y.Z., Zhang, Q.S., 2014. Measurement of flow around a cactus-analogue grooved cylinder at $Re_D = 5.4 \times 10^4$: wall-pressure fluctuations and flow pattern. J. Fluids Struct. 50, 120–136.

Wang, W., Mao, Z., Song, B., Tian, W., 2021. Suppression of vortex-induced vibration of a cactus-inspired cylinder near a free surface. Phys. Fluids 33 (6), 067103.

White, F.M., 2011. Fluid Mechanics, 7th ed. McGraw Hill, New York, NY.

Yan, X., Carriveau, R., Ting, D.S-K., 2018. On laminar to turbulent buoyant vortex ring regime in terms of Reynolds number, Bond number, and Weber number. J. Fluids Eng. 140 (5) 054502-(1-5).

Yong, E., 2013. Why a jellyfish is the ocean's most efficient swimmer. Nature doi:10.1038/nature.2013.13895.

Young, A.M., Gómez-Ruiz, P.A., Peña, J.A., Uno, H., Jaffe, R., 2018. Wind speed affects pollination success in blackberries. Sociobiology 65 (2), 225–231.

Zhdanov, O., Green, R., Busse, A., 2021. Experimental investigation of the angle of attach dependence of the flow past a cactus-shaped cylinder with four ribs. J. Wind Eng. Ind. Aerodyn. 208, 104400.

Zhu, H., Zhao, H., Zhang, L., Xu, J., Zhu, C., Zhao, H., Lv, G., 2017. Dandelion root extract suppressed gastric cancer cells proliferation and migration through targeting lncRNA-CCAT1. Biomed. Pharmacother. 93, 1010–1017.

Part 4

Ecophysiology-flavored Engineering Heat Transfer

Chapter 13

Steady conduction of thermal energy

"Heat, like gravity, penetrates every substance of the universe, its rays occupy all part of space."

Jean Baptiste Joseph Fourier

Chapter Objectives

- Understand Fourier's law of heat conduction.
- Fathom the inverse relationship between thermal conductivity and temperature gradient.
- Appreciate the parallel between electric current flow and heat flow, and between electric resistance and thermal resistance.
- Master one-dimensional heat conduction in planar, cylindrical, and spherical coordinates.
- Differentiate the parallel-path method from the isothermal-plane method.

Nomenclature

A   area; A_{avg} is the average area, A_{insul} is the area of the path containing insulation, A_j is the area correspond to the j path, A_{stud} is the area of the stud, A_{tot} is the total heat transfer area

E   energy; E_{wall} is the energy of the wall

h_{conv} convection heat transfer coefficient; h_{in} is the internal heat convection coefficient, h_{out} is the external convection heat transfer coefficient

k   thermal conductivity; k_{1cm} is the thermal conductivity of a 1-cm-thick material, k_{3cm} is the thermal conductivity of a 3-cm-thick material, $k_{contact}$ is the contact thermal conductance

L   length

Q   thermal energy; Q' is the heat transfer rate, Q'_{cond} is the conduction heat transfer rate, Q'_{cyl} is the heat transfer rate through the cylinder, Q'_{gap} is the heat transfer rate through the gap or void, Q'_{in} is the entering heat transfer rate, Q'_{insul} is the heat transfer rate through the path containing insulation, Q'_{join} is the heat transfer rate through the (perfectly) joined section, Q'_{out} is the outgoing heat transfer rate, Q'_{stud} is the heat transfer rate through the path containing a stud, Q'_{tot} is the total heat transfer rate

q'' heat flux

Thermofluids: From Nature to Engineering. DOI: https://doi.org/10.1016/B978-0-323-90626-5.00017-3

R   resistance; R_1 is the thermal resistance of layer 1, R_2 is the thermal resistance of layer 2, R_{1cm} is the thermal resistance of a 1-cm-thick material, R_{3cm} is the thermal resistance of a 3-cm-thick material, $R_{contact}$ is the contact resistance, $R_{conv,in}$ is the inner convection resistance, $R_{conv,out}$ is the outer convection resistance, R_{cyl} is the thermal resistance of the cylinder, R_{in} is the inner thermal resistance, R_{insul} is the thermal resistance of the path with insulation, R_{out} is the outer thermal resistance, R_{sph} is the thermal resistance of a sphere, R_{stud} is the thermal resistance of the path with a stud, $R_{stud,insul}$ is the stud-insulation thermal resistance, R_{tot} is the total thermal resistance, $R_{tot,avg}$ is average total resistance, $R_{tot,insul}$ is the total thermal resistance along the path with insulation, $R_{tot,stud}$ is the total thermal resistance along the path with a stud

r   radius; r_{in} is the inner radius, r_{out} is the outer radius

T   temperature; $T_{1,R}$ is the actual temperature of the right edge of Layer 1 in the presence of voids, $T_{2,L}$ is the actual temperature of the left edge of Layer 2 in the presence of voids, T_H is the higher temperature, T_{in} is the inner temperature, T_L is the lower temperature, T_{out} is the outer temperature, $T_{\infty,in}$ is the inside fluid temperature, $T_{\infty,out}$ is the ambient fluid temperature, ΔT is the temperature difference, $\Delta T_{contact}$ is the temperature drop due to imperfect contact.

t   time

U   U-value; U_{avg} is the average thermal conductance, U_j is the thermal conductance along the j path

V   voltage; ΔV is the voltage difference, V_H is the higher voltage, V_L is the lower voltage

x   distance; Δx is the thickness, Δx_{1cm} is 1 cm thick, Δx_{3cm} is 3 cm thick

y   lateral direction or distance

13.1   Fourier's law of heat conduction

Fourier was among the first inquisitive minds who made some significant strides in heat conduction. Envision steady-state heat conduction through a layer of homogeneous, material as shown in Fig. 13.1. It is apprehensible that the heat transfer rate increases with the available surface area for heat transfer, temperature difference across the layer, and decreasing thickness of the layer. We could picture it like hot lava moving down the slope of a mountain. The steeper the slope (representing the temperature gradient) and/or the wider the path (analogous to the available heat transfer area), the faster the lava (or heat

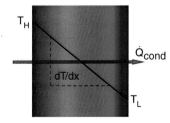

FIGURE 13.1   Steady heat conduction through a homogeneous solid, Fourier's law of heat conduction (created by D. Ting). We may envision the flow of heat similar to lava falling down the mountain, the steeper the slope, the higher the flow rate.

transfer rate) pours down the mountain (or temperature slope). The rate of heat conduction can thus be expressed mathematically as follows:

$$Q'_{cond} \propto A\Delta T/\Delta x, \tag{13.1}$$

where A is the available heat transfer area, ΔT is the temperature difference, and Δx is the thickness or distance. We can replace the proportionality with an equal sign by introducing a proportionality constant, k, such that

$$Q'_{cond} = -kA\Delta T/\Delta x = -kA(T_L - T_H)/\Delta x = kA(T_H - T_L)/\Delta x, \tag{13.2}$$

where T_L is the lower temperature and T_H is the higher temperature. A negative sign is added as the thermal conductivity, k, has a positive value, whereas the second law of thermodynamics entails that heat naturally travels in the decreasing temperature direction, that is, negative $\Delta T/\Delta x$ direction. In the limit that the thickness, Δx, approaches zero, we have

$$Q'_{cond} = -kAdT/dx. \tag{13.3}$$

This is called *Fourier's law of heat conduction*.

Concerning the thermal conductivity, k, consider an everyday example of feeling colder when we touch a piece of metallic furniture compared to one made of fabric. In this case, the heat transfer surface area, A; the higher temperature of our body that keeps our hands warm, T_H; the lower temperature of the room where the furniture is equilibrating, T_L; and the distance between the high and low temperatures, Δx, remains the same whether we touch the metallic or fabric furniture. We feel colder when touching the metallic furniture because of the higher heat transfer rate from our body, via our hand, to the furniture into the room. This higher heat transfer rate is due to the higher thermal conductivity of the metallic furniture.

Example 13.1 Quality of winter coat and thermal conductivity.

Given: Dr. JAS wears a bulky 3-cm thick jacket to keep her warm through the winter, where the ambient is at -20°C; see Fig. 13.2. A student in the Turbulence and Energy Lab, T–E Banyak, on the other hand, feels equally warm with her 1-cm thick jacket. To put it another way, the rate of heat transfer through Dr. JAS' bulky jacket is the same as that through T–E Banyak's tenuous jacket.

Find: What can you say about the thermal conductivity of the two jackets?

Solution: The thermal conductivity, k, signifies how well a material conducts heat. Assume that the outer surface of the coat is at -18°C (2°C warmer than the ambient air due to the film thermal resistance), and the inner coat surface is at 18°C. For equal heat transfer rates through both winter coats, the conduction heat transfer equation, Eq. (13.2), gives

$$Q' = kA(T_H - T_L)/\Delta x = k_{3cm}A(18 + 18)°C/3cm = k_{1cm}A(18 + 18)°C/1cm.$$

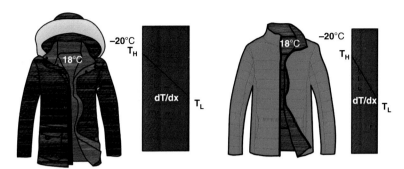

FIGURE 13.2 Thermal conductivity and thermal quality of winter coats (created by X. Wang). If both winter coats resist heat transfer so that the resulting heat transfer rate is the same, then the thinner coat with a lower thermal conductivity is a better-quality overcoat.

We see that the thermal conductivity of the 1-cm thick jacket, k_{1cm}, is 1/3 that of the 3-cm thick jacket, k_{3cm}. The lower thermal conductivity makes it possible for T–E Banyak to have a thinner and lighter jacket for shielding against the same temperature difference as Dr. JAS' three times thicker jacket.

To put in another way, T–E Banyak's 1-cm thick coat has the same thermal resistance as Dr. JAS' 3-cm coat. Namely,

$$R_{1cm} = \Delta x_{1cm}/(k_{1cm}A) = R_{3cm} = \Delta x_{3cm}/(k_{3cm}A).$$

Canceling out the heat transfer area, A, we get

$$\Delta x_{1cm}/k_{1cm} = \Delta x_{3cm}/k_{3cm}.$$

Substituting for the respective thicknesses results in

$$k_{1cm} = k_{3cm}/3.$$

In short, the thermal conductivity of T–E Banyak's 1-cm thick coat is one-third of that of Dr. JAS' 3-cm thick coat.

The winter coat can be envisioned as a wall separating the nasty Mr. Freeze from getting inside and freezes everything. Even with the wall standing in the way, Mr. Freeze presses on to freeze everything inside the wall. In other words, the barricade acts as a shield, resisting the coldness spurting from Mr. Freeze's ice gun and freezing those inside the fortification. A quality shield is a thin bulwark that can keep the inside warm, that is, a winter coat that safeguards the body it encompasses from freezing, as effectively as its thicker or bulkier counterparts. The thin suit of armor maintains a warm environment inside, in spite of intimidating "cold pressure" on the outside. The more efficacious temperature shield is the one that can uphold a larger temperature gradient. This is illustrated in Fig. 13.2, where, for restricting the heat (or cold) transfer at the same rate, the thinner coat has a larger thermal resistance, or smaller thermal conductance, per unit thickness, and thereupon a larger temperature gradient. To

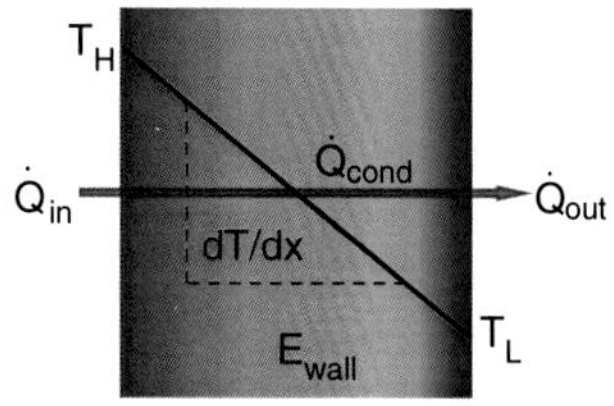

FIGURE 13.3 One-dimensional, steady heat conduction through a passive homogeneous wall (created by D. Ting). The wall is passive because it is neither generating heat (not a heat source) nor is it absorbing heat (not a heat sink). It is homogeneous in the sense that the thermal properties, such as the thermal conductivity, are uniform throughout the wall.

put it another way, the thinner coat can resist the larger temperature gradient from enforcing too much heat transfer by its higher thermal resistance compared to its thicker counterpart. We may envisage the thinner coat as a water (representing heat) chute at a steeper slope (resembling a larger temperature gradient). To permit the same rate of water (heat) going down the steeper chute as that going down the more gradual chute, there must be more resistance against water (heat) flowing through the steeper chute.

The aforementioned winter jackets, which quality that is made based heavily on sound conduction heat transfer knowledge, are typically under-appreciated by the general public. An obvious exception is in winter sports. For example, skiers lean on outfits that are thin and light with excellent conduction heat transfer characteristics to excel. When fighting fires, heat transfer is everything. For this reason, the research on firefighter clothing continues to be of high priority. For instance, Deng et al. (2021) revealed many details concerning the importance of heterogeneous air gaps in firefighter clothing on heat transfer and skin burn. Aerogels, ultra-light solids with a thermal conductivity lower than air, come to our mind when talking about enhancing thermal resistance. The promise of using aerogels in clothing is comprehensively reviewed by Greszta et al. (2021) very recently. Some of the major outstanding challenges include their delicate and brittle structure and their tendency to dusting.

We have thus far assumed steady, 1-dimensional conduction; see Fig. 13.3. It is steady because the high and low temperatures remain unchanged with respect to time. As a consequence, the heat transfer rate remains unchanged with respect to time. Applying conservation of energy to the wall, we have the rate of change of energy within the wall,

$$dE_{wall}/dt = Q'_{in} - Q'_{out}, \tag{13.4}$$

where Q'_{in} is the rate of heat entering the wall and Q'_{out} is the rate of heat exiting the wall. Under steady-state conditions, there is no net change in the energy content of the wall, that is, $dE_{wall}/dt = 0$ and, hence, we have

$$Q'_{in} = Q'_{out} = Q'_{cond}. \tag{13.5}$$

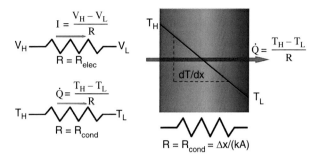

FIGURE 13.4 Thermal resistant concept of heat transfer (created by D. Ting). Temperature potential is analogous to electric voltage, and the thermal energy flow rate is akin to the current flow rate.

This says that the rate of thermal energy entering the wall is equal to that exiting the wall and, thus, equal to the rate of heat being conducted through the wall.

13.2 From electric resistance to thermal resistance

Learning from existing knowledge that is more mature in another field has proven to be potent for progress. By the time the heat transfer pioneers were making ground with heat transfer analysis, electric circuit concepts were already well established. It was the realization that heat flow is analogous to current flow that significantly advanced the field of heat transfer. Following Fig. 13.4, the flow of current is driven by a potential or voltage difference,

$$\Delta V = V_H - V_L, \tag{13.6}$$

where V_H denotes the higher voltage and V_L is the lower voltage. Consonantly, thermal energy is caused to flow in the presence of a temperature gradient. When covering fluid mechanics, we have been educated by Bernoulli that fluid flows when there is a pressure head gradient. To this end, a temperature gradient is akin to a voltage potential and an elevation head difference, each result in heat, current, and fluid flow, respectively. Electric resistance resists the flow of electric current, while friction limits fluid flow. In a similar manner, thermal resistance keeps the heat transfer rate in check. For steady, one-dimensional conduction heat transfer, such as that portrayed in Fig. 13.4, the heat transfer rate,

$$Q' = (T_H - T_L)/(\Delta x/kA). \tag{13.7}$$

We note that the denominator is the *heat transfer resistance*, R, which is analogous to the electric resistance, that is,

$$R = \Delta x/(kA). \tag{13.8}$$

As expected, the thermal or heat transfer resistance increases with increasing material thickness, decreasing thermal conductivity, and decreasing available

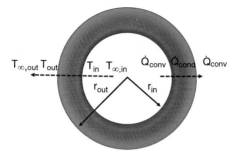

FIGURE 13.5 One-dimensional radial heat conduction through the wall of a long cylindrical pipe (created by D. Ting). Note that the available heat transfer area increases with radius. Therefore, insulating at a smaller radial circumference is more effective compared to insulating at a larger radial location.

surface area for heat transfer. With this heat transfer resistance defined, the one-dimensional heat conduction rate can be expressed as

$$Q' = (T_H - T_L)/R. \tag{13.9}$$

13.3 One-dimensional heat conduction in cylindrical coordinates

Cylindrical pipes and ducts are an integral part of engineering and natural systems. Consider a long cylindrical pipe whose cross-section is depicted in Fig. 13.5. With negligible end effects, the heat conduction is by-and-large one-dimensional radially. Fourier's law of the rate of heat conduction through the cylindrical wall can be expressed as

$$Q'_{cyl} = -kAdT/dr. \tag{13.10}$$

Call to mind that the negative sign is present because heat transfer in the negative temperature gradient, dT/dr, direction. The available heat transfer area for a section of a cylinder of length L,

$$A = 2\pi rL = f(r), \tag{13.11}$$

where r is the radius of the cylinder. We note that the available heat transfer area, A, increases with radius, r. Eq. (13.10) can be rearranged into

$$\left(Q'_{cyl}/A\right)dr = -kdT. \tag{13.10a}$$

Under steady-state conditions, the rate of heat transfer through the cylinder, Q'_{cyl}, is a constant. On that account, we can take the integration according to

$$Q'_{cyl} \int (1/A)dr = -k \int dT, \tag{13.12}$$

noting that the thermal conductivity, k, is also assumed to be a constant. Replacing area, A, with cylinder radius, r, and length, L, we have

$$Q'_{cyl} \int [1/(2\pi rL)] \, dr = \left[Q'_{cyl}/(2\pi L) \right] \int (1/r) dr = -k \int dT. \qquad (13.12a)$$

Integrating from the inner radius, r_{in}, which is at temperature, T_{in}, to the outer radius, r_{out}, which is at temperature, T_{out}, gives

$$\left[Q'_{cyl}/(2\pi L) \right] \ln(r_{out}/r_{in}) = -k(T_{out} - T_{in}). \qquad (13.13)$$

This can be rearranged into

$$Q'_{cyl} = 2\pi Lk(T_{in} - T_{out})/\ln(r_{out}/r_{in}). \qquad (13.13a)$$

Using the electric circuit analogy expounded in the previous section, the thermal (conduction) resistance across the cylinder wall is

$$R_{cyl} = \ln(r_{out}/r_{in})/(2\pi Lk). \qquad (13.14)$$

With this radial thermal conduction resistance, we can rewrite the heat transfer rate through the cylindrical shell as

$$Q' = (T_{in} - T_{out})/R_{cyl} = (T_{in} - T_{out})/[\ln(r_{out}/r_{in})/(2\pi kL)]. \qquad (13.13b)$$

If we include the convection between the fluid inside the pipe and the inner pipe surface and also the convection from the pipe exterior to the ambient fluid, we have

$$Q' = (T_{\infty,in} - T_{\infty,out})/R_{tot}. \qquad (13.14)$$

Here, $T_{\infty,in}$ signifies the fluid temperature inside the pipe and $T_{\infty,out}$ denotes the ambient temperature. The total thermal resistance is

$$R_{tot} = R_{conv,in} + R_{cyl} + R_{conv,out}$$
$$= 1/(2\pi r_{in}Lh_{in}) + \ln(r_{out}/r_{in})/(2\pi kL) + 1/(2\pi r_{out}Lh_{out}), \qquad (13.15)$$

where h_{in} is the internal heat convection coefficient and h_{out} is the convection heat transfer coefficient of the external surface.

Example 13.2 Heat conduction through a homogeneous leg.
Given: Heat stress is a common ailment and thermal treatment is an effective therapeutic. As an illustration, our body responds to heat stress by increasing blood flow and cardiac output (Chiesa et al, 2016). A marathon runner soaks his 80 mm diameter legs in an ice bath after a rough competition. Assume that the thermal energy carried by the blood vessels at 37°C is conducted radially through 31.5-mm thick flesh with k = 0.49 W/(m·°C) into the ice bath at 4°C.
Find: The rate of heat loss from each of his 400-mm-long legs.

Solution: The heat conduction rate, in cylindrical coordinate, can be cast as

$$Q' = (T_{in} - T_{out})/R_{cyl} = (T_{in} - T_{out})/[\ln(r_{out}/r_{in})/(2\pi kL)].$$

The outer radius, $r_{out} = 40$ mm and the inner radius, $r_{in} = 40 - 31.5 = 8.5$ mm. The thermal resistance,

$$R_{cyl} = \ln(r_{out}/r_{in})/(2\pi Lk) = \ln(40/8.5)/(2\pi(0.4)(0.49)) = 1.26\,^\circ C/W.$$

With that in mind, the heat conduction rate,

$$Q' = (T_{in} - T_{out})/R_{cyl} = (37 - 4)/1.26 = 26.24 \text{ W}.$$

This is quite significant when contrasted with a resting adult dissipating a total of about 80 W through the entire body, including that through breathing. Admittedly, we have neglected the convective thermal resistance between the skin and the ice water. Including this thermal resistance would raise the skin temperature somewhat.

13.4   Heat conduction radially through a sphere

Radial heat conduction of a sphere is similar to that of a cylindrical pipe except that the increase in available heat transfer area with increasing radius is faster than that of a cylinder. The rate of heat conduction through the spherical shell can be described as

$$Q' = (T_{in} - T_{out})/R_{sph}. \tag{13.16}$$

Here, T_{in} is the temperature of the inner wall and T_{out} is the temperature of the outer surface. The thermal resistance of the spherical shell,

$$R_{sph} = (r_{out} - r_{in})/(4\pi r_{in}r_{out}k). \tag{13.17}$$

As the heat is assumed to transfer from fluid inside the spherical shell, at $T_{\infty,in}$, to the outside fluid, at $T_{\infty,out}$, we can write

$$Q' = \left(T_{\infty,in} - T_{\infty,out}\right)/R_{tot}. \tag{13.18}$$

If the outside is hotter, the heat will transfer from outside in. Under those circumstances, the heat transfer rate according to Eq. 13.18 will be negative. In other words, a negative heat transfer or heat transfer rate value signifies the flow of thermal energy in the opposite direction to that is assumed while formulating the equation. The total thermal resistance consists of convection inside the sphere, conduction through the spherical shell, and convection to the outside fluid. Namely,

$$R_{tot} = R_{conv,in} + R_{sph} + R_{conv,out}$$
$$= 1/\left(4\pi r_{in}^2 h_{in}\right) + (r_{out} - r_{in})/(4\pi r_{in}r_{out}k) + 1/\left(4\pi r_{out}^2 h_{out}\right). \tag{13.19}$$

Unlike pipe flow, the fluid inside a sphere is enclosed and, thus, there is not a constant supply of hot (or cold) fluid. Therefore, for steady-state operation, there has to be a source (or sink) inside the sphere. This can be realized in practice via a slow chemical reaction or similar; for example, a nuclear reaction where a cherry-size amount of uranium is enough to power a passenger car over its entire life span of twenty or so years.

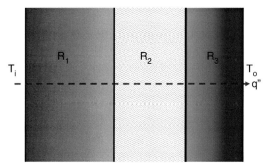

FIGURE 13.6 One-dimensional heat conduction through a three-layer wall (created by D. Ting). Under steady-state conditions, the heat that is conducted through the first layer passes the second and third layers. It follows that the heat transfer rate at any point along the path is the same.

13.5 Steady conduction through multilayered walls

Fig. 13.6 illustrates one-dimensional steady heat conduction through a multi-layered wall. An everyday example is the transfer of heat through the wall of a building. In hot climates, outdoor thermal energy penetrates through multilayered walls into the nicely cooled indoors. The direction of heat flow is reversed for those in colder climates during the heating season. Consider the case where heat is driven from the warm indoors, through a multilayered wall, to a cold ambient. Through the three different layers of materials making up the wall, as shown in Fig. 13.6, thermal energy is conducted all the way through. The total conductive resistance of the wall is the sum of the resistance of the three layers, that is,

$$R_{tot} = R_1 + R_2 + R_3. \tag{13.20}$$

It follows that the conduction heat transfer rate,

$$Q' = (T_{in} - T_{out})/R_{tot}. \tag{13.21}$$

The corresponding heat transfer rate per unit area of the wall, that is, the heat flux, can be described as

$$q'' = (T_{in} - T_o)/(R_{tot}A) = U(T_i - T_o). \tag{13.22}$$

The U-value is no more than the inverse of the total thermal resistance, that is, $U = A/R_{tot}$. The thermal conductivity of commercial thermal materials such as building insulations is often given in terms of the U-value, in $W/(m^2 \cdot {}^\circ C)$ or $Btu/(h \cdot ft^2 \cdot {}^\circ C)$, and/or the R-value, in $m^2 \cdot {}^\circ C/W$ or $h \cdot ft^2 \cdot {}^\circ F/Btu$.

Example 13.3 Heat conduction through a muscle-fat-skin-layered leg.
Given: Our legs are layered from inside out with muscles, fat, and skin. Reconsider the marathon runner who soaks his 80-mm-diameter legs in an ice

bath after a rough competition. Assume that the thermal energy carried by the blood vessels at 37°C is conducted via 25 mm of muscle (k = 0.49 W/(m·°C)), followed by 5 mm of fat (k = 0.21 W/(m·°C), and finally, 1.5 mm of skin (k = 0.37 W/(m·°C), into the ice bath at 4°C.

Find: The rate of heat loss from each of his 400-mm-long legs.

Solution: The heat conduction rate, in cylindrical coordinates, can be cast as:

$$Q' = (T_{in} - T_{out})/R_{cyl} = (T_{in} - T_{out})/[\ln(r_{out}/r_{in})/(2\pi kL)].$$

For a typical adult human leg modeled as a multilayered 80-mm-diameter cylinder, the outer radius, r_{out} = 40 mm and the inner radius, r_{in} = 40 - 1.5 - 5 - 25 = 8.5 mm. For the inner layer composed of muscles, the thermal resistance, according to $R_{cyl} = \ln(r_{out}/r_{in})/(2\pi Lk)$, is

$$R_1 = \ln(33.5/8.5)/(2\pi(0.4)(0.49)) = 1.11°C/W.$$

For the layer of fat above the muscle, the corresponding thermal resistance,

$$R_2 = \ln(38.5/33.5)/(2\pi(0.4)(0.21)) = 0.26°C/W.$$

The thermal resistance of the outer layer of skin is

$$R_3 = \ln(40/38.5)/(2\pi(0.4)(0.37)) = 0.04°C/W.$$

The total thermal resistance is the sum of the three layers of thermal resistors, that is,

$$R_{cyl} = R_1 + R_2 + R_3 = 1.38°C/W.$$

With that, the heat conduction rate,

$$Q' = (T_{in} - T_{out})/R_{cyl} = (37 - 4)/1.38 = 23.96W.$$

We note that the layer of fat with the lowest thermal conductivity, k of 0.21 W/(m·°C), that covers the muscle particularly helps to reduce the risk of hypothermia.

13.5.1 Thermal contact resistance

We have assumed perfect contact between consecutive layers of the wall in Fig. 13.6. As it happens, in reality, the contact between adjacent layers is never perfect, that is, each contact introduces some additional resistance. Imagine you ride on Dr. JAS' Starship T&E, which is shrunk into an infinitely small ship that rides on "an element of thermal energy" as it is being conducted through the three-layer wall. Layer 1, in Fig. 13.7, is typically drywall, in North American houses, and its thermal conductivity is moderate, at approximately 0.3 W/(m·K). Let us imagine this to be equivalent to riding at 300 km/h on an element of thermal energy through Layer 1 alone, for the given indoor and outdoor temperatures. Layer 2 is largely insulation, with a thermal conductivity

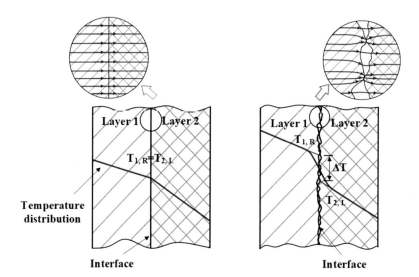

FIGURE 13.7   Thermal contact resistance (created by X. Wang). The imperfect contact has voids that are typically of significantly lower thermal conductivity. For that reason, the transfer of heat is slowed down, and this is manifested as a temperature drop, $\Delta T_{contact} = T_{1,R} - T_{2,L}$, from the warmer layer to the cooler one.

an order of magnitude smaller, that is, in the neighborhood of 0.03 W/(m·K). Accordingly, Starship T&E moves 10 times slower through Layer 2 of the same thickness and subjected to the same temperature difference. A zoomed-in illustration of the intersection is portrayed in Fig. 13.7. We see that there are many gaps or voids filled with air, where the thermal resistance is much higher than that of Layer 1 or Layer 2; recall that atmospheric air has a rather low thermal conductivity of roughly 0.025 W/(m·K). For this reason, Starship T&E is slowed down further to 25 km/h by the bottleneck imposed by the gap. This additional layer of resistance imposed by the interface is called *thermal contact resistance*. Mathematically, the total heat transfer rate from Layer 1 into Layer 2 can be expressed as:

$$Q' = Q'_{join} + Q'_{gap} = R_{contact}\Delta T_{contact} = R_{contact}\left(T_{1,R} - T_{2,L}\right)$$
$$= k_{contact}A\left(T_{1,R} - T_{2,L}\right), \tag{13.23}$$

where Q'_{join} is the heat transfer rate through the (perfectly) joined section, Q'_{gap} is the slower heat transfer rate through the gap, $R_{contact}$ is the contact resistance, $T_{1,R}$ is the actual temperature of the right edge of Layer 1 in the presence of voids, $T_{2,L}$ is the actual temperature of the left edge of Layer 2, and $k_{contact}$ is the thermal contact conductance. Note that, in the ideal case without contact resistance caused by voids, $T_{1,R}$ is equal to $T_{2,L}$, that is, $\Delta T_{contact} = 0$.

Example 13.4 Fire walking.

Given: Following the Turbulence and Energy Lab tradition, Dr. JAS brought a group of graduate researchers to Pelee Island, Ontario, Canada, for a field trip. Imagine you had the opportunity to tag along. In the evening, the host welcomed all of you with a campfire, barbeque, singing, and storytelling. The highlight of the evening was fire walking. Just when you were trying to figure out how the magicians managed to walk on fire, you witnessed Dr. JAS walk briskly through a 4-m stretch of fire with her bare feet.

Find: How did Dr. JAS do it without burning her feet? Assume Dr. JAS knows no magic, that is, she knows only science and its applications, engineering.

Solution: This is clearly a heat transfer conundrum. We first warn the readers that firewalking is unsafe, and it can definitely result in serious burns. But how do people do it? Within the context of heat transfer, or more specifically conduction heat transfer, here are some insights into how it works; see, for example, Brain (2021) and Kanchwala (2021).

First of all, the fire is lit for a sufficiently long time that the wood is burned down to non-flaming coals. To put it another way, it is a bed of ashes or embers acting as a layer of thermal insulation on top of extinguishing coals, which are also poor conductors of heat. To enhance the appearance of fire and to fade the visibility of the insulating ashes, firewalking is performed when it is dark. On top of that, the added contact thermal resistance between our feet and the ashes and, most of all, the very short contact time between the feet and the bed, due to brisk walking, minimize the possibility your feet will reach a high enough temperature to be damaged.

Note that our bodies can sustain several minutes of near-water-boiling temperatures of sauna air, but not 50°C water. This is mainly because the thermal conductivity of air is more than 20 times lower than that of water, roughly 0.025 $W/(m \cdot {}^\circ C)$ as compared to 0.6 $W/(m \cdot {}^\circ C)$. We can infer from this that placing a steel or aluminum plate over a bed of burning coals is not practiced in fire walking, for this will surely burn even the best fire walkers. With all that being said, it only takes a couple of seconds to get a third-degree burn in 60°C water (Moritz and Henriques, Jr, 1947). For this reason, domestic hot water is limited to a maximum temperature of 50°C. As importantly, firewalking is hazardous and, thus, we should exclude it from our bucket list or replace it with competing in a marathon, for example.

13.6 Multilayered inhomogeneous walls

More often, in practice, the layers involved in conduction heat transfer have distinct thermal properties and not all layers are homogeneous. For building envelopes, for example, the insulation layer needs structural supports. These

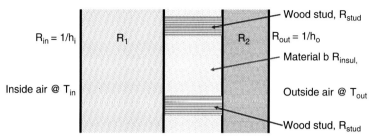

FIGURE 13.8 Heat transfer via the parallel-path method (created by Y. Yang). The parallel-path method assumes perfectly one-dimensional heat flow, resulting in lower and higher heat transfer rates through the passage consisting of less and more conductive materials, respectively. For the case shown, Q'_{stud}/A_{stud} is larger than Q'_{insul}/A_{insul}.

structural supports, in most residential houses, are wood studs. The values of the thermal conductivity of these supports are often significantly larger than that of the insulation. Fig. 13.8 shows a simple two-dimensional section of the three-layer wall with wood studs along the horizontal direction. In reality, there are also wood frames in the vertical direction. Let us stay with the two-dimensional case depicted in Fig. 13.8. Suppose Dr. JAS lines up the entire engineering cohort of 1600 students along the interior wall, say in 16 rows of a hundred students each. She shrinks the students into infinitely small particles that ride on the "thermal slides" from the highest point (temperature), the warm indoors, all the way to the lowest altitude (on the temperature scale), the cold outdoors. As the rows of students slide down the temperature slope[1] from the warm indoor air, the first thermal resistance is the convection heat transfer resistance between the indoor air and the inner wall. The next thermal resistance is the conduction resistance of the first layer of the building envelope, the drywall. Keep in mind that the thermal resistance sets the limit of the temperature slope to be finite, that is, more than zero, the higher the thermal resistance, the steeper the temperature slope. In other words, a higher thermal-resistance material can manage a larger temperature difference across a shorter distance. Between the inner wall (drywall) and the outside wall, typically brick or concrete, is a two-path passage. At this junction between the drywall and the outside wall, the students can either go through the easier, less thermally resistive path, which is typically the studs used for supporting the wall, or the notably more resistive "pink" insulating material. Analytically, there are two engineering approximations to resolve this problem: (1) the parallel-path method and (2) the isothermal-plane method.

---

1. More correctly, the speed of heat flow corresponds to the inverse of the temperature gradient. For layers of different materials of the same thickness, Δx, the material with a larger temperature gradient has a higher thermal resistance. A zero-temperature gradient means zero thermal resistance, that is, it takes no time for you to reach the destination temperature, for you are already at that temperature to start with.

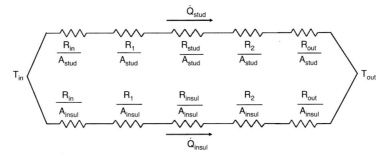

FIGURE 13.9 A schematic of a wood-frame wall section of a typical residential building in North America (created by D. Ting). Thermal energy travels considerably faster through the supporting structures, such as studs, as compared to the "pink" insulation. For this reason, some of the thermal energy in the proximity of the studs travels through the studs instead of the more restrictive path made of insulation.

13.6.1 Parallel-path method

The parallel-path method assumes that the portion of the heat transmitted through the less resistive path involving the relatively higher conductive stud or other materials remains parallel from the indoor wall to the outside surface. Imagine the lineup of Dr. JAS' students, each carrying Q amount of heat, starting before the inside wall of Fig. 13.9. Assume Rows 1 and 2, line up above the top stud, Rows 3 to 6 between the horizontal lines bounding the top stud, Rows 7 to 10 between the two studs, Rows 11 to 14 between the horizontal lines bounding the lower stud, and Rows 15 and 16 below the bottom stud. In this way, we would expect Rows 3 to 6 and 11 to 14, which encounter a section of the lower-resistance stud path, to travel faster than Rows 1, 2, 7 to 10, 15, and 16. This is portrayed in Fig. 13.9, where Q'_{stud} is the heat transfer rate along the parallel horizontal path, which consists of a section of lower resistance stud, and the lower heat transfer rate, Q'_{insul}, through the parallel path which includes a section of high-resistance insulation path. The total heat transfer rate per unit of stud-insulation section is the sum of the heat transfer rate through the parallel path encompassing the stud and the path that comprises the insulation, that is,

$$Q'_{tot} = Q'_{stud} + Q'_{insul}$$
$$= [A_{stud}(T_{in} - T_{out})]/R_{tot,stud} + [A_{insul}(T_{in} - T_{out})]/R_{tot,insul}, \quad (13.24)$$

where subscript "insul" denotes the board-insulation pathway. The total average thermal resistance, $R_{tot,avg}$, can be deduced from

$$A_{tot}/R_{tot,avg} = A_{stud}/R_{tot,stud} + A_{insul}/R_{tot,insul}. \quad (13.25)$$

This can be rearranged to give the average total conductance,

$$\frac{1}{R_{tot,avg}} = \frac{A_{stud}}{A_{tot}} \frac{1}{R_{tot,stud}} + \frac{A_{insul}}{A_{tot}} \frac{1}{R_{tot,insul}}. \quad (13.26)$$

The underlying assumption here is that every row of students stays on their respective course faithfully. In other words, the students do not deviate from their respective path and move into a faster lane, even when they are confronted with much slower traffic in their lane. In reality, however, some of these students, especially those next to a less congested passage, would take advantage of the speedway, rendering the heat transfer two- or three-dimensional. It is sometimes more convenient to use the U-value for the parallel-path method. The average conductance, the U-value, of an inhomogeneous wall can be deduced from

$$U_{avg}A_{avg} = \sum (U_jA_j). \tag{13.27}$$

13.6.2 Isothermal-plane method

Another way to solve the inhomogeneous wall problem, without resorting to tedious and costly numerical modeling, is to invoke the isothermal-plane method. The isothermal-plane method assumes excellent or perfect lateral heat transfer. The fallout of this is a constant temperature along every lateral plane, that is,

$$T = T(x)\,\text{only, i.e.,} \neq f(y), \tag{13.28}$$

where y is the lateral direction. Following Dr. JAS 16 rows of one hundred students' metaphor, all the students are arrayed according to their respective row and column. The array of 16 rows by one hundred columns move in unison from the inner wall indoors to the outer wall that is exposed to the outdoors. It follows that the students moving through the more-resistive passage are towed, to a certain extent, by the faster-moving ones on the less-resistive route. As a consequence, the slower ones speed up while the faster ones slow down to the same speed. To put it another way, all students along the same initial column move through the wall shoulder by shoulder. Because of that, the temperature is uniform over any vertical plane. The corresponding thermal circuit is shown in Fig. 13.10. According to this thermal circuit, the total average resistance,

$$R_{tot,avg} = R_{in} + R_1 + R_{stud,insul} + R_2 + R_{out}, \tag{13.29}$$

where

$$A_{tot}/R_{stud,insul} = A_{stud}/R_{stud} + A_{insul}/R_{insul}, \tag{13.30}$$

that is,

$$R_{stud,insul} = \left(\frac{A_{stud}}{A_{tot}}R_{stud} + \frac{A_{insul}}{A_{tot}}R_{insul}\right)^{-1}. \tag{13.31}$$

It follows that the total heat transfer rate is

$$Q'_{tot} = A_{tot}(T_{in} - T_{out})/R_{tot,avg} = U_{avg}A_{tot}(T_{in} - T_{out}). \tag{13.32}$$

We note that it is easier to use the R-value when applying the isothermal-plane method.

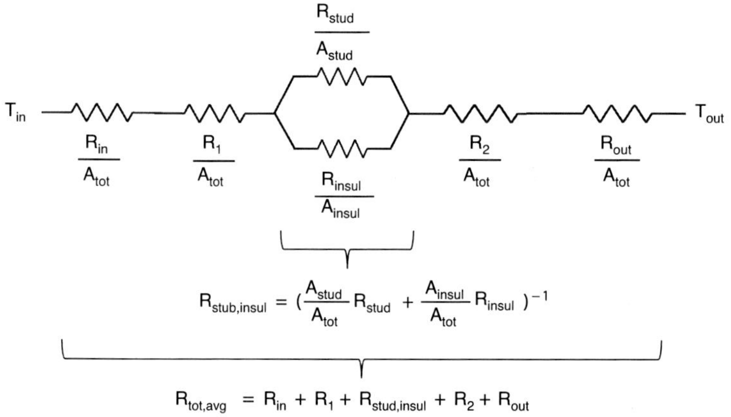

FIGURE 13.10 Equivalent-circuit diagram for the isothermal-plane method (created by Y. Yang). The isothermal-plane method assumes that the temperature at any cross-section along the decreasing temperature heat-transfer path is at the same temperature.

Both parallel-path and isothermal-plane methods are good enough for most engineering problems. Comparing the calculated results from both methods would provide an obvious indication of the validity of these methods. A noticeable difference is a cue that the heat transfer is highly two- or three-dimensional. For such cases, a detailed computational fluid dynamic simulation is required.

Problems

13.1 Thermal conductivity of an unknown material.

The Turbulence and Energy Laboratory came up with a promising insulator. To determine the thermal conductivity of the new material, Dr. JAS placed a large 0.02-m-thick block of the new material on top of boiling water and wet ice was placed on top of the block. A heat flux, q'', of 1500 W/m^2 was measured. What is the thermal conductivity? Assume the bottom surface is at 100°C and the top surface is at 0°C, while being mindful that these are approximations, as there is another thermal resistance between the boiling water and the lower surface and also between the wet ice and the upper surface. Note, however, the thermal resistance imposed by the boiling water, in particular, is very small, that is, the convection heat transfer coefficient is very high, ranging from 3000 to 100,000 W/(m$^2 \cdot$°C).

13.2 Energy-saving windows.

You are thinking about upgrading the windows of your house to save some heating costs. The two options available correspond to windows with U-values of 2.7 W/(m$^2 \cdot$°C) and 1.2 W/(m$^2 \cdot$°C). The windows cover a total building envelope area of 50 m^2 and the indoor temperature is fixed at

20°C. How much energy can be saved over a 10-h night? Assume that the indoor and outdoor surfaces of the windows are at 17°C and 0°C, respectively; the drop in surface temperature from the surrounding fluid is due to convection heat transfer resistance.

13.3 Defrosting.

The windshield of a typical car is about 5 mm thick, with a thermal conductivity of 1 W/(m·°C). What is the minimum heating power from hot air at 45°C required to just defrost the ice over a 0.5 m^2 area, under ideal conditions? If there is a total of 1 kg of snow over this 0.5 m^2 area, what is the minimum duration to melt the snow? We emphasize "minimum" because we have omitted contact and convection resistance. More significantly, much of the thermal energy is lost but, thankfully, mostly within the compartment and, thus, is being utilized to keep the driver and passenger from freezing.

13.4 Insulating a hot-water pipe.

Hot water at 50°C is delivered via a 1.59-cm outside diameter copper (k = 400 W/(m·°C)) pipe with a thickness of 0.07 cm. The ambient temperature is 20°C. Determine the required thickness of the insulation, with a thermal conductivity of 0.03 W/(m·°C), required to cover the pipe, if the heat loss rate is to be 50% lower than that of the naked pipe. How would your answer change if the copper pipe is doubled in diameter?

13.5 Heat loss through a wall.

Between the 0.635-cm inner drywall (k = 0.17 W/(m·°C)) and the outer brick wall (0.8 W/(m·°C)) is 10-cm-thick insulation (k = 0.03 W/(m·°C)) alternated by 4.6-cm-thick wood supports (k = 0.1 W/(m·°C)) every 60 cm of height. The indoor surface of the drywall is at 17°C and the outer surface of the brick wall at -12°C. Calculate the heat transfer rate through a 60-cm by 60-cm cross-section of the wall. This 60-cm by 60-cm cross-section is a representative wall section enclosing the 55.4-cm by 55.4-cm insulation with 2.3-cm wood borders sandwiched between the drywall and the brick wall.

13.6 Garage temperature.

A 3-m-high and 7-m-deep attached garage has a 8-m-span exterior wall with U = 2.5 W/m^2·°C. The other three walls (U = 0.7 W/m^2·°C) separate it from the indoors, where the indoor wall surface is at 18°C. The garage has a ceiling plus attic plus roof with U = 8.8 W/m^2·°C. The outside air is at -5°C. What is the temperature of the garage?

13.7 Freezing location.

A three-layer flat roof has its top layer of shingles laying on top of plywood, with a total R-value of 0.2 m^2·°C/W. The 10-cm gap between the plywood and the ceiling is filled with fiberglass insulation, with an R-value of 0.7 m^2·°C/W. The ceiling has an R-value of 0.15 m^2·°C/W. For indoor and outdoor temperatures of 18°C and -5°C, respectively, find the location at which the temperature is 0°C.

13.8 Green dome.

Communities living in a row of attached houses can save significant heating and cooling energy. The main reason behind the energy savings is the reduction in heat exchange area between indoors and outdoors. Building shape, like that of an igloo, can be even more effective in energy conservation. How much conduction heat loss can be reduced if a dome of the same indoor space (volume) is used instead of a 6-m-tall, 8-m-wide, and 150-m-long building? Assume the interior surface is at 17°C and the outer wall is at -15°C. Note that actual savings are likely more as infiltration is also reduced with building envelope area. In real life, however, solar radiation is a pivotal factor for both heating and cooling. The idea is to maximize solar heat gain during the cold season while minimizing it during the warm season.

References

Brain M., 2021. "How firewalking works," howstuffworks, 2021 https://entertainment. howstuffworks.com/arts/circus-arts/firewalking.htm, (accessed June 8, 2021).

Chiesa, S.T., Trangmar, S.J., González-Alonso, J., 2016. Temperature and blood flow distribution in the human leg during passive heat stress. J. Appl. Physiol. 120, 1047–1058.

Deng, M., Psikuta, A., Wang, Y., Annaheim, S., Rossi, R.M., 2021. Numerical investigation of the effects of heterogeneous air gaps during high heat exposure for application in firefighter clothing. Int. J. Heat Mass Transfer 181, 121813.

Greszta A., Krzmenińska S., Bartkowiak G., Dąbrowska A., "Development of high-insulating materials with aerogel for protective clothing applications—an overview," 112(2): 164-172, 2021.

Kanchwala, H., 2021. Walking on hot coal: How do people firewalk? Science ABC 2021. https://www.scienceabc.com/humans/how-do-some-people-firewalk.html. (accessed June 8, 2021).

Moritz, A.R., Henriques Jr., F.C., 1947. Studies of thermal injury. II. The relative importance of time and surface temperature in the causation of cutaneous burns. Am. J. Pathol. 23 (5), 695–720.

Chapter 14

Transient conduction of thermal energy

"Knowledge of the soul is the only universal truth and the only wisdom – all other knowledge is transient."

Plato

Chapter Objectives

- Understand what a lumped system is and when it is applicable.
- Appreciate the Biot number and its application in transient conduction.
- Master one-dimensional transient problems for a large plate, a long cylinder, and a sphere.
- Comprehend the solution approach for a semi-infinite wall.

Nomenclature

A	area
A_n	constants
Bi	Biot number, the ratio of internal conductance resistance within the body over the external convective resistance at the surface of the body
C	capacitance or heat capacity; c is specific heat capacity, c_P is specific heat capacity at constant pressure
$C_1, C_2, \ldots$	constants
E	energy
e	the base of natural logarithms
erf	error function
erfc	complementary error function
F	a function
Fo	Fourier number, the diffusive transport rate with respect to the storage rate
G	a function
h	(convection) heat transfer coefficient
J_0, J_1	Bessel functions of the first kind
k	thermal conductivity
L	length
m	mass, or number, 1, 2, 3 …
n	number, 1, 2, 3 …

The essentials of heat transfer in chemical engineering. DOI: hidden/doi.org/10.1016/B978-0-323-90626-5.00002-1

Q   thermal energy; Q' is heat transfer rate, Q_{in}' is the heat transfer rate into the system
r   radius; r_C is radius of a cylinder, r_S is radius of a sphere
T   temperature; T_0 is initial temperature at time $t = 0$, T_{in} is indoor temperature, T_{out}
             is outdoor temperature, T_∞ is surrounding temperature
t   time
u   a variable
X   non-dimensional distance
x   distance

Greek and other symbols
α thermal diffusivity
θ non-dimensional temperature
λ a constant; λ_n are constants for n = 1, 2, 3, …
ρ density
τ time constant

14.1   A lumped system with homogeneous temperature

A cold-blooded creature can adjust its body temperature to be in equilibrium
with its environment; see Fig. 14.1. To this end, the creature and the environment
that it is in are at the same temperature. As the temperature of an environment
typically changes slowly, the system, the lizard in this case, and the environment
are always at the same temperature. Let us concern ourselves with the system,
that is, the cold-blooded lizard. If you poke a tiny meat thermometer into the
lizard, you will find that the temperature throughout the body of the lizard is
approximately uniform. As discussed earlier in this book, there have also been
experimental studies performed by placing a frog, an equally familiar cold-
blooded creature, in a pot of water and slowly heating up the water. The entire
body of the frog adjusts to the temperature of the water as it is heated up. In a

FIGURE 14.1   A cold-blooded creature such as a lizard is at the same temperature as its environment (created by O. Imafidon). Treating the lizard as a lumped system implies that the entire body
of the lizard is at a homogeneous temperature as the lizard adjusts to changes in its environment.
This lumped system assumption is valid only when the change in the environmental temperature is
sufficiently slow. This is true in real life, except in situations such as when the environment becomes
unbearably hot such as under scorching sun in a desert, the lizard dives into the sand to avoid being
toasted.

desert when the sand is scorched by the sun, a lizard performs a special dance by lifting two legs (one foreleg and one hide leg on opposite sides) at a time alternately so that the legs do not get burned by the baking hot sand; see (BBC Studios, 2008), for example. When the "cool heat sand dance" is unable to spare its legs and the rest of its body from the sizzling sun, the lizard dives into the sand to cool off.

A system in which the temperature of the entire lump varies with time, that is, the temperature is the same throughout the system at any instant, is a *lumped system*. There are two scenarios in which this is the case. First, the temperature of the environment in which the system is in changes sufficiently slow that the entire system adjusts uniformly to the change. The frog in a slow cooking pot is such an example. Secondly, the thermal conductivity within the system is high; any temperature change of the surface exposed to the surroundings is swiftly transmitted and shared by the entire "community." This is the situation when a piece of ingot is subjected to a change in temperature. The high conductivity allows the prompt transfer of heat throughout the ingot, maintaining an even temperature for the lump of ingot.

Imagine a one-dimensional scenario, say a 100-foot-tall bamboo tree[1], as shown in Fig. 14.2. After a brief drought season, a light afternoon shower brings torrential moisture to the soil. If the "water conductivity" of the bamboo trunk is high, the much-needed moisture absorbed by the roots will be immediately distributed all the way up the 100-foot height. Every section of the bamboo will be evenly quenched. On the other hand, if the "water conductivity" is low, only the lower portion near the roots will be doused, while the top of the bamboo tree remains parched. Analogously, a highly (thermally) conductive, 10-foot-long copper rod with its base connected to an incinerator will have roughly the same temperature throughout the rod. The temperature of the rod varies with the temperature of the incinerator. The lumped system approximation improves significantly when the dimensions, specifically the dimensions through which the thermal energy must diffuse through, are smaller. For example, when we are dealing with a 1-cm silver ball exposed to the same environment. Any alteration in the environment is felt by the surface of the ball and quickly conducted through the 0.5-cm radius.

It should be noted that it is more than just thermal conductivity, k, that is important, the thermal diffusivity is also of importance. Thermal diffusivity, $\alpha = k/(\rho c_P)$, represents the portion of thermal energy conducted through the material with respect to that retained by the medium, where ρ is density and c_P is the specific heat capacity. Let us invoke the 100-foot-tall bamboo tree metaphor where water diffusivity is analogous to thermal diffusivity. If the water diffusivity

---

1. Since we are on the topic of a 100-foot bamboo tree, the ancient oriental proverb, "One hundred foot bamboo, progress one step," is appropriate. The moral is that even after you have excelled one hundred steps, strive yet the next step of progress. In other words, keep striving!

FIGURE 14.2   Due to its excellent water conductivity, a bamboo tree can be considered as a one-dimensional (vertically) lumped system in terms of water distribution (created by N. Bhoopal). Likewise, we can approximate a copper rod as a lumped system in terms of heat distribution, because of its superior thermal conductivity.

is low, most of the water will be retained by the water-greedy lower section of the tree. On the other hand, a high water diffusivity implies generosity, sharing most of the water with its siblings, that is, the sections of the bamboo tree that are above it. At room temperature, copper has a relatively high thermal diffusivity of 1×10^{-4} m²/s, whereas that of wood is almost three orders of magnitude lower, around 3×10^{-7} m²/s. Accordingly, for transient conduction, it is generally sound to consider a piece of copper as a lumped system, but not a piece of wood.

Example 14.1   A thermocouple junction as a lumped system.

Given: A thermocouple with a time constant, τ, of 0.03 s initially at 22°C is suddenly placed into a hot water bath at 88°C, as shown in Fig. 14.3. Note that a time constant is the time required to respond to approximately $(e - 1)/e$ of a step change, where e is the base of natural logarithms.

Find: The temperature sensed by the thermocouple as a function of time.

Solution: Invoking the first law of thermodynamics, conservation of energy states that the rate of energy change of the thermocouple junction or bead is equal to its rate of energy gain. Namely,

$$dE/dt = Q'_{in}. \tag{E14.1.1}$$

Here E is the energy of the system, t is time, and Q_{in}' is the heat transfer rate into the system of concern, that is, the thermocouple bead. For the fixed-mass system, the bead, with an approximately constant heat capacity over the

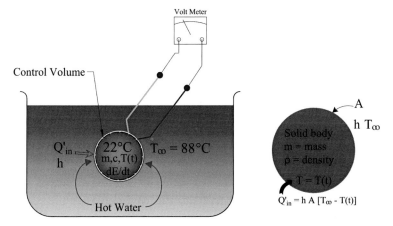

FIGURE 14.3 The energy balance of a thermocouple junction in the form of a bead (created by F. Kermanshahi). The rate of energy increase in the bead is equal to the net rate of heat gain. This increase in energy content, dE/dt, is reflected in temperature rise according to dE/dt = m c_P dT/dt.

range of temperature change, the rate of energy content change with respect to time is

$$dE/dt = mc_P dT/dt, \qquad (E14.1.2)$$

where m is the mass, c_P is the heat capacity at constant pressure, and T is the temperature. This equation states that the rate of energy change of the bead is directly manifested by the change in the temperature of the bead. The entire junction is treated as a lumped system at temperature T.

Let the heat transfer coefficient between the hot water and the bead be h. Then, the rate of heat transfer from the hot water to the bead,

$$Q'_{in} = h A[T_\infty - T(t)], \qquad (E14.1.3)$$

where A is the surface area of the bead, T_∞ is the temperature of the hot water, which is fixed at 88°C, and T(t) is the temperature of the bead, which is a function of time t.

From Eqs. (E14.1.2) and (E14.1.3), we have

$$mc_P dT/dt = h A[T_\infty - T(t)]. \qquad (E14.1.4)$$

This can be rearranged into

$$[mc_P/(hA)]dT/dt + T(t) = T_\infty, \qquad (E14.1.5)$$

which is a first-order differential equation. The term in square brackets, $mc_P/(hA)$, is the time constant, τ. Rearranging, we can express it as

$$1/[T(t) - T_\infty]dT = -[hA/(mc_P)] dt, \qquad (E14.1.6)$$

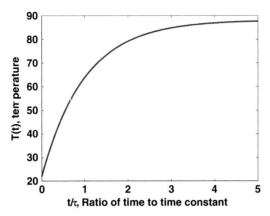

FIGURE 14.4   Plot of the response, $T(t) - T_\infty = [T_0 - T_\infty] \exp\{-h\,At/(mc_P)\} = [T_0 - T_\infty] \exp(-t/\tau)$, of a thermocouple junction subjected to a step change from $T_0 = 22°C$ to $T_\infty = 88°C$ (created by X. Wang).

Integrating this from $t = 0$, where $T_0 = T(t = 0) = 22°C$, to $t = t$, where $T(t = t) = T(t)$, we have

$$\ln[T - T_\infty] - \ln[T - T_\infty] = \ln\{[T - T_\infty]/[T - T_\infty]\} = -h\,At/(mc_P),$$
(E14.1.7)

where T_0 is the initial temperature at time zero, that is, $T_0 = 22°C$. Taking the anti-ln, we get

$$[T(t) - T_\infty]/[T_0 - T_\infty] = \exp\{-h\,At/(mc_P)\} = \exp(-t/\tau).$$ (E14.1.8)

This can be rearranged into

$$T(t) - T_\infty = [T_0 - T_\infty]\exp\{-h\,At/(mc_P)\} = [T_0 - T_\infty]\exp(-t/\tau).$$
(E14.1.8a)

Substituting for the known values, including the initial temperature and the hot water temperature, where T is in °C, we have

$$T(t) - 88 = -66\exp(-t/0.03).$$

This is plotted in Fig. 14.4. Note that the time constant, τ, is defined as the time required for the thermocouple to respond to 63.2% of the step change. The time constant of a thermocouple, τ, is equal to $mc_P/(hA)$. That being the case, an increase in the mass (size) or heat capacity would lead to an increase in the time constant, that is, it takes more time to heat up a larger mass and, hence, a slower response. Because of that, a small and exposed (for promoting heat transfer) thermocouple (bead or junction) is needed for sensing rapidly fluctuating temperatures. The drawback is its fragility. Another way to improve the response is by increasing its surface area (for heat exchange). Practically, a fine junction is fragile and, thus, is not conducive to flattening to enlarge its surface area. Another way to lower the response time is to augment the heat

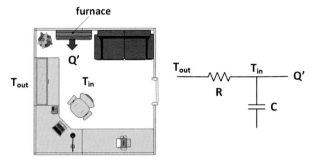

FIGURE 14.5 A single zone building modeled in terms of a capacitor and a resistor (created by X. Wang). The capacitance, C, describes the thermal mass of the building, while the resistance, R, represents the overall heat transmission resistance of the building envelope. In the summertime, Q' is negative, indicating the rate of cooling, that is, heat removal rate.

transfer coefficient. In the current example, this can be done by stirring the water bath.

In Example 14.1, the system of concern is a small thermocouple junction in the form of a tiny bead. Its smallness, along with the fact that it is made of highly conductive metals, grant the validity of the lumped system assumption. For a much larger system, such as a building, a single temperature can also be used to set the indoors at a comfortable temperature. This is particularly true for a single-zone building, such as a residential dwelling (Ting, 2020). Under those circumstances, we can model the dynamic response of a thermostat set-back, used for reducing the heating bill in the winter, as a simple electric system, as shown in Fig. 14.5. The rate of heat input into the system (indoors) is designated as Q', and this heat supply balances with the heat loss to the cold ambient. Typically, the indoor is maintained at T_i and the outdoor is at T_∞. The heat loss rate from the warmer indoor to the cold outdoor is dictated by the overall heat transfer resistance, R. The thermal mass of the building is represented by the capacitance, C. The heat balance of the building can be expressed mathematically as

$$CdT_{in}/dt = [T_{out} - T_{in}]/R + Q', \qquad (14.1)$$

where T_{in} is the (instantaneous) indoor temperature and T_{out} is the outdoor temperature, which is assumed to be fixed in this context. To save energy, the occupants of a house can lower the thermostat from 20°C to around 10°C before they vacate the house for the day for work and/or school. Practically, 10°C is roughly the lower limit to prevent possible damage such as pipes bursting due to freezing. Eq. (14.1) can be rearranged into

$$\tau dT_{in}/dt + T_{in} = T_{out} + RQ', \qquad (14.2)$$

where the time constant, $\tau = RC$. Letting $T(t) = T_{in} - T_{out} - RQ'$, we see that $dT(t)/dt = dT_{in}/dt$, since the other terms are not a function of time. With this in mind, we can write

$$\tau dT/dt + T = 0. \tag{14.3}$$

Following Example 14.1, the solution to this first-order equation is

$$T = T_{out}\exp(-t/\tau). \tag{14.4}$$

After lowering the thermostat at time zero, the indoor temperature, T, will drop exponentially, as illustrated in Fig. 14.4 but with decreasing temperature from 20 to 10°C. The larger the time constant, the slower the decrease. This is expected, as a larger time constant implies a larger thermal mass and/or overall heat transfer resistance, which both slow down the decrease in temperature. The indoor temperature will reach the setback temperature, for example, 10°C, asymptotically.

Note that the reverse occurs in the summertime when we allow our house to heat up to the equilibrium warm and perspiring temperature set by the hot environment, at the mercy of the scorching sun. An automated or remote controller is typically employed to turn the cooling system on some time before the first occupant returns home. This is because we want our home to be a sweet (cool) and not sweat home, after a long day at work or school. The desirable pre-cooling time can be determined from the temperature versus time plot, such as that shown in Fig. 14.4. For this summer cooling case, the plot is reversed, that is, it goes from an uncomfortably high temperature to a cool one, asymptotically.

14.2   Biot number

Jean-Baptiste Biot (1774–1862) is credited for the dimensionless quantity denoting the thermal resistance inside a body with respect to that at the surface of the body. The thermal resistance inside a body of dimension L is equal to L/k, where k is the thermal conductance; see Fig. 14.6. Convection takes place at the surface of the body and, thus, the corresponding thermal resistance is 1/h. It follows that the Biot number,

$$Bi = (L/k)/(1/h) = hL/k. \tag{14.5}$$

We note that the Biot number, Bi, becomes very small when the convection heat transfer coefficient, h, is very small and/or the thermal conductivity, k, is very large. Under this condition, the temperature within the body is homogeneous. To put it another way, any change in the surface temperature via convection is felt all the way through the entire body, via rapid conduction. This is the case for a highly conductive metal ball subjected to a varying ambient air temperature, for example. Conversely, for a large Biot number, where the convective heat transfer at the surface is fast while heat conduction through the body is slow, changes in the surface temperature are not immediately appreciated, especially by the inner parts of the body. Thanksgiving turkey is a conspicuous real-life example, that is, it takes a long time for the inner part of the turkey to be heated up enough for it to be cooked, even when the oven environment it is in is at 200°C or more. It follows that there is a definite

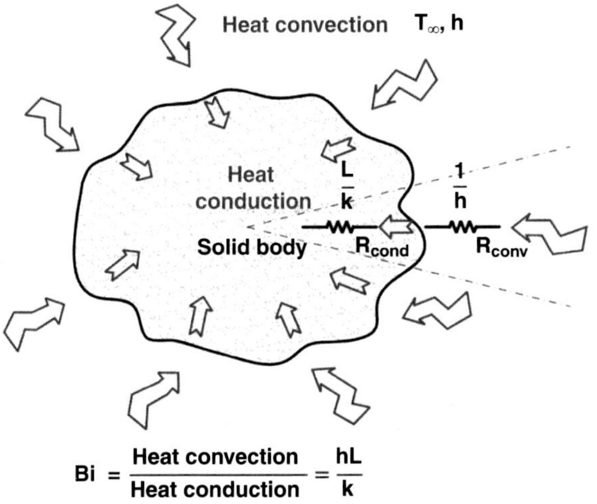

$$Bi = \frac{\text{Heat convection}}{\text{Heat conduction}} = \frac{hL}{k}$$

FIGURE 14.6 Biot number illustrated (created by X. Wang). When the within-the-system thermal conduction resistance is small compared to the surface convective resistance, any change in surface temperature is promptly transmitted through the entire system and, hence, the system has a homogeneous temperature. With that in mind, this small-Bi condition grants the utilization of the lumped parameter assumption.

temperature gradient after any change in the environment and the rate at which this temperature gradient diminishes decreases with increasing Biot number.

Example 14.2 Shipping food to a starving island.
Given: An evenly populated island is suddenly hit by severe hail, destroying crops and livestock alike. The good neighboring nations respond by sending fleets of cargo ships, dropping off food along the shoreline enveloping the island.
 Find: What would happen if the island's Biot number, with food symbolizing thermal energy, is large? How would the situation change to represent a small Biot number?

Solution: For a large Bi, the internal, or inland in the starving-island case, resistance is large with respect to shoreline resistance. Much food may be dropped off along the shoreline, but the distribution of the food inland is relatively slow. The near-shore residents may be saved from starvation but not those residing far inland.

Analogously, with thermal energy, the distribution is similar; the abundant supply of heat may be immediately received along the surface. Imagine the island representing the cross-section of a circular cylinder. The peripheral of the cylinder (circle) will be heated up, but the temperature at the inner core will take some time to change.

A small Biot number, in contrast, resembles well-established land transportation. Every piece of cargo is immediately distributed throughout the island. In this case, the survival of the entire population is at the mercy of the incoming shipments.

If the corresponding cylinder is made of a highly thermally conductive material, say graphene[2] with a thermal conductivity of around 5000 W/(m·K) (Sang et al., 2019) compared to approximately 400 W/(m·K) for copper, we have a small Biot number situation. With such grand thermal conductivity, any change in the surface temperature imposed by convection heat transfer is immediately felt by, and distributed throughout, the entire cylinder. Under these circumstances, the entire cylinder can be lumped together represented by a single temperature.

14.3 One-dimensional transient problems

The exposure of a long plate to a sudden temperature change is a common encounter in sheet metal forming. Since the length and width are long with respect to the thickness, this problem can be approximated as a one-dimensional problem in the thickness direction. Namely, the temperature change starts from the two large surfaces and propagates into the center of the plate thickness. To put it another way, at any instant in time, the temperature inside the solid plate or wall varies only with the distance with respect to the surface, in the thickness direction, which responds promptly to changes in the ambient temperature, due to a large convection coefficient. This is depicted in Fig. 14.7, where the thickness of the plate is set to be 2L. Due to the fact that the temperature distribution is symmetrical, we only need to consider the right half of the plate thickness, from $x = 0$ to $x = L$. We see that the temperature distribution is also a function of the thermal conductivity, k, the heat capacity, c_P, and the density, ρ. The effects of heat capacity and density can be expressed in terms of the thermal diffusivity, $\alpha = k/(\rho c_P)$, which signifies the amount of thermal energy a layer of material allows to pass through with respect to that it retains. To rephrase it, the temperature, T, is a function of x, L, t, k, and α. The change in the temperature of the surroundings is the driver. As a consequence, the temperature at any spatial location and instant in time, T, is also affected by the temperature gradient caused by the change in the surrounding temperature, that is, T_∞ and T_0, and also the effectiveness of heat convection at the surface, that is, h. To put these into a mathematical function, we have

$$T = f(x, L, t, k, \alpha, T_\infty, T_0, h). \tag{14.6}$$

The heat diffusion equation for the one-dimensional transient heat conduction problem without any internal heat sink or source can be expressed as

$$\frac{\partial}{\partial x}\left(k\frac{\partial T}{\partial x}\right) = \rho c_P \frac{\partial T}{\partial t} \tag{14.7}$$

---

2. Graphene is a new material classified as a metal. It is an allotrope, consisting of a single layer and, thus, the thinnest known material of carbon arranged in a hexagonal lattice; see Colapinto (2014) and Graphenea (2021), for example.

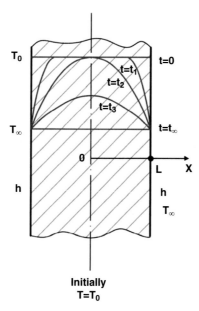

FIGURE 14.7 One-dimensional transient heat conduction through the thickness of a large plate with both sides exposed to a change in the surrounding temperature (created by X. Wang). For the particular case shown, the plate is initially at equilibrium with the ambient at T_0, and at time $t = 0$, the ambient temperature drops to T_∞, and the plate starts to respond to that step change. The change in the temperature distribution inside the plate with increasing time, t, is of significant practical importance.

For constant thermophysical properties, this can be rearranged into

$$\partial^2 T/\partial x^2 = (1/\alpha)\partial T/\partial x. \tag{14.8}$$

Initially, the entire plate is at T_0, that is, $T(x, t{=}0) = T_0$. Due to symmetry in the x direction, the temperature gradient at $x = 0$ is always zero. For the case that the object of interest is initially hot, the heat is conducted from the inner part of the object to the surface, which is losing heat via convection to the cooler ambient. At any instant in time, the rate of heat conducted to the surface is equal to that convected from the surface, namely, $-k\ \partial T(L, t)/\partial x = h\ [T(L, t) - T_\infty]$. In summary, the initial and boundary conditions are:

$$T(x, 0) = T_0, \tag{14.8a}$$

$$\partial T(0, t)/\partial x = 0, \text{ and} \tag{14.8b}$$

$$-k\partial T(L, t)/\partial x = h[T(L, t) - T_\infty]. \tag{14.8c}$$

The first boundary condition equation states that the entire object is at the initial temperature, T_0, initially at time, $t = 0$. The second boundary equation conveys that the temperature gradient at $x = 0$ is always zero because of symmetry in the temperature distribution. The third boundary condition expression bespeaks that at the surface, that is, at $x = L$, the rate of heat conduction from the object to its surface is equal to the rate of heat convection from the surface to the surroundings. It is often more convenient to express these equations in a normalized or nondimensional form. We can define a nondimensional temperature as

$$\theta = (T - T_\infty)/(T_0 - T_\infty), \tag{14.9}$$

where temperature, T, is a function of both space, x, and time, t. The nondimensional distance from the mid-plane, or middle or a cylinder or sphere, is

$$X = x/L. \tag{14.10}$$

Recall that L is one-half the plate thickness, or the corresponding radius for a cylinder or sphere. The remaining term in Eq. (14.8) is the thermal diffusivity, and it can be normalized as:

$$Fo = \alpha t/L^2. \tag{14.11}$$

French mathematician Jean–Baptiste Joseph Fourier (1768–1830) is credited for this non-dimensional parameter. This is why Fo is called the *Fourier number*; it denotes the diffusive transport rate with respect to the storage rate. It can also be regarded as a (storage) time scale; the larger the value of Fo, the slower the storage and the faster heat propagates (conducts) through the body. With that in mind, we can express the one-dimensional transient heat conduction equation in nondimensional form as

$$\partial^2\theta/\partial X^2 = \partial\theta/\partial Fo. \tag{14.12}$$

Initially, at time, $t = 0$, $Fo = 0$ and, hence, we have

$$\theta(X, 0) = 1. \tag{14.12a}$$

This initial condition conveys that the temperature is initially homogeneous and equal to the ambient, that is, $T = T_0$ and, thus, $\theta = (T - T_\infty) / (T_0 - T_\infty) = 1$. The inner boundary condition at the middle of the wall, cylinder, or sphere, that is at $X = 0$, is

$$\partial\theta(0, Fo)/\partial X = 0. \tag{14.12b}$$

This says that the temperature gradient at $x = 0$ and, hence, $X = 0$, is zero. That being the case, this problem is the same as a wall of thickness L (one-half that of a 2L-thick wall) with a well-insulated surface, at $x = 0$, and the other surface is exposed to the changing environment. The external boundary condition, at $X = 1$, is

$$\partial\theta(1, Fo)/\partial X = -Bi\,\theta(1, Fo), \tag{14.12c}$$

stating that heat conduction at the surface is equal to heat convection from it. In a nutshell, $\theta = f(Fo, X, Bi)$.

We can apply the separation of variables, that is, $\theta(X, Fo) = F(X) G(Fo)$, to obtain the solution. Substituting $\theta(X, Fo) = F(X) G(Fo)$ into Eq. (14.12), we obtain

$$(1/F)d^2F/dX^2 = (1/G)dG/dFo. \tag{14.13}$$

Setting this equal to $-\lambda^2$, where λ is a constant, leads to

$$d^2F/dX^2 + \lambda^2F = 0, \tag{14.14}$$

and

$$dG/dFo + \lambda^2G = 0. \tag{14.15}$$

The general solutions for these two equations are

$$F = C_1\cos(\lambda X) + C_2\sin(\lambda X), \tag{14.16}$$

and

$$G = C_3\exp(-\lambda^2Fo). \tag{14.17}$$

Therefore, we have

$$\theta = FG = \exp(-\lambda^2Fo)[C_3 \cos(\lambda X) + C_4\sin(\lambda X)]. \tag{14.18}$$

There are an infinite number of solutions of the form,

$$\theta = \sum A_n\exp(-\lambda_n^2Fo)\cos(\lambda_n X), \tag{14.19}$$

where the summation is from $n = 1$ to infinity. The constants, A_n, can be deduced from the initial condition, $\theta(X, 0) = 1$, and, thus,

$$1 = \sum A_n\cos(\lambda_n X), \tag{14.20}$$

which is a Fourier series describing a constant in terms of an infinite series of cosine functions. Multiplying both sides by $\cos(\lambda_m X)$ and integrating from $X = 0$ to 1 gives

$$\int \cos(\lambda_m X)dX = \int \sum A_n\cos(\lambda_m X)\cos(\lambda_n X)dX. \tag{14.21}$$

It can be shown that all integrals vanish except when $m = n$, that is,

$$\int \cos(\lambda_n X)dX = A_n \int \cos^2(\lambda_n X)dX. \tag{14.22}$$

Accordingly, the coefficient, A_n, becomes

$$A_n = 4\sin\lambda_n/[2\lambda_n + \sin(2\lambda_n)]. \tag{14.23}$$

The solutions for the large plate (plane wall), along with those for a long cylinder and a sphere, are tabulated in Table 14.1. The reader can refer to Weisstein (2002) regarding the Bessel functions. Remember that a plane wall of thickness L with one side (surface) perfectly insulated is analogous to that of the large plate of thickness 2L with both surfaces exposed to the same convective environment.

TABLE 14.1 Solutions for one-dimensional transient conduction for a large plate, a cylinder, and a sphere.

Geometry	Solution	λ_n are the roots of
Large Plate (plane wall)	$\theta = \sum\limits_{n=1}^{\infty} \dfrac{4\sin\lambda_n}{2\lambda_n + \sin(2\lambda_n)} e^{-\lambda_n^2 Fo} \cos\left(\dfrac{\lambda_n x}{L}\right)$	$\lambda_n \tan \lambda_n = Bi$
Long cylinder	$\theta = \sum\limits_{n=1}^{\infty} \dfrac{2}{\lambda_n} \dfrac{J_1(\lambda_n)}{J_0^2(\lambda_n)+J_1^2(\lambda_n)} e^{-\lambda_n^2 Fo} J_0\left(\dfrac{\lambda_n r}{r_C}\right)$	$\lambda_n J_1(\lambda_n)/J_0(\lambda_n) = Bi$
Sphere	$\theta = \sum\limits_{n=1}^{\infty} \dfrac{4(\sin\lambda_n - \lambda_n \cos\lambda_n)}{2\lambda_n + \sin(2\lambda_n)} e^{-\lambda_n^2 Fo} \dfrac{\sin(\lambda_n r/r_S)}{\lambda_n r/r_S}$	$1 - \lambda_n \cot \lambda_n = Bi$

The plate has a thickness of 2L. The cylinder has a radius of r_C. The sphere has a radius of r_S. The object is subjected to convection at the surface(s). Also, J_0 and J_1 are the Bessel functions of the first kind.

TABLE 14.2 Approximate solutions for one-dimensional transient conduction for a large plate, a cylinder, and a sphere, for Fourier number, Fo, larger than 0.2.

Geometry	Approximate solution
Large plate	$\theta = A_1 e^{-\lambda_1^2 Fo} \cos\left(\dfrac{\lambda_1 x}{L}\right)$
Long cylinder	$\theta = A_1 e^{-\lambda_1^2 Fo} J_0(\lambda_1 r/r_C)$
Sphere	$\theta = A_1 e^{-\lambda_1^2 Fo} \dfrac{\sin(\lambda_1 r/r_S)}{\lambda_1 r/r_S}$

The plate has a thickness of 2L. The cylinder has a radius of r_C. The sphere has a radius of r_S. The object is subjected to convection at the surface(s). Also, J_0 and J_1 are the Bessel functions of the first kind.

As with Taylor series, the terms in the summation decrease quickly with increasing n. This is especially true for large Fourier number, Fo. It is found that for Fo larger than 0.2, keeping the first term alone leads to an error of less than 2 percent. For this reason, in practice, we can often resort to one-term approximations, as tabulated in Table 14.2. Note that A_1 and λ_1 are functions of the Biot number, Bi, only. Sample values of A_1 and λ_1 are listed in Table 14.3, whereas sample values of J_0 are tabulated in Table 14.4. These values are based on Çengel et al. (2016).

Example 14.3 Broiling a Thanksgiving Turkey.

Given: Butterball is just about the most appropriate name for a thanksgiving turkey, that is, it can be treated like a ball. Twenty minutes before her students arrive at her house, Dr. JAS places a 10-kg turkey ($\rho = 960$ kg/m^3, k = 0.7 W/(m·°C), c = 6000 J/(kg·°C)) at 15°C into a preheated oven (heat transfer coefficient, h = 114 W/(m^2·°C)) at 205°C.

Find: The temperature at a one-half radius of the turkey ball when the students arrived at Dr. JAS' house.

TABLE 14.3 Sample values of A_1 and λ_1 for a large flat plate, a long cylinder, and a sphere.

	Large plate		Cylinder		Sphere	
Bi	A_1	λ_1	A_1	λ_1	A_1	λ_1
0.01	1.0017	0.0998	1.0025	0.1412	1.0030	0.1730
0.02	1.0033	0.1410	1.0050	0.1995	1.0060	0.2445
0.04	1.0066	0.1987	1.0099	0.2814	1.0120	0.3450
0.06	1.0098	0.2425	1.0148	0.3438	1.0179	0.4217
0.08	1.0130	0.2791	1.0197	0.3960	1.0239	0.4860
0.2	1.0311	0.4328	1.0483	0.6170	1.0592	0.7593
0.4	1.0580	0.5932	1.0931	0.8516	1.1164	1.0528
0.8	1.1016	0.7910	1.1724	1.1490	1.2236	1.4320
2.0	1.1785	1.0769	1.3384	1.5995	1.4793	2.0288
4.0	1.2287	1.2646	1.4698	1.9081	1.7870	2.5704
6.0	1.2479	1.3496	1.5253	2.0490	1.8338	2.6537
8.0	1.2570	1.3978	1.5526	2.1286	1.9106	2.8044
10	1.2620	1.4289	1.5677	2.1795	1.9249	2.8363
20	1.2699	1.4961	1.5919	2.2880	1.9781	2.9857
40	1.2723	1.5325	1.5993	2.3455	1.9942	3.0632
100	1.2731	1.5552	1.6015	2.3809	1.9990	3.1102
∞	1.2732	1.5708	1.6021	2.4048	2.0000	3.1416

TABLE 14.4 Sample values of the zeroth-order Bessel function of the first kind, J_0, as a function of $\lambda_1 r / r_C$.

$\lambda_1 r / r_C$	0.0	0.1	0.2	0.4	0.6	0.8
$J_0(\lambda_1 r / r_C)$	1.0000	0.9975	0.9900	0.9604	0.9120	0.8463
$\lambda_1 r / r_C$	1.0	1.2	1.4	1.6	1.8	2.0
$J_0(\lambda_1 r / r_C)$	0.7652	0.6711	0.5669	0.4554	0.3400	0.2239
$\lambda_1 r / r_C$	2.2	2.4	2.6	2.8	3.0	3.2
$J_0(\lambda_1 r / r_C)$	0.1104	0.0025	-0.0968	-0.1850	-0.2601	-0.3202

Solution: Assume the turkey as a sphere, that is, $\rho\,(4/3)\pi r_S^3 = 10$ kg. From which we get the radius, $r_S = 13.55$ cm.

The solution for this one-dimensional transient heat conduction problem, as expressed by Eq. (14.19), is

$$\theta = \sum A_n \exp\left(-\lambda_n^2 Fo\right)\cos(\lambda_n X).$$

Invoking one-term approximation by employing the first term. From Table 14.2, we have

$$\theta = (T - T_\infty)/(T - T_\infty) = A_1 \exp\left(-\lambda_1^2 Fo\right)\sin(\lambda_1 r/r_S)/(\lambda_1 r/r_S)$$

Substituting the corresponding values into the Fourier number or time, we get

$$Fo = \alpha t/L^2 = [k/(\rho c_P)]t/r_S^2 = 0.008$$

Similarly, the Biot number, with known values of the respective variables, is

$$Bi = (L/k)/(1/h) = hL/k = 114(0.1355)/0.7 = 22.06$$

From Table 14.3, for $Bi = 20$, we get $A_1 = 1.9781$ and $\lambda_1 = 2.9857$.
  Substituting the values into the one-term approximation formula leads to

$$(T - 205)/(15 - 205)$$
$$= 1.9781\exp\left[(2.9857)^2(0.008)\right]\sin[(2.9857)(0.5)]/[(2.9857)(0.5)] = 0.037$$

We get $T = 198°C$ at one-half radius, 20 min after the turkey is placed in the preheated oven. This tells us that most part of the turkey is ready for Dr. JAS' students to enjoy after only 20 min of broiling at 205°C. A much longer broiling time, however, is required so that the most interior part of the turkey is cooked. After all, we do not wish to be contending with campylobacteriosis, such as salmonella infection (Webmd, 2021).

14.4 Semi-infinite solid

A thick plate that requires a relatively long time for the changing surrounding temperature to penetrate can be treated as a semi-infinite solid. The term "relative" is the key. In material processing applications such as welding, the rather-conductive metal is only subjected to short periods of temperature change. In this case, the temperature change does not propagate all the way into the metal before the environment it is in changes its temperature again. On that account, the piece of metal can also be analyzed as a semi-infinite solid (Ghoshdastidar, 2012). It is worth mentioning another everyday semi-infinite solid, that is, the floor in our basement, or ground floor that laid on top of the ground or soil.

  Let us consider a semi-infinite solid or wall, as shown in Fig. 14.8. The initial condition is $T(x, 0) = T_0$, that is, the entire solid is at the same temperature as the surroundings at time, $t = 0$. This occurs when the solid and the environment have been "sitting" at the same temperature for a sufficiently long time that they reach the equilibrium condition. With this in mind, we also appreciate that far enough into the wall, the temperature will remain at this initial value, that is, $T(x\rightarrow\infty, t) = T_0$. The other boundary condition is at the surface; it is at the ambient temperature, that is, $T(0, t) = T_\infty$. We note that $x = 0$ corresponds to the surface, and the positive x direction is into the semi-infinite solid. The

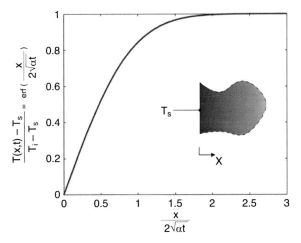

FIGURE 14.8 A semi-infinite wall subjected to a change in ambient and, thus, surface temperature (created by X. Wang). The wall is so thick that the temperature at large x remains fixed at T_0.

appropriate governing equation for describing heat conduction in a semi-infinite solid in normalized form can be pressed as

$$\partial^2\theta/\partial x^2 = (1/\alpha)\partial\theta/\partial X, \qquad (14.24)$$

where, as defined earlier, the nondimensional temperature, $\theta = (T - T_\infty)/(T_0 - T_\infty)$, and $X = x/L$. The question here is what is L for a semi-infinite wall? It has been found that an appropriate characteristic length can be defined as $L = \sqrt{(4\alpha t)}$, where α is the thermal diffisivity of the material making up the wall and t is simply the time elapsed since the ambient and, hence, the surface changes its temperature from T_0. With that in mind, we can write the initial condition at time, t = 0, as

$$\theta(X, 0) = 1, \qquad (14.24a)$$

where $X = x/\sqrt{(4\alpha t)}$. The deep-into-the-wall boundary condition is

$$\theta(X \rightarrow \infty, t) = 1. \qquad (14.24b)$$

This conveys that the temperature deep into the wall remains unchanged. The solution describing the temperature with respect to distance from the surface and time has been found to be

$$\theta = [T(X, t) - T_\infty]/[T_0 - T_\infty] = \text{erf}\left[x/\sqrt{4\alpha t}\right], \qquad (14.25)$$

where erf designates the error function and $x/\sqrt{(4\alpha t)}$ is called the similarity variable. Fig. 14.9 shows the plot of the error function, erf, versus the similarity variable, $x/\sqrt{(4\alpha t)}$. Also shown is the non-dimensional temperature, θ, as a function of the similarity variable, $x/\sqrt{(4\alpha t)}$, according to the error function, erf.

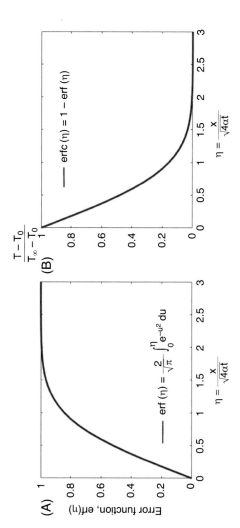

FIGURE 14.9   Solution for a semi-infinite wall subjected to a change in surface temperature (created by X. Wang). (A) Error function, erf, versus similarity variable, $x/\sqrt{(4\alpha t)}$, where u is a variable. (B) Nondimensional temperature, θ, as a function of the similarity variable, $x/\sqrt{(4\alpha t)}$, according to the complementary error function, erfc.

Problems

14.1 Time of death of a rhino.

You are a co-op student working for Dr. JAS' conservation-enforcement team in Borneo. Specifically, your group is assigned to protect rhinoceroses. You come across a rhino that has been shot dead and you immediately notify Dr. JAS via your handheld transceiver. She asks you to estimate the time of death so that she can appropriately deploy her enforcement force to apprehend the poacher(s). With a good engineering thermofluids education under your belt, you draw out a slim and long thermometer and probe it through the rhinoceros' body. The hot tropical rainforest is at a steady ambient temperature of 34°C, just right for warm-blooded rhinoceroses, with body temperatures of 37.8°C, to thrive. At 5 cm, 10 cm, 20 cm, 40 cm into the rhino's body, the temperatures are 34.5°C, 35.0°C, 35.5°C, and 36.0°C, respectively. Estimate the time of death using a lumped parameter analysis with a body temperature of 35.5°C. Compare this with the estimate based on a one-dimensional cylinder with a radius of 1.3 m.

14.2 Predict the initial temperature based on current temperature.

A large, 2-cm-thick, hot metal ($\rho = 8000$ kg/m^3, $k = 18$ W/(m·°C), $c = 500$ J/(kg·°C)) plate is drawn out of a furnace and set to cool in a room at 27°C. The heat transfer coefficient is estimated to be around 21 W/(m^2·°C). After 5 min, the plate temperature is measured to be 455°C. What is the temperature of the furnace?

14.3 Heating time for billets.

Long billets at a room temperature of 25°C are to reach 270°C at half its radius. The properties of the metal are: $\rho = 8000$ kg/m^3, $k = 18$ W/(m·°C), and $c = 500$ J/(kg·°C). How long should you place billets of 5-cm radius in a furnace at 500°C? How much longer do you need for billets of 10-cm radius?

14.4 Cooling time for a brick.

How long does it take a 225-mm by 112.5-mm by 75-mm brick from a kiln at 1100°C to cool to 50°C? The properties of the brick are: $\rho \approx 1800$ kg/m^3, $k \approx 0.72$ W/(m·°C), $c \approx 829$ J/(kg·°C), and $\alpha \approx 1.7 \times 10^{-3}$ m^2/s. Hint: use an equivalent length defined as the volume divided by the surface area. Deduce the Biot number first to see if the lumped-parameter assumption is valid.

14.5 Cooking an egg, a tomato, and a potato.

Compare the cooking time required to cook an egg, a tomato, and a potato in boiling water, assuming all are spheres with a radius of 2.5 cm, have been sitting in a refrigerator at 8°C for a few days, and the cooking temperature for all is 71°C. For the egg, $\rho \approx 1000$ kg/m^3, $k \approx 0.34$ W/(m·°C), $c \approx 3300$ J/(kg·°C). For the tomato, $\rho \approx 1000$ kg/m^3, $k \approx 0.59$ W/(m·°C), $c \approx 3990$ J/(kg·°C), and/or $\alpha \approx 1.41 \times 10^{-7}$ m^2/s. For the potato,

$\rho \approx 1080$ kg/m^3, k ≈ 0.52 W/(m·°C), c ≈ 3650 J/(kg·°C), and/or α $\approx 8.9 \times 10^{-7}$ m^2/s. Note that the values are approximates and, therefore, some discrepancies are expected. What is the value of the convective heat transfer coefficient?

14.6 Thermal energy storage of a concrete base.
The floor of a greenhouse is made of a very thick layer of concrete ($\rho = 2500$ kg/m^3, k $= 2.7$ W/(m·°C), and c $= 1000$ J/(kg·°C). With the rising sun, the greenhouse temperature increases from 15°C to 34°C in 4 h. What are the temperatures at 2 cm, 5 cm, and 10 cm deep? How much thermal energy has been stored within the top 12-cm layer of the concrete for a 7-m by 21-m greenhouse?

14.7 Minimum depth to prevent water pipe from freezing.
In a city in Canada, the ground surface can be assumed to stay around -15°C in January and February. How deep should a water pipe be buried in the ground to ensure it does not freeze?

14.8 Estimate the heat transfer coefficient from a cooling steel ball.
A 2.5-cm diameter steel ($\rho = 7880$ kg/m^3, k $= 19$ W/(m·°C), and c $= 450$ J/(kg·°C)) ball at 100°C is left to cool in a room at 21°C. The temperature sensor at 0.5 cm from the center reads 72°C and 57°C after 50 s and 100 s, respectively. What is the heat transfer coefficient acting on the sphere in the room?

References

BBC Studios, https://www.youtube.com/watch?v=1rkkKyYCxio, October 13, 2008, (accessed October 11, 2021).

Çengel, Y.A., Cimbala, J.M., Turner, R.H., 2016. Fundamentals of Thermal-Fluid Sciences, 5th ed McGraw-Hill, New York, NY.

Webmd, 2021. https://www.webmd.com/food-recipes/food-poisoning/what-is-salmonella, (accessed October 12, 2021).

Colapinto, J., 2014. Material question. Graphene may be the most remarkable substance ever discovered. But what's it for? Annals Innovation. https://www.newyorker.com/magazine/2014/12/22/material-question. (accessed October 11, 2021).

Ghoshdastidar, P.S., 2012. Heat Transfer, 2nd ed. Oxford University Press, New Delhi.

Graphenea, https://www.graphenea.com/pages/graphene, (accessed October 11, 2021), 2021.

Sang, M., Shin, J., Kim, K., Yu, K.J., 2019. Electronic and thermal properties of graphene and recent advances in graphene based electronic applications. Nanomaterials (Basel) 9 (3), 374.

Ting, D.S-K., 2020. Lecture Notes on Engineering Human Thermal Comfort. World Scientific, Singapore.

Weisstein E.W., "Bessel function of the first kind," MathWorld—A Wolfram Web Resource, 2002. https://mathworld.wolfram.com/BesselFunctionoftheFirstKind.html, (accessed June 2, 2021).

Chapter 15

Natural convection

"Reading about nature is fine, but if a person walks in the woods and listens carefully, he can learn more than what is in books, for they speak with the voice of God."

George Washington Carver

Chapter Objectives

- Appreciate everyday natural convection and thermals.
- Be familiar with the underlying mechanisms of natural convection.
- Identify the relevant nondimensional parameters of natural convection.
- Recognize classical Rayleigh–Bernard convection and natural convection inside a cavity.
- Comprehend natural convection along a vertical plate.
- Fathom continuous thermal plumes and buoyant jets.

Nomenclature

A	area
a	acceleration
Ar	Archimedes number; $Ar \equiv$ gravitational/viscous force, $Ar = Gr/Re^2$
C	coefficient; C_0, C_1 are empirical coefficients
c_P	specific heat capacity at constant pressure
F	force; F_B is buoyancy force, F_g is gravity force, $F_{z,net}$ is the net force in the z (vertical) direction
g	gravity
Gr	Grashof number; $Gr \equiv$ buoyancy force/viscous force, Gr_z is the Grashof number at z
h_{conv}	convection heat transfer coefficient
k	thermal conductivity
L	length; L_L is the longer length, L_S is the shorter length
Nu	Nusselt number; $Nu \equiv$ convection/conduction heat transfer, $Nu_L = h_{conv}L/k$
P	pressure
Q	heat transfer; Q' is heat transfer rate
Pr	Prandtl number; $Pr \equiv$ momentum/thermal diffusivity
R	gas constant
Ra	Rayleigh number; $Ra \equiv$ gravity/thermal diffusivity, $Ra \equiv Gr\ Pr$, Ra_L is Rayleigh number for the entire length L, $Ra_{L,c}$ is critical Rayleigh number from stagnant fluid to laminar flow, Ra_z is the Rayleigh number at z

Thermofluids: From Nature to Engineering. DOI: https://doi.org/10.1016/B978-0-323-90626-5.00015-X

Re   Reynolds number, $Re \equiv$ inertia/viscous force

SATP   standard atmospheric temperature and pressure

T   temperature; T_{bulk} is the bulk temperature of the fluid, T_{cold} is the cold (surface) temperature, T_H is the higher temperature, T_{hot} is the hot (surface) temperature, T_L is the lower temperature, T_{max} is the maximum temperature, T_S is the surface temperature, T_{wall} is the temperature of the wall, T_∞ is the ambient temperature

U   velocity in the (stream-wise) x-direction

W   velocity in the (vertical) z-direction; W_0 is the vertical velocity at $z = 0$, W_{max} is the maximum velocity in the z-direction, W_z is the vertical velocity at z, W_∞ is the vertical velocity of the surrounding fluid

x   (distance in) the stream-wise direction

z   (distance in) the vertical direction, z_c is the critical z distance.

Greek and other symbols

α thermal diffissivity, $\alpha = k/(\rho c_P)$

Δ difference

β thermal expansion coefficient, $\beta = -(1/\rho)\,(\partial\rho/\partial T)_P$

δ thickness or small change; δ_T is the thermal boundary layer thickness, δ_v is the momentum boundary layer thickness

μ dynamic or absolute viscosity

ν kinematic viscosity, $\nu = \mu/\rho$

ρ density; ρ_H is the density of the higher temperature fluid, ρ_L is the density of the lower temperature fluid, ρ_∞ is the density of the ambient fluid

$\forall$ volume

15.1 Natural convection and thermals

Natural convection is the convection of thermal energy made possible by fluid motion driven by the buoyancy force due to density differences associated with temperature variations. It is also called free convection, that is, it is free from (the need of) other fluid motion such as those induced by a fan or atmospheric wind. Natural convection is so common that "hot air rises" is a household phrase, though often uttered without appreciating the beauty behind this essential natural phenomenon. Do you know that hot air only rises in the presence of gravity?

Many birds appreciate and live on natural convection. For example, turkey vultures soaring in a circular path is a familiar afternoon sight. This is a clear natural display of natural convection, except the thermals that the birds ride on are invisible to our naked eyes. It is true that predators such as vultures soar in circles looking for prey. As importantly, they orbit around the thermals so that they do not exhaust their energy by flapping their wings to keep their relatively weighty bodies afloat. Without the oresence of this natural convection called thermals, birds such as the Andean condor (Vulture gryphus), the heaviest soaring bird, would have gone extinct before the great flood. Shepard et al. (2011) tracked the three-dimensional traveling path of Andean condors. They found that

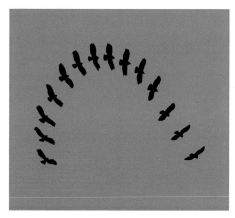

FIGURE 15.1 A flock of birds riding on a thermal (created by S. Akhand). The figure equally illustrates multiple snapshots of one vulture rising on a thermal in a roughly circular motion.

while hunting for food, the big birds' soaring behavior appeared to be determined by the need to ascend only to the extent to gain their position in acquiring food on the ground. To put it another way, they get the free ride up to the appropriate altitude and then, apply the minimum needed energy in a rapid descent to catch their prey.

Fig. 15.1 makes visible the thermal that a flock of vultures ride on. This captivating scene is seen typically[1] from late morning until midafternoon. This is because the vultures are too big for their muscles and/or they do not have enough stamina to sustain flying beyond a few flaps at a time (Pennycuick, 1973), especially early in the morning when the muscles are not yet warmed up. Moreover, most thermals are only formed after the sun has risen and has been heating the ground until it is sufficiently warm to form thermals. The plume of hot air involves a circulating motion as it rises. The warmest air in the middle of the plume rises most speedily, setting the adjoining air spinning into a vortex ring circumscribing the thermal. The upward movement of an element of the inner part of the vortex ring also enhances the upward propagation of the adjacent hot air, furthering the climb of the rising plume. The outer boundary of the vortex ring induces a downward movement of the nearby atmospheric air. Under those circumstances, hot air rises and the "vacuum" created by it is filled by the descending cooler air.

The question is, how do these thermals come into existence in the first place? Envision the ground basking under the rising Sun. It soon warms up and, thus,

---

1. It is worth noting that thermals are not the only natural rides on which birds rely. Wind shear is known to sustain birds such as albatrosses hundreds of kilometers at one time (Richardson, 2011; Sachs et al., 2013).

FIGURE 15.2   A beautiful display of cumulus clouds above the Detroit river (photo taken by X. Wang).

heats up the near-ground layer of air. As the air is heated, it becomes more "thirsty," evaporating the near-surface water into vapor and carrying it in volumes of hot and humid air. When a volume of air is sufficiently heated, it rises. An aftereffect that is clearly visible is the "cotton candy clouds," cumulus clouds, formed by the rising hot and moist air as it reaches a sufficient elevation and is cooled and equilibrates to the much lower temperature at the respective elevation. In other words, the water vapor condenses into tiny water droplets, exhibiting as a puffy, white cloud. Fig. 15.2 is a splendid display of cumulus clouds above the Detroit River. The abundant moisture supplied by the river eases the formation of the rich cumulus clouds.

Other than natural thermals, there are also man-made and/or man-caused thermals. A landfill in eastern Pennsylvania that flares off methane has been found to provide thermals for turkey vultures (Mandel and Bildstein, 2007), in addition to providing carrion and rotten vegetables on which they feast. Thermal power plants in Manaus, Central Amazon, Brazil also furnish artificial thermals for turkey vultures, as well as black vultures (Freire et al., 2015). It is interesting to note that these birds take advantage of man-made thermals when natural thermals are weak or scarce; for example, in the late afternoon after feeding, they get a free ride on man-made thermals to a safe and cozy place to roost.

It is fascinating to learn that thermals are not only harnessed by heavy birds, they are also exploited by many small, insect-feeding common swifts. Hedrick et al. (2018) discovered that swifts that have exquisite maneuvering capabilities spend only 25% of their time flapping and over 70% gliding with extended wings.

While gliding, the average power they utilize for changing speed or elevation is next to zero, at about 0.84 W/kg.

Less alluring, but indispensable in terms of their utility, are common radiators or site heaters for keeping indoor occupants warm during the cold season. Natural convection is also essential for boiling water and cooking. Many residential furnaces and fireplaces lean heavily on natural convection to safely remove the toxic flue gases. Cooling of a cup of hot coffee, a hard-boiled egg, or a hot potato is primarily via natural convection of heat from the hot object into the cooler surroundings.

15.2 Thermal expansion and buoyancy force

Ever wonder why we call those bodies of rising hot air *thermals*? A necessary ingredient for natural convection is thermal expansion empowered by thermal energy. It is the expansion of a fluid such as air when heated that leads to a decrease in its density. The thermal expansion coefficient of a fluid is

$$\beta = -(1/\rho)(\partial\rho/\partial T)_P, \tag{15.1}$$

where ρ is density, T is temperature, and P is pressure. The subscript P signifies that the partial derivative is taken at a constant pressure P. The negative sign indicates expansion with increasing temperature, that is, a decrease in density with increasing temperature. For an ideal gas, such as air at standard atmospheric temperature and pressure (SATP) conditions, the gas density is a function of pressure and temperature according to

$$\rho = P/(RT), \tag{15.2}$$

where R is gas constant. Substituting this into the thermal expansion equation, we get

$$\beta = -(RT/P)\{\partial[P/(RT)]/\partial T\}_P. \tag{15.3}$$

This can be simplified into

$$\beta = -T\partial(1/T)/\partial T = -T\partial T^{-1}/\partial T = -T(-1/T^2) = 1/T. \tag{15.4}$$

We note that, here, temperature T is the absolute temperature in degrees Kelvin. The equation conveys that the thermal expansion coefficient decreases with temperature, that is, an ideal gas expands less per unit increase in temperature at higher temperatures.

Let us consider a spherical volume of air at temperature T_H in a large environment that is at a lower temperature T_L, as depicted in Fig. 15.3. We assume free boundary flow, where the only flow motion is associated with that of the relatively small volume of rising hot air. To put it another way, the blob of hot air is in an infinite backdrop of quiescent cooler fluid. Explicitly, the main assumptions are:

1. An infinitely large environment of stagnant air at T_L.
2. The volume of hot air is at a uniform temperature T_H.

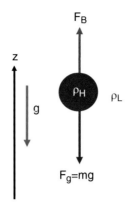

FIGURE 15.3   A spherical volume of hot air in a cooler, stagnant environment (created by D. Ting). In the ideal case, we assume the volume of hot air stays as a sphere with a uniform temperature.

3. There is no heat transfer between the hot air and the cooler environment.
4. There is no mass exchange between the volume of hot air and the ambient air.
5. The volume of hot air remains intact as a perfect sphere.

The weight of the volume of hot air is

$$F_g = mg = \rho_H \forall g, \tag{15.5}$$

where m is the mass of the hot air, g is gravity, ρ_H is the density of the hot air at T_H, and $\forall$ is the volume of the hot air. Taking the vertically upward direction as the positive z direction, the net upward force, as illustrated in Fig. 15.3, is equal to the upward buoyancy force minus the downward gravity force, that is,

$$F_{z,net} = F_B - F_g = F_B - mg = (\rho_L - \rho_H)\forall g. \tag{15.6}$$

Here, ρ_L is the density of the lower-temperature surrounding fluid, and ρ_H is the density of the higher-temperature volume. We note that the upward-directed buoyancy force,

$$F_B = \rho_L \forall g, \tag{15.7}$$

where ρ_L is the density of the surrounding air at a lower temperature T_L. It is clear that the volume of air of interest will rise only when ρ_H is less than ρ_L, that is, when ρ_H minus ρ_L is positive; it will sink if ρ_H minus ρ_L is negative. Eq. (15.6) can be recast as

$$F_{z,net} = \Delta\rho\forall g. \tag{15.8}$$

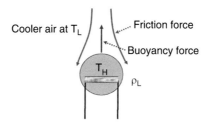

FIGURE 15.4 Buoyant hot air, heated via an electric resistor, being opposed by its viscous stagnant surroundings (created by D. Ting). The buoyant force of the heated lighter air has to be greater than the viscous force imposed by the surrounding stagnant air before the volume of hot air can rise.

The density difference is proportional to the temperature difference, that is,

$$\Delta\rho = \rho_L - \rho_H \propto \Delta T = T_H - T_L. \tag{15.9}$$

In other words, we are only considering the variation in density caused by temperature differences. We can change the proportional sign into an equal sign by introducing the thermal expansion coefficient, β. Mathematically, we have

$$\Delta\rho = \rho_L - \rho_H = \beta\Delta T = \beta(T_H - T_L). \tag{15.10}$$

15.3 Nondimensional parameters in natural convection

Recall from fluid mechanics that all real fluids are viscous. Air is a relatively thin fluid, but it still has finite viscosity, $\mu \approx 2 \times 10^{-5}$ kg/(m·s) at SATP, compared to 10 kg/(m·s) for honey. Viscosity resists any motion and there is no exception for the volume of hot air at T_H that is trying to rise against the stagnant surrounding air at T_L, in Fig. 15.4. With that in mind, the buoyancy force has to be larger than the impeding viscous force before the volume of hot air starts to rise. The importance of this buoyant-viscous force ratio was first disclosed by a German engineer, Franz Grashof. The Grashof number can be expressed as

$$Gr \equiv buoyancy\,force/viscous\,force = g\beta(T_H - T_L)\forall/\nu^2, \tag{15.11}$$

where ν is the kinematic viscosity. Due to the tug of war between buoyancy force and viscous force, there is some minimum Gr before the buoyant volume can overcome the imposed resistance by viscosity. The specific value of this critical Grashof number depends on the conditions involved. The volume term is typically replaced by an appropriate characteristic length, L, to the power of three. Namely,

$$Gr = g\beta(T_H - T_L)L^3/\nu^2. \tag{15.12}$$

Let us look at an example illustrating the balancing of buoyancy force and viscous force in terms of drag.

Example 15.1 Approximate natural convection velocity of a hot plume.

Given: An environment of stagnant cool air with a heating coil in the midst, as shown in Fig. 15.4.

Find: An approximate expression for the upward velocity of the rising volume of heated air.

Solution: When the volume of air immediately surrounding the heating coil in Fig. 15.4 is heated, it undergoes thermal expansion. A thermally expanded fluid element such as this will experience a buoyancy force per unit mass of $g\Delta\rho/\rho$.

The generic expression for the upward velocity is

$$W = dz/dt,$$

where z is the vertical distance. The corresponding upward acceleration,

$$a = dW/dt = -g = \text{constant}.$$

Substituting for $dt = W/dz$ from the previous equation, we get

$$a = dW/(dz/W),$$

which can be rearranged into

$$adz = WdW.$$

We can integrate this from $z = 0$ to $z = z$,

$$\int adz = a \int dz = \int WdW.$$

Initially at $z = 0$, $W_0 = 0$, and, therefore, the integration leads to

$$az = \frac{1}{2}\left(W_z^2 - W_0^2\right) = \frac{1}{2}W_z^2.$$

The vertical velocity W at any elevation z is thus

$$W = W_z = \sqrt{(2az)}.$$

The acceleration is a function of gravity and density difference, that is,

$$a = g\Delta\rho/\rho = g\beta\Delta T.$$

Substituting this acceleration into the vertical velocity equation leads to

$$W = \sqrt{(2g\beta\Delta Tz)}.$$

Thermal diffusion spreads out the heat and, thus, reduces the temperature gradient, resulting in a lower than ideal buoyancy force. Furthermore, viscosity drags the upward moving blob of hot air and, thus, slows it down. Accordingly, the actual upward velocity is less than $\sqrt{(2\ g\ \beta\ \Delta T\ z)}$. To account for these, we can remove the square root of two, reducing the velocity estimate from the ideal upper limit by roughly 40%, that is,

$$W \approx \sqrt{(g\beta\Delta Tz)}.$$

More accurate results, if needed, and the substantially more effort required is justifiable, can be achieved by conducting detailed computational fluid dynamic analysis.

Example 15.1 furnishes an estimate of the rising velocity of a hot volume of thermally expanded fluid. The upward velocity,

$$W \approx \sqrt{(g\beta\Delta Tz)}, \tag{15.13}$$

where z is the distance above the heat source. For the ideal case where there is no heat transfer between the volume of hot fluid and its surroundings, the blob of rising hot air is analogous to a freely falling object under gravity. In other words, the hot blob will accelerate upward until it reaches its terminal velocity, when the buoyancy force is equal to the opposing drag force. The actual happenings are significantly more complex; nevertheless, Eq. (15.13) provides a first estimate of the rising velocity not too far above the heat source. Let us exploit this approximation in a real-life application.

Example 15.2 Ball-part velocity of hot air from a home radiator.
Given: A home radiator in stagnant air in a room of 3 m height is turned on. The radiator reaches a temperature of 70°C while the room air is around 20°C.
Find: The approximate velocity of the natural convection of hot air in the room.
Solution: Assume the mean velocity occurs at roughly mid-height, that is, 1.5 m above the radiator, then,

$$W \approx \sqrt{(g\beta\Delta Tz)} = \sqrt{(9.81 \times 3.4 \times 10^{-3} \times 20 \times 1.5)} = 1 \text{ m/s}.$$

Here, we assume that, due to heat transfer in the presence of a temperature difference or gradient, volumes of heated air rising above the radiator are at roughly 40°C. Furthermore, the volume expansion coefficient of air at atmospheric pressure is estimated to be 3.4×10^{-3} per °C.

It is interesting to note that thermoelectric generators can be employed to upgrade part of the thermal energy from a (hydronic) radiator into high-quality energy, electricity, before it is dissipated as heat; see, for example, Al-Widyan et al. (2021).

As electricity passes through the resistor shown in Fig. 15.4, the coil becomes hot. Prior to the onset of fluid motion, the thermal energy is transmitted to the nearby fluid via diffusion, that is, conduction heat transfer. This is characterized by thermal diffusivity, $\alpha = k/(\rho c_P)$, where k is the thermal conductivity. Thermal diffusivity denotes the amount of thermal energy diffusing through the fluid with respect to what it withholds. The portion that it retains warms up the fluid; the extent of the temperature rise is defined by the specific heat capacity, c_P. The remaining thermal energy is passed to the subsequent fluid. It is clear that when the thermal diffusivity is high, the resulting heat is spread out and, thus, the temperature gradient is low. For that reason, an increase in thermal diffusivity tends to diminish natural convection.

Ultimately, fluid motion is needed to effectuate natural convection. The momentum diffusivity indicates the rate at which momentum spreads. The rate of momentum diffusion with respect to thermal diffusion is described by the Prandtl number,

$$\text{Pr} = \nu/\alpha = (\mu/\rho)/[k/(c_P\rho)] = c_P\mu/k, \tag{15.14}$$

where μ is the dynamic viscosity. The Prandtl number for air at SATP is about 0.7, that is, flow momentum in air diffuses at the same order of magnitude but somewhat slower than thermal energy. For liquid mercury at room temperature, the corresponding Prandtl number is approximately 0.025 (Engineering ToolBox, 2021). This implies that it is significantly easier to induce natural convection in air than in mercury. In other words, when applying heat to a local region in a pool of mercury, the heat is readily conducted from the heated zone outward throughout the pool of mercury. For that reason, there is not enough a temperature gradient to cause natural convection unless the heating rate is faster than the outward conduction rate.

It follows that both Grashaf number and Prandtl number are important in natural convection. This was first solidified by Lord Rayleigh. The Rayleigh number is the product of Grashaf number and Prandtl number, that is,

$$\text{Ra} \equiv \text{gravity/thermal diffusivity} = \text{Gr Pr} = g\beta(T_H - T_L)L^3/(\nu\alpha). \tag{15.15}$$

As such, we can look at Rayleigh number as the timescale for thermal transport via diffusion with respect to the timescale for thermal transport via convection. The larger the timescale, the weaker the corresponding term. For example, a large timescale for thermal transport via diffusion relative to that via convection implies that thermal diffusion is slower, that is, it takes a longer time to realize. It follows that the convection becomes progressively more dominant with increasing Ra. For low-Pr fluids such as liquid mercury, the corresponding Grashaf number has to be notably larger, compared to high-Pr fluids, in order to achieve the same Rayleigh number for significant natural convection. Interested readers should check out (Cioni et al., 1997; Kaminski and Jaupart, 2003; Davaille et al., 2011; Wang et al., 2018), best after going through the next section on the classical Rayleigh–Bernard convection.

15.4   The classical Rayleigh–Bernard convection

Some of the earliest systematic studies on natural convection were those carried out by Rayleigh and Bérnard. Fig. 15.5 shows a computer simulation of Rayleigh–Bérnard convection. A compressible, or expandable, fluid is enclosed in the rectangular cavity. A two-dimensional case is used here because of its simplicity and clarity. With the two side walls perfectly insulated, heating the bottom plane leads to thermal energy diffusion into the fluid above the hot plane. The fluid in the vicinity of the hot plane expands as it is heated, lowering density with respect to the surrounding cold air. When the resulting buoyancy

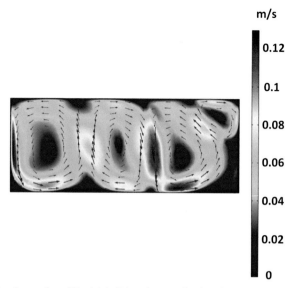

m/s

0.12

0.1

0.08

0.06

0.04

0.02

0

FIGURE 15.5 A snapshot of Rayleigh–Bérnard convection based on a two-dimensional simulation of air in a 0.2 m high and 0.5 m wide cavity (created by X. Wang). The base is at 20°C while the air is at 10°C.

force is sufficiently large, the natural convection currents, such as those shown in Fig. 15.5, set in.

For a narrow rectangular enclosure, as shown in Fig. 15.6, the flow is significantly more limited by its shorter dimension, L_S, than its longer length, L_L. Let us first consider the low-horizontal-enclosure case with the shorter dimension L_S being the height, and base horizontal area of L_L by L_L. Studies have found that the critical Rayleigh number below which the heated fluid layer is thermally stable, that is, not moving, is

$$Ra_{L,c} \approx 1700. \qquad (15.16)$$

The characteristic length, L, in this context is the height or vertical gap, L_S. We note that below this critical Rayleigh number, the fluid in the enclosure remains stagnant and heat transfer from the hot bottom plane to the top plane occurs via conduction. The corresponding heat conduction rate from the bottom hot plane at T_{hot} to the cooler top plane at T_{cold} is

$$Q' = h_{conv}A(T_{hot} - T_{cold}) = kA(T_{hot} - T_{cold})/L, \qquad (15.17)$$

where A is the area of the horizontal plane. This is the exact same heat transfer that takes place if we reverse the enclosure so that the top plane is hot and the bottom plane is cold. As such, the average Nusselt number,

$$Nu_L = h_{conv}L/k = 1, \qquad (15.18)$$

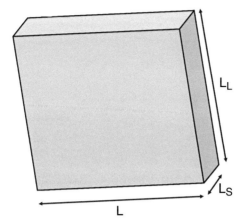

FIGURE 15.6 A rectangular enclosure (or cavity) with shorter dimension L_S and longer dimension L_L (created by D. Ting). The lower horizontal plane is heated until the Rayleigh number is larger than the critical value to form laminar Rayleigh–Bérnard cells. Further increase in the Rayleigh number beyond the laminar-to-turbulent critical value leads to random turbulent motion driven by intense natural convection.

for Ra less than 1700. The convection heat transfer coefficient, h_{conv}, in W/(m²·K) in SI units, is equal to the "conduction heat transfer coefficient," k/L, where thermal conductivity, k, is in W/(m·K). To put it another way, the heat convection takes place in its limiting mode without any movement of fluid, that is, in conduction-only convection, when Ra is less than 1700. Turbulent motion sets in when the Rayleigh number is larger than about 3×10^5.

Increasing the Rayleigh number above 1700 will form and set into motion regular Rayleigh–Bernard cells; see Fig. 15.5. As long as Ra is no more than 5×10^4, these regular motions are laminar. Naturally, the Nusselt number is more than unity and it increases as these Rayleigh–Bernard cells become more intense with Ra. The convective motions become unsteady above Ra of 5×10^4, and transition into fully turbulent ones at Ra of 3×10^5. The average Nusselt number for the turbulent case can be expressed as

$$Nu_L = 0.069Ra_L^{1/3}Pr^{0.074}. \qquad (15.19)$$

Vertical cavities, with the shorter dimension L_S in Fig. 15.6 being the width, are common in engineering. Among others, many building envelopes have enclosed vertical air gaps. The performance of double-pane and triple-pane windows leans on minimizing the formation of Rayleigh–Bernard cells within the air gap. The heat transfer rate is lowest in the absence of any air motions. With the height much larger than the gap size, there is more room to rise along the longer dimension, L_L, for a given cavity or gap of L_S. Under those circumstances, L_S is the more appropriate characteristic length, L, for defining the Rayleigh number. It follows that the critical Rayleigh number below which the fluid remains motionless is lower than that of the low-horizontal-cavity case at

$$Ra_{L,c} \approx 1000. \qquad (15.20)$$

Increasing Ra above 1000 gives rise to a single-cell cellular flow that becomes progressively less quiescent. Secondary cells spawn at the corners with increasing Ra. The Nusselt number is also influenced by the aspect ratio, L_L/L_S. The following equations apply to the various conditions. For $1 \lesssim L_L/L_S \lesssim 2$, $10^{-3} \lesssim Pr \lesssim 10^5$, $10^3 \lesssim RaPr/(0.2+Pr)$,

$$Nu = 0.18[Pr\,Ra/(0.2 + Pr)]^{0.29}. \qquad (15.21)$$

For $2 \lesssim L_L/L_S \lesssim 10$, $Pr \lesssim 10^5$, $10^3 \lesssim Ra \lesssim 10^{10}$,

$$Nu = 0.22[Pr\,Ra/(0.2 + Pr)]^{0.28}(L_L/L_S)^{-1/4}. \qquad (15.22)$$

Example 15.3 Rayleigh number of a double-pane window.

Given: A 2-m high double-pane window has an 1-cm air gap. For occupant thermal comfort, the indoor air temperature is kept at 20°C. The surface of the inner pane in contact with the air gap may be assumed to be at 17°C.

Find: If the convection heat transfer through the air gap involves more than conduction alone, when the surface of the outer pane facing the air gap is at (1) 0°C, (2) −15°C, and (3) −30°C.

Solution: An appropriate expression for the Rayleigh number in this vertical-cavity case is

$$Ra = g\beta(T_H - T_L)L^3/(\nu\alpha),$$

where $g = 9.81$ m/s and $L = 0.01$ m.

For atmospheric air at 0°C, $\beta = 0.0037$/K, $\nu = 1.3 \times 10^{-5}$ m²/s, $\alpha = 1.9 \times 10^{-5}$ m²/s and, thus,

$$Ra = g\beta(T_H - T_L)L^3/(\nu\alpha) = 2498.$$

At −15°C, $\beta = 0.0039$/K, $\nu = 1.2 \times 10^{-5}$ m²/s, and $\alpha = 1.7 \times 10^{-5}$ m²/s and, therefore,

$$Ra = g\beta(T_H - T_L)L^3/(\nu\alpha) = 6001.$$

At −30°C, $\beta = 0.0042$/K, $\nu = 1.1 \times 10^{-5}$ m²/s, and $\alpha = 1.5 \times 10^{-5}$ m²/s and, hence,

$$Ra = g\beta(T_H - T_L)L^3/(\nu\alpha) = 11,736.$$

We see that when the temperature difference across the gap is larger, the natural convection within the air gap is more significant. The temperature difference between the two inner surfaces has to be less than approximately 12°C to keep Ra below the critical value of 1700, where there is no natural convection.

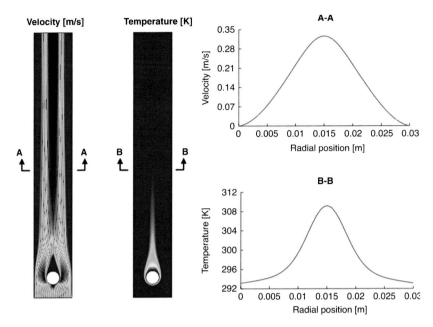

FIGURE 15.7 A continuous thermal plume of air (created by X. Wang) generated from a 0.01 m diameter cylinder at 380 K in 293 K ambient air moving upward at 0.15 m/s. For air at typical conditions, $Pr \approx 0.7$ and, thus, a wider velocity wake than the thermal wake.

15.5 Continuous thermal plumes and buoyant jets

A sufficiently heated horizontal wire, such as that shown in Fig. 15.7, can lead to a continuous plume rising up in an open and otherwise stagnant environment. The temperature at the middle of the plume, at x = 0.015 m (middle of the wire) and z = 0^+, is the highest. The superscript plus sign associated with the z-coordinate (vertically upward) accounts for the very thin wire, which in the ideal case has zero thickness. Along the vertical path upward, the same amount of thermal energy is spread over a progressively larger x-span. Because of this, the peak temperature along x = 0.015 m decreases with increasing z. An appropriate way to normalize the temperature profile is to employ

$$(T - T_\infty)/(T_{max} - T_\infty), \tag{15.23}$$

where T is the temperature, T_∞ is the ambient temperature, and the maximum temperature, T_{max}, corresponds to the temperature at x = 0.015 m and z=0^+. Similarly, the upward velocity can be normalized as

$$(W - W_\infty)/(W_{max} - W_\infty), \tag{15.24}$$

where W is the velocity in the upward z-direction, W_∞ is the ambient velocity in the vertical direction, and W_{max} is the maximum upward velocity. We note that the maximum velocity occurs where the temperature is the highest, as the flow

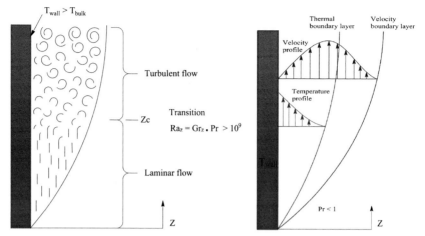

FIGURE 15.8 Free convection of air along a vertical hot plate (created by F. Kermanshahi). Note that for air at typical near-atmospheric conditions, Pr $\approx$ 0.7 and, hence the momentum boundary layer is thicker than the thermal boundary layer.

is induced by buoyancy, which is a function of temperature. Also, the ambient fluid is stagnant, that is, $W_\infty = 0$. Therefore, the normalized velocity is simply W/W_{max}.

Unsurprisingly, the velocity profile development is similar to that of the temperature profile development. As it happens, they are exactly the same when the Prandtl number is one. For air at SATP, Pr = 0.7, and this gives a wider velocity wake (plume) because it takes a longer distance for the momentum to diffuse than the thermal plume.

Thermal jets can be viewed as thermal plumes with an induced momentum. The induced momentum enables them to be directed in any direction, unlike thermal plumes that can rise upward only. Both hair dryers and heat air guns produce a thermal jet, with the thermal jet from the heat gun at a much higher temperature, up to $\sim 1200°C$ compared to up to $\sim 100°C$ for hair dryers. It is important to stress that thermal jets are enforced by an external force and, therefore, are not natural but forced convection.

15.6 Free convection along a vertical plate

The heat transfer rate of a surface undergoing natural convection is a function of the surface orientation, among other parameters. Let us consider a hot vertical plate, as shown in Fig. 15.8. Recall the general expression for the average Nusselt number,

$$Nu = C_0 (Gr\,Pr)^{C_1} = C_0 Ra^{C_1}. \qquad (15.25)$$

For a heated vertical plate, C_0 is usually less than one, C_1 is typically around 1/4 for the laminar flow portion, and approximately one-third for the turbulent flow region. Note that transition to turbulent flow occurs when the local Rayleigh number, $Ra_z = Gr_z \, Pr$, reaches approximately one billion, that is, 10^9.

As we move vertically upward along the heated plate portrayed in Fig. 15.8, the amount of thermal energy in the fluid next to the hot surface increases. This is because, in addition to the locally generated heat, there is also the heat convected from the lower part of the hot plate. That being so, the induced buoyancy force and the extent of energy transfer in the direction perpendicular to the plate, that is, x-direction, also increase. We can see that the temperature of the fluid in the boundary layer decreases exponentially away from the hot plate in the x-direction. With no-slip conditions, both at the solid boundary and the velocity boundary at the interface with the stagnant fluid, the peak velocity at any elevation takes place near the midpoint between these two boundaries. The velocity profile is skewed, with a depression on the side nearer to the stagnant surrounding fluid. Or, more correctly, the slowing down imposed by the solid boundary is lessened by the larger buoyant force, due to its higher temperature.

Let us analyze the laminar portion in more detail by following Ghoshdastidar (2012). To simplify the analysis, the following assumptions are invoked:

1. The flow is steady and two-dimensional.
2. The fluid properties remain unchanged in the flow region.
3. The fluid is incompressible apart from the density difference driving the buoyancy force.

The z-momentum equation in the vertical direction is

$$W\frac{\partial W}{\partial z} + U\frac{\partial W}{\partial x} = -\frac{1}{\rho}\frac{\partial P}{\partial z} + \nu\frac{\partial^2 W}{\partial x^2} - g. \tag{15.26}$$

Applying the boundary layer approximation, which says that the pressure is only a function of z, that is, it is not a function of the x direction normal to the plate. Namely,

$$\partial P/\partial x = 0. \tag{15.27}$$

Further, assume that the z-pressure gradient in the boundary layer is equal to that outside the boundary layer. For the case under consideration, the surrounding fluid is stagnant, that is, $W = U = 0$. The corresponding pressure gradient can be deduced from the above z-momentum equation to be

$$\partial P/\partial z = -\rho_\infty g. \tag{15.28}$$

This indicates that the pressure is hydrostatic, that is, $P = -\rho_\infty gz$. Substituting this back into the z-momentum equation (for the momentum boundary layer), we get

$$W\frac{\partial W}{\partial z} + U\frac{\partial W}{\partial x} = \frac{g}{\rho}(\rho_\infty - \rho) + \nu\frac{\partial^2 W}{\partial x^2}. \tag{15.29}$$

It is clear that the first term on the right is the buoyancy force, which drives the fluid upward. Recall that the volumetric thermal expansion coefficient is $\beta = -(1/\rho)\,(\partial\rho/\partial T)_P$. Following Boussinesq (1903), this can be approximated as

$$\beta \approx -(1/\rho)(\rho_\infty - \rho)/(T_\infty - T). \tag{15.30}$$

Substituting this into the z-momentum equation gives

$$W\frac{\partial W}{\partial z} + U\frac{\partial W}{\partial x} = g\beta(T - T_\infty) + \nu\frac{\partial^2 W}{\partial x^2}. \tag{15.31}$$

The left side entails the inertia terms. The first term on the right side is the buoyancy term, and the last term is the friction term.

In short, the three fundamental equations are as follows.
The continuity equation,

$$\frac{\partial W}{\partial z} + \frac{\partial U}{\partial x} = 0. \tag{15.32}$$

The z-momentum equation,

$$W\frac{\partial W}{\partial z} + U\frac{\partial W}{\partial x} = g\beta(T - T_\infty) + \nu\frac{\partial^2 W}{\partial x^2}. \tag{15.33}$$

The energy equation,

$$W\frac{\partial T}{\partial z} + U\frac{\partial T}{\partial x} = \alpha\frac{\partial^2 T}{\partial x^2}, \tag{15.34}$$

where the viscous dissipation term has been neglected.

It is more convenient and versatile to use nondimensional parameters. The fitting nondimensional parameters are:

$$x* = x/L, \tag{15.35}$$

$$z* = z/L, \tag{15.36}$$

$$U* = U/W_o, \tag{15.37}$$

$$W* = W/W_o, \tag{15.38}$$

and

$$T* = (T - T_\infty)/(T_S - T_\infty), \tag{15.39}$$

where W_o is the reference vertical velocity, and T_S is the surface temperature of the vertical plate. With the above nondimensional parameters, the z-momentum equation can be re-expressed in the normalized form,

$$W*\frac{\partial W*}{\partial z*} + U*\frac{\partial W*}{\partial x*} = \frac{g\beta(T_S - T_\infty)L}{W_o^2}T* + \frac{1}{Re_L}\frac{\partial^2 W*}{\partial x*^2}. \tag{15.40}$$

The first term on the right-hand side is the buoyancy term and it can be described as:

$$g\beta(T_S - T_\infty)L/W_o{}^2 = Gr_L/Re_L{}^2, \tag{15.41}$$

where $Re_L = W_o L/\nu$. Similarly, the energy equation can be normalized as:

$$W * \frac{\partial T*}{\partial z*} + U * \frac{\partial T*}{\partial x*} = \frac{1}{Re_L \, Pr} \frac{\partial^2 T*}{\partial x*^2}. \tag{15.42}$$

The term, $Gr_L/Re_L{}^2$, represents the buoyancy to inertia force ratio, and it is referred to as the Archimedes number. We see that the normalized heat transfer coefficient, the Nusselt number,

$$Nu_L = f(Re_L, Gr_L, Pr). \tag{15.43}$$

We have free convection when the Archimedes number, $Gr_L/Re_L{}^2$, is much larger than one, that is, when the buoyant influence is much larger than the inertia effect. Under this condition, the inertia term, the last term in the z-momentum equation, can be neglected with respect to the buoyant term, the second last term in the z-momentum equation. Without the inertia or Re_L term, the Nusselt number is only a function of the Grasholf number and the Prandtl number. Namely,

$$Nu_L = f(Gr_L, Pr). \tag{15.44}$$

At the other end, the buoyant term is much less than the inertia term, that is, $Gr_L/Re_L{}^2 \ll 1$. In this case, the role of natural convection can be neglected, we have forced convection, where

$$Nu_L = f(Re_L, Pr). \tag{15.45}$$

Both free and forced convections are important when $Gr_L/Re_L{}^2$ is of the order of one. This in-between regime is called *Mixed Convection*, where

$$Nu_L = f(Re_L, Gr_L, Pr) = f(Re_L, Ra_L). \tag{15.46}$$

As such, Gr/Re^2 represents the importance of natural convection/forced convection. Plots of $Nu/Re^{\frac{1}{2}}$ versus Gr/Re^2 are commonly utilized to separate natural convection from forced convection, with mixed convection marrying the two. In general, forced convection may be neglected when Gr/Re^2 is greater than 10, that is, there is natural convection only. For $0.1 < Gr/Re^2 < 10$, both natural and forced convection are important. When $Gr/Re^2 < 0.1$, forced convection prevails and, thus, natural convection needs not be considered.

Churchill and Chu (1975) expression may be used to describe the entire range of natural convection along a vertical wall, that is,

$$Nu_L = \left\{ 0.825 + \frac{0.387 \, Ra_L^{1/6}}{\left[1 + (0.492/Pr)^{9/16}\right]^{8/27}} \right\}. \tag{15.47}$$

A better fit for the more commonly encountered laminar regime, when the vertical plate is not too long or the temperature difference is not too large, that is, when $Ra_L < 10^9$, is

$$Nu_L = 0.68 + \frac{0.670 Ra_L^{1/4}}{\left[1 + (0.492/\,Pr)^{9/16}\right]^{4/9}}. \tag{15.48}$$

These two expressions are strictly for the constant-surface-temperature case. Nonetheless, they also grant a reasonable approximation for the uniform-heat-flux condition.

Multiple vertical plates packed close to one another are commonly seen in engineering systems for dissipating heat. These heat fins promote heat transfer by furnishing a large surface area for heat convection. Packing them too close together, however, will limit convection. The optimum spacing between consecutive heat fins is defined by factors such as the temperature difference between the fin and the ambient, and the dimensions of the fins. Fans are added to force the convection, allowing more closely packed fins than those that rely only on natural convection.

15.7   Other free convection cases

There are many other kinds of free or partially free convection. In engineering, hot cylinders are common in many heat exchange systems, such as tubes and shell heat exchangers. Natural convection of a hot sphere is also encountered in practice. For an isothermal cylinder, as depicted in Fig. 15.9, the boundary layer thickens from the lower point around the cylinder upward. This is because of increasing thermal energy being transferred into the fluid around the cylinder and causing the hot fluid to rise. As far as the local heat transfer rate is concerned, the Nusselt number, Nu, is largest at the lowest physical point, where the boundary layer thickness is the smallest, that is, temperature gradient between the hot cylinder surface and the cool ambient fluid is the largest due to the shortest distance separating them. The value of Nu stays relatively high away and upward from the lowest physical point. This is presumably due to the balance between the thickening boundary layer, which decreases the heat transfer rate, and the progressively more disturbed (turbulent) boundary layer. Once the boundary layer is fully turbulent, the local Nu decreases rapidly with increasing thickness, and it reaches its minimum at the upper stagnation point.

The average Nusselt number for a Rayleigh number, based on the diameter of the cylinder, of less than 10^{13} can be described by Churchill and Chu (1975) correlation,

$$Nu = \left\{0.60 + 0.387\,Ra^{1/6}/\left[1 + (0.559/Pr)^{9/16}\right]^{8/27}\right\}^2. \tag{15.49}$$

300.0   300.5   301.0   301.5   302.0   302.5   303.0

FIGURE 15.9 Free convection from a 2 m diameter horizontal circular cylinder at 600 K in a 300 K gas with Prandtl number of 0.7, specific heat of 712 J/(kg·K), dynamic viscosity of 1×10^{-5} kg/(m·s), and molar mass 28.9 g/mol (created by R. Wang).

It is easier to employ

$$\text{Nu} = 0.36 + 0.518\text{Ra}^{1/6}/\left[1 + (0.559/\text{Pr})^{9/16}\right]^{4/9}. \tag{15.50}$$

For a horizontal isothermal circular cylinder in a medium with $\text{Pr} \approx 0.7$, air being the most typical example, the expression suggested by Morgan (2007) is probably best in terms of both convenience and accuracy. Specifically,

$$\text{Nu} = C_0\text{Ra}^{C_1}, \tag{15.51}$$

where values of C_0 and C_1 for different ranges of Rayleigh number, Ra, are tabulated in Table 15.1.

For an isothermal sphere, Churchill (1983) suggests,

$$\text{Nu} = 2 + 0.587\,\text{Ra}^{1/4}/\left[1 + (0.469/\text{Pr})^{9/16}\right]^{4/9}\}. \tag{15.52}$$

TABLE 15.1 Nusselt–Rayleigh Correlation for Pr = 0.7, Nu = C_0 Ra^{C_1}.

Ra	C_0	C_1
$10^{-2} \sim 10^2$	1.02	0.148
$10^2 \sim 10^4$	0.850	0.188
$10^4 \sim 10^7$	0.480	0.250
$10^7 \sim 10^{12}$	0.125	0.333

FIGURE 15.10 Hot air balloons (created by X. Wang). One of the most popular hot air balloon rides is at Cappadocia, Turkey; see Planet (2021), for example.

This is valid for Ra less than 10^{11} and Pr greater or equal to 0.7. Concerning the local Nusselt number around the sphere, readers can refer to Chiang et al. (1964), where the solutions for both isothermal and uniform-heat-flux cases are disclosed. Other than circular cylinder and sphere in isolation and in group, heated objects of other shapes have also been extensively studied. Also studied thoroughly are concentric cylinders and concentric spheres.

You may be wondering why soul-soothing hot air balloons were not introduced earlier. Here it is, as depicted in Fig. 15.10, free convection has been harnessed for recreation and sightseeing for many years. Fig. 15.10 illustrates that the hot air rises after being heated by the flame; once it reaches the top of the balloon, it is cooled, and further cooling takes place on its way down the side of the balloon. More intense heating is required to make the balloon ascend, and heating is reduced or stopped for a safe descent. Before we think that hot

FIGURE 15.11 A classic airship (created by X. Wang). Traditionally, airships have been employed for scientific exploration and battles.

air balloons were invented specifically for peacetime leisure, we are reminded that a key driver leading to their maturity was scientific exploration or, more specifically, atmospheric air monitoring and weather studies. In fact, the earliest airships, such as the Graf Zeppelin shown in Fig. 15.11, were used for battles and wars.

Even more common is cooling of a cup of coffee or tea via natural convection and, conversely, the warming up of a can or bottle of cold drink. We would have to consume raw food without natural convection cooking and boiling. Ostensibly, free convection is also critical for sparing us from having heat stroke, especially when we are mad. Don't you see the fiery hot plume convecting from someone's head when the individual loses control?

Problems

15.1 Cooling a hot potato on Mars.

A group of engineering students who are addicted to sweet potatoes, for their taste and for playing the "hot potato" game, went on a mission with Dr. JAS to Mars. Does it take longer, shorter, or the same time to cool a sweet potato via natural convection on Mars (where the gravity is 3.721 m/s^2) than on Earth? Approximate a typical sweet potato as a 0.07-m-diameter sphere. Estimate the difference in the cooling rate from 95°C to 50°C, at which temperature it is thrilling, yet safe, to play the hot potato game in a 20°C room temperature.

15.2 Bluff body natural convection.

An irregular bluff body at 70°C, with an equivalent diameter of 1 cm, is placed in a room where the otherwise stagnant air is at 20°C. The corresponding average heat flux caused by natural convection is measured to be 125 W/m^2. Another bluff body of the same shape, with an equivalent diameter of 3 cm, at the same temperature, is placed in the same environment. What is the convection heat transfer rate associated with the 3-cm body?

15.3 Natural convection of a building façade.

The facade of a 30-m tall building is made of two panes of glass separated by a 1-cm air gap. On a typical winter night, the inner surfaces of the two panes of glass are at 12°C and 2°C. What is the rate of heat loss through a 1-m wide section of the facade?

15.4 Cooling a vertical plate via natural convection.

A 2-m by 2-m vertical plate at 75°C is left in a room at 25°C to cool off, that is, both surfaces are exposed to the still room air. What is the initial heat transfer rate? What is the rate of heat transfer when the plate is cooled to 50°C?

15.5 Cooling a short cylinder.

Your absent-minded friend left seven dozen 6-cm diameter by 12-cm long beverage cans in the car. The cans were at a temperature 45°C when you were getting ready for the guests to come for a picnic in 25 min. Would you place every one of them horizontally or vertically to cool off via natural convection by ambient air at 21°C? Calculate the Nusselt number to back up your answer. Estimate the temperature of the cans by the time the guests show up, assuming they are punctual.

15.6 Cool head.

How much heat is lost via natural convection from an average human head in a cool room at 18°C? Assume the head to be a 0.2-m diameter sphere at 37°C.

References

Al-Widyan, M., Al-Nimr, M., Al-Oweiti, Q., 2021. A hybrid TEG/thermal radiator system for space heating and electric power generation. J. Build. Eng. 41, 102364.

Boussinesq, J., 1903. Théorie Analytique de la Chaleur, II. Gauthier-Villars, Paris.

Chiang, T., Ossin, A., Tien, C.L., 1964. Laminar free convection from a sphere. J. Heat Transfer 86 (4), 537–541.

Churchill, S.W., 1983. Comprehensive, theoretically based, correlating equations for free convection from isothermal spheres. Chem. Eng. Commun. 24, 339–352.

Churchill, S.W., Chu, H.H.S., 1975. Correlating equations for laminar and turbulent free convection from a horizontal cylinder. Int. J. Heat Mass Transfer 18 (9), 1049–1053.

Cioni, S., Ciliberto, S., Sommeria, J., 1997. Strongly turbulent Rayleigh-Bénard convection in mercury: comparison with results at moderate Prandtl number. J. Fluid Mech. 335, 111–140.

Davaille, A., Limare, A., Touitou, F., Kumagai, I., Vatteville, J., 2011. Anatomy of a laminar starting thermal plume at high Prandtl number. Exp. Fluids 50, 285–300.

Engineering ToolBox, 2021. https://www.engineeringtoolbox.com/mercury-d_1002.html, (accessed October 13, 2021).

Freire, D.A., Gomes, F.B.R., Cintra, R., Novaes, W.G., 2015. Use of thermal power plants by new world vultures (Cathartidae) as an artifice to gain lift. Wilson J. Ornithol. 127 (1), 119–123.

Ghoshdastidar, P.S., 2012. Heat Transfer, 2nd ed. Oxford University Press, New Delhi.

Hedrick, T.L., Pichot, C., de Margerie, E., 2018. Gliding for a free lunch: biomechanics of foraging flight in common swifts (Apus apus). J. Exp. Biol. 221, 186270 jeb.

Kaminski, E., Jaupart, C., 2003. Laminar starting plumes in high-Prandtl-number fluids. J. Fluid Mech. 478, 287–298.

Mandel, J.T., Bildstein, K.L., 2007. Turkey vultures use anthropogenic thermals to extend their daily activity period. Wilson J. Ornithol. 119 (1), 102–105.

Morgan, V.T., 2007. Heat transfer by natural convection from a horizontal isothermal circular cylinder in air. Heat Transfer Eng. 18 (1), 25–33.

Pennycuick, C.J., 1973. The soaring flight of vultures. Sci. Am. 229 (6), 102–109.

Planet, 2021. https://www.planetware.com/world/best-hot-air-balloon-rides-in-the-world-us-ut-198. htm, (accessed October 13, 2021).

Richardson, P.L., 2011. How do albatrosses fly around the world without flapping their wings? Prog. Oceanogr. 88, 46–58.

Sachs, G., Traugott, J., Nesterova, A.P., Bonadonna, F., 2013. Experimental verification of dynamic soaring in albatrosses. J. Exp. Biol. 216, 4222–4232.

Shepard, E.L.C., Lambertucci, S.A., Vallmitjana, D., Wilson, R.P., 2011. Energy beyond food: foraging theory informs time spent in thermals by a large soaring bird. PLoS ONE 6 (11), e27375.

Wang, X., Xu, F., Zhai, H., 2018. An experimental study of a starting plume on a mountain. Int. Commun. Heat Mass Transfer 97, 1–8.

Chapter 16

Forced convection

"Heat is required to forge anything. Every great accomplishment is the story of a flaming heart."

Mary Lou Retton

Chapter Objectives

- Differentiate forced convection from natural convection.
- Appreciate the convection heat transfer coefficient.
- Be familiar with the forced convection of a uniform flow over a flat plate.
- Know the key parameters underlying forced convection.
- Recognize mixed convection that marries forced convection and natural convection.
- Identify the principal nondimensional parameters in forced convection.
- Comprehend the dependence of Nu on Re and Pr.
- Distinguish the constant-temperature from uniform-heat-flux boundary condition.
- Understand the relation between forced convection and wall shear.
- Fathom forced convection around a cylinder.
- Get acquainted with forced convection inside a conduit.

Nomenclature

A	area; A_c is cross-sectional area, A_s is surface area
a	dimension
b	dimension
C	coefficient; C_o, C_1, ... are coefficients with empirically-obtained values
C_f	friction or drag coefficient; $C_{f,x}$ is the local friction coefficient at distance x
c_P	specific heat capacity at constant pressure
D	diameter; D_h is the hydraulic diameter
f	friction factor
F_D	drag force
g	gravity
Gr	Grashof number, Gr ≡ buoyancy force/viscous force
Gz	Graetz number, Gz ≡ thermal capacity/convective heat transfer
h	(convection) heat transfer coefficient; h_{conv} is the convection heat transfer coefficient, h_x is the local convection heat transfer coefficient
k	thermal conductivity
L	length; L_{hydro} is the hydrodynamic entrance length, $L_{thermal}$ is the thermal entrance length

Nu	Nusselt number, Nu $\equiv$ convection/conduction heat transfer; Nu_x is the local Nusselt number, Nu_θ is the Nusselt number at angle θ
p_r	perimeter.
Pe	Péclet number, Pe $\equiv$ rate of advection/rate of diffusion
Pr	Prandtl number, Pr $\equiv$ momentum/thermal diffusivity
Q	heat; Q' is heat transfer rate, Q'_{conv} is the convection heat transfer rate
q'	heat flux; q'_{cond} is conduction heat flux, q'_{conv} is convection heat flux
Re	Reynolds number, Re $\equiv$ inertia/viscous force; Re_C is the critical Reynolds number, Re_x is the Reynolds number based on distance x
SATP	standard atmospheric temperature and pressure, 20°C and 1 atm
St	Stanton number, St $\equiv$ heat transfer with respect to heat capacity
T	temperature; T_b is the bulk mean temperature, $T_b = \frac{1}{2}(T_i + T_e)$, T_e is the exiting temperature, T_f is the film temperature, T_H is the higher temperature, T_i is the inlet temperature, T_L is the lower temperature, T_s is the surface temperature, $T_{s,x}$ is the surface temperature at x, T_∞ is the surrounding temperature, $T_{\infty,x}$ is the freestream temperature at x, ΔT is the temperature difference, ΔT_{ln} is the logarithmic mean temperature difference, T^* is a nondimensional temperature, $T^* \equiv (T_s - T)/(T_s - T_\infty)$
U	velocity; U_{air} is the velocity of air, U_{water} is the velocity of water, U_∞ is the freestream velocity
x	distance in the stream-wise direction; x_{cr} is the critical distance corresponding to the critical Reynolds number
y	distance in the direction normal to the wall; y^* is a nondimensional distance, $y^* \equiv y/L$
m	mass; m' is the mass flow rate
P	pressure

Greek and other symbols

α	thermal diffusivity, $\alpha = k/(\rho c_P)$
Δ	difference
β	thermal expansion coefficient, $\beta = -(1/\rho)(\partial\rho/\partial T)_P$
δ	thickness or small change; δ_T is the thermal boundary layer thickness, δ is the momentum boundary layer thickness
θ	angle, from the stagnation point
μ	dynamic or absolute viscosity; μ_b is the dynamic viscosity at the bulk mean temperature, μ_s is the dynamic viscosity at the surface temperature, μ_∞ is the dynamic viscosity at the freestream temperature
ν	kinematic viscosity, $\nu = \mu/\rho$
ρ	density
τ	shear stress; τ_w is shear stress at the wall
$\forall$	volume

16.1 What is the force behind forced convection?

Forced convection is the convection of thermal energy by a flow that is not driven by the buoyancy force associated with the temperature gradient or the density variation of the fluid of concern. In plain English, we can define forced convection in the following manner.

FIGURE 16.1 Fanning to cool, an everyday application of forced convection (created by S. Akhand). A working folding fan can forcefully convect away hot flashes.

Forced convection is the transfer of heat from a hot expanse to a cooler region due to flow motions induced by an external source.

Since the fluid involved is subjected to a temperature gradient and, therefore, a density gradient, free or natural convection is always present. For that reason, forced convection invariably entails natural convection. The transport of thermal energy via fluid motion is called forced convection when natural convection is inconsequential compared to forced convection.

An everyday example of forced convection is fanning ourselves to keep us cool in moderately hot surroundings, as shown in Fig. 16.1[1]. The term "moderately" is used to highlight forced convection without significant sweating; sweating, which involves evaporation (latent heat of vaporization), is a much more effective mode of heat transfer than single-phase convection. Unless when phase change is explicitly spelled out, single-phase convection is in context. When the ambient temperature is in the neighborhood of 20°C, and the relative humidity is in the ballpark of 50%, natural convection is just about right for keeping a person with a healthy body temperature of approximately 37°C cool and, hence, thermally comfortable; see Ting (2020) for more on engineering human thermal comfort. The additional cooling brought about via increasing fanning is enough when the ambient temperature rises to around 28°C, beyond that temperature, intelligent design calls upon sweating, where evaporative cooling involving phase change from liquid sweat to water vapor, to keep us from having heat stroke.

When the ambient temperature is greater than 36°C, it becomes too hot even for honeybees. To survive, they fan to force convective cooling of their colony (Muth, 2013; Peters et al., 2019); see Fig. 16.2. It would be pretty difficult to evaporate honey to augment the heat transfer rate beyond what forced convection

---

1. Forced convection brought about by an old-fashioned folding fan can be effective in managing unpredictable hot flashes (Oemi̇ev, 2020).

FIGURE 16.2 Honeybees fanning to force convective cooling of their colony (created by X. Wang). The collective effort of the many small but reverberating wings keep a bee colony cool.

FIGURE 16.3 Igloos, an efficacious means for lessening forced convection by the bone-chilling Arctic wind (created by S. Akhand). Snow is also an excellent insulator with a thermal conductivity of the order of 0.1 W/(m·K), an order of magnitude less conductive than concrete with a thermal conductivity of around 1 W/(m·K).

can effectuate. Instead, honeybees collect water to invoke evaporative cooling under forced convection, made possible by fanning their powerful wings at high frequencies. The point here is collaborative corporate fanning by a multitude of worker honeybees.

On the other side of the temperature scale, many creatures have to cope with bone-chilling winter temperatures. Among the ample opportunities nature grants us to grow stronger, frigid wind-chill is definitely one means for building one's character. The resilient Arctic people, the Inuit, engineer literally air-tight igloos, as depicted in Fig. 16.3, to shield themselves from forced convection by the merciless arctic wind. A channel-shaped doorway is often designed to further block convective heat loss from the relatively warm and cozy occupant space to the shivery outdoors. This doorway is also made of blocks of snow "fused" together, with the opening facing away from the prevailing bitter wind.

There may not be homo sapiens living around the south pole, but there are many other intelligent and intrepid creatures. In terms of the ability to fight

FIGURE 16.4 Emperor penguins huddling to shield the bone-chilling blast via a dynamically interchanging windbreak (created by X. Wang). Hazardous forced convection is kept at the outskirt of the huddle, where any penguin is exposed to but only momentarily in rotation.

against losing too much body heat to the chill of Antarctica, the emperor penguins stand up. Being the only vertebrate species that breeds during the harshest part of the Antarctic winter, they have been intelligently designed to thrive even in −50°C wind at up to 200 km/h (Gerum et al., 2013; Ancel et al., 2015). Fig. 16.4 shows that the adults shield their babies by forming a windbreak around them in a closely knitted congregation. While their bodies may not provide as much insulation as the snow, they are heat sources themselves, providing heat to keep their youngsters warm. Most intriguingly, they move around in a unique sequence within the fold, called huddling (Gilbert et al., 2008). The intelligently devised huddling movement ensures that no penguins remain on the outer circle for too long, equalizing each adult's exposure to the nasty wind with proper "warming" breaks inside the cozy huddle. Another huddling creature worth mentioning is the Inca Dove. They form a huddle pyramid to keep one another warm. It appears that the ones on top are exploiting the cozy natural convection plume produced by their comrades on the base to fight the cold.

16.2 The convection heat transfer coefficient

Fundamentally, the heat transfer rate increases with the temperature gradient,

$$Q' \propto \Delta T = T_s - T_\infty, \tag{16.1}$$

where ΔT is the temperature difference, T_s is the surface temperature of the object of concern, and T_∞ is the surrounding fluid temperature. We will assume by default that the surface is at a higher temperature than its surroundings unless we state the reverse, that is, the surface is cooler than its surroundings, explicitly. It is also intuitive that the larger the surface area, the larger the heat transfer rate, that is,

$$Q' \propto A_s. \tag{16.2}$$

Combining Eqs. (16.1) and (16.2), and introducing a proportionality constant, h_{conv}, we can write a general expression for the convection heat transfer rate,

$$Q'_{conv} = h_{conv}A_s(T_s - T_\infty). \tag{16.3}$$

The proportionality constant, h_{conv}, is called the *convection heat transfer coefficient*. It is clear that the larger the value of h_{conv}, the higher the heat transfer rate, for a given heat transfer area and a temperature difference between the surface and its surrounding fluid. Let us press ahead and see what dictates the magnitude of this convection heat transfer coefficient, h_{conv}.

Example 16.1 Heat fins for dinosaurs and elephants?

Given: Dinosaurs such as stegosauruses are equipped with a beautiful display of sails and plates. Note that surface-area-to-volume ratio decreases as a body becomes larger, whereas creature metabolism increases with body size.

Find: A possible purpose of the sails and/or plates within the context of forced convection; see, for example, Farlow et al. (1976) and Wright (1984).

Solution: The intimidating sails and plates of stegosauruses have been postulated to fence off the enemies of the relatively mild-mannered dinosaurs. It is equally likely that the extra surface area associated with these unique features enhances heat convection, according to Eq. (16.3). It is interesting to note that these beautiful creatures (and possibly also the good planet Earth that we once shared with them) may be a lot younger than we thought; see Anderson (2021), for example.

In addition to providing a large heat dissipating area, elephants also flap their ears to augment the convection heat transfer coefficient, h_{conv}, of Eq. (16.3) (Wright, 1984).

16.3 Forcing heat to convect from a flat surface

The most elementary forced convection is plausibly forced convection induced by a uniform flow over a horizontal flat plate with a homogeneous surface temperature that is higher than that of the free stream. We will discuss the reverse case, where the plate is at a temperature lower than that of the fluid, later. Recall that convection is preponderantly a flow phenomenon; therefore, a brief recap of the underlying fluid mechanics is warranted. The familiar uniform flow over a flat plate is illustrated in Fig. 16.5. The highlight is the boundary layer, which develops from the leading edge of the plate. For the originally smooth and uniform incoming flow, the primary parameter is the free-stream velocity. Put

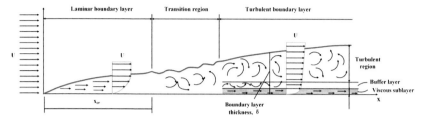

FIGURE 16.5 Uniform, parallel flow over a flat plate (created by X. Wang). Due to finite viscosity, the velocity is reduced to zero at the plate and, thus, also forced convection, leaving only conduction through the stagnant fluid at the solid surface.

simply, the larger the velocity, the larger the rate of heat convection. If the viscosity of the involved fluid is zero, the fluid is inviscid (frictionless) and the free-stream velocity is practically the only powering fluid mechanics parameter. For this idealized inviscid flow, thermal energy is convected downstream at the free-stream velocity, with thermal diffusion (conduction) spreading the heat upward, perpendicular to the plate. We note that conduction heat transfer perpendicular to the hot plate is dictated by the thermal conductivity of the fluid. Only under very special conditions, such as when a highly conductive fluid like mercury is involved, is conduction heat transfer considerable in forced convection. More typically, the convection heat transfer coefficient is predominantly a function of the velocity, U, that is,

$$h_{conv} = f(U). \tag{16.4}$$

In reality, all fluids have finite viscosity and, therefore, the boundary layer more-or-less dictates the convection heat transfer. This is because the boundary layer is the bottleneck for the thermal energy transfer path between the plate and the freestream. We will expound on this shortly. As discussed in an earlier chapter, the viscosity causes the fluid next to the solid plate to "stick" to the plate, creating the "no-slip" boundary condition where the velocity is zero; see Fig. 16.5. That being so, the convection heat transfer at the plate surface is purely conduction, through the thin stagnant fluid layer along the plate. Specifically, the convection heat flux at the plate,

$$q'_{conv,y=0} = q'_{cond} = -k\, \partial T/\partial y|_{y=0}, \tag{16.5}$$

where q'_{cond} is the conduction heat flux (heat transfer rate per unit area), k is the thermal conductivity of the fluid, T is temperature, and y is the direction perpendicular to and pointing upward with respect to the plate, with y=0 at the plate. This is depicted in Fig. 16.6. To keep the discussion simple, we take on the case where the temperature boundary layer[2] is in the order of the velocity

2. The velocity boundary layer is the region of fluid from the solid surface to the location where the local velocity reaches 99% free-stream velocity. Similarly, the temperature boundary layer

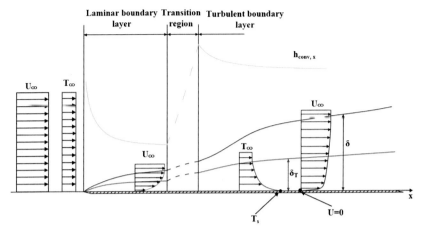

FIGURE 16.6 Velocity and temperature profiles for a uniform flow over a hot flat plate (created by X. Wang). In general, the thermal and momentum diffusivities are not equal and, thus, also their respective thicknesses.

boundary layer. This implies that the diffusivity of heat is equal to the diffusivity of the momentum of the fluid. For the warmer plate with respect to the colder fluid condition, as shown in Fig. 16.6, the temperature profile is "reversed" with respect to the velocity profile.

Recall that convection is the transfer of thermal energy between a solid surface and the adjacent moving fluid. To that end, convection involves the collective effects of conduction and fluid motion. Eq. (16.5) tells us that right at the plate, due to no-slip condition, there is no fluid motion and, thus, only conduction takes place. To put it another way, the local convection at the plate, y=0, is equal to conduction. As a consequence, the heat "convects" most slowly through the stagnant layer of fluid, forming the bottleneck for convection heat transfer from the hot plate to the flowing fluid. This bottleneck can be envisioned as the smallest section of the hourglass, as illustrated in Fig. 16.7.

As far as forced convection is concerned, we cannot alter the size of the bottleneck, which signifies the thermal conduction resistor at the solid surface, as illustrated in Fig. 16.7. This is because thermal conductivity and thermal diffusivity are properties of the fluid involved and, hence, their values are set by the thermodynamic conditions (temperature and pressure), and have nothing to do with convection per se. To speed up the flow of sand, or heat, the passage before the bottleneck should be widened and the stretch of the bottleneck shortened. When we speed up the flow, the velocity gradient and, thus, the temperature gradient next to the plate, $\partial T/\partial y|_{y=0}$, also increase; this augments the heat transfer next to the surface, $q_{conv,y=0}'$, see Eq. (16.5), and, hence, the

---

is the region bounded by $T_s - T = 0.99(T_s - T_\infty)$, where T_s is the surface temperature, T is the local temperature, and T_∞ is the free-stream temperature.

FIGURE 16.7   Falling sand in an hourglass represents heat convection from a hot flat plate (created by X. Wang). The rate of sand falling is predominantly dictated by the bottleneck, which signifies conduction-only heat transfer at the no-slip boundary.

convection heat transfer coefficient h_{conv}. Under the steady-state condition, no heat is accumulated and, thus, the amount of heat conducted from the surface to the fluid next to the surface must be convected away. From Eq. (16.5), we can write

$$h_x(T_s - T_\infty) = -k \, \partial T / \partial y|_{y=0}, \qquad (16.6)$$

where h_x is the local convection heat transfer coefficient, T_s is the surface temperature, and T_∞ is the temperature of the freestream fluid. We can rearrange the equation to give

$$h_x = -k \, \partial T / \partial y|_{y=0} / (T_s - T_\infty). \qquad (16.7)$$

With a known thermal conductivity, the local convection heat transfer coefficient, h_x, can be determined from the temperature gradient next to the surface.

Let us start from the leading edge of the flat plate by following Fig. 16.8. The boundary layer initiates from zero thickness at the leading edge; nevertheless, the finite viscosity imposes no-slip even at this point. Note that right at the leading edge there is virtually zero heat transfer area and, hence, the first measurable-heat-transfer point is at a small but finite distance from the leading edge. At this measurable-heat-transfer point, the boundary layer thickness is close to, but not exactly, zero. For these reasons, we do not have next-to-infinite heat transfer from the leading edge in real life. At the same time, because the boundary layer is so thin, the local convection heat transfer coefficient is very large at x=0. In terms of the analogous shape of the hourglass, this can be visualized as a cross-section corresponding to the magnitude of the free-stream velocity just before a very short bottleneck. Accordingly, the heat transfer rate is very high, as there is no slowdown except at the short bottleneck. The ultimate 'choking' heat transfer

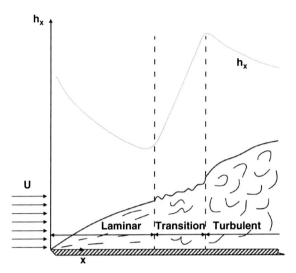

FIGURE 16.8   Stream-wise convection heat transfer coefficient variation for a uniform flow over a hot plate (created by X. Wang). The initial decrease is due to increasing laminar boundary layer thickness, transitioning into a turbulent boundary layer substantially augments the heat transfer coefficient, and the subsequent decrease is caused by the thickening turbulent boundary layer.

rate that prescribes the thermal energy traffic hinges on the thermal conductivity of the fluid (and the fluid viscosity and flow that dictate the thickness of this layer, along with the temperature difference).

As we move downstream, the boundary layer thickness increases, (and thereupon the velocity and temperature gradients decrease), and this can be represented by an elongating narrow converging passage to the bottleneck. Resultantly, the local convection heat transfer coefficient, h_x, decreases as the laminar boundary layer thickens with distance from the leading edge. When the boundary layer enters the transitional phase, the flow within the boundary layer becomes progressively more three-dimensional and both the velocity and temperature profiles become fuller with a larger gradient next to the no-slip film. The velocity (component) in the direction perpendicular to the plate brings pockets of warmer fluid away from the hot plate and cooler fluid from the free stream toward the hot plate. As a result of that, the local convection heat transfer coefficient, h_x, begins its rapidly increasing trend. It is worth noting that the boundary layer continues to grow; however, the enhancement in h_x caused by the increasing velocity in the perpendicular direction outshines what the thickening boundary layer tries to dampen. This may be conceived as small sticks moving up and down, promoting the falling sand through the bottleneck of the hourglass, so also analogously is the transport of thermal energy from the hot plate. Another important mechanism is also worth mentioning, the velocity in the direction perpendicular to the plate results in significant momentum transport in that

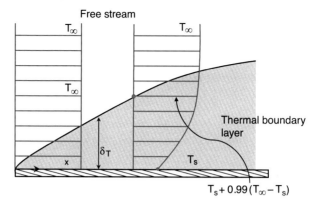

FIGURE 16.9 Uniform flow over a cold plate (created by O. Imafidon). The relatively inactive boundary layer is further inactivated by the "numbing" cold wall.

direction. In other words, the up and down movements of the small sticks also widen the converging section of the passage leading to the bottleneck. At the fluid mechanics' end, this is exhibited by the stream-wise velocity in the boundary layer that approaches the free-stream value incrementally closer to the plate, that is, the velocity profile becomes increasingly fuller. Likewise, the temperature profile becomes fuller with a progressively larger temperature gradient next to the plate, promoting the heat transfer.

The aforementioned two mechanisms, increasing up and down transport and larger near-plate stream-wise velocity, overpower the continuous slowdown due to boundary layer growth, leading to a substantial climb in the convective heat transfer coefficient. This continues until the boundary layer becomes fully turbulent, beyond which the slowing-down of h_x with growing boundary layer thickness dominates the control of the local heat transfer rate. But the question is, how can the maximum value of h_x at the point when transition to fully developed turbulent exceed[3] that at the leading edge where there is essentially no boundary layer? This is speculated to be due to a thinner viscous sublayer when the boundary layer is fully turbulent, compared to the boundary layer in the vicinity of the leading edge.

Convecting heat from a uniform flow onto a cooler flat plate. The reverse situation, where the incoming parallel flow is warmer than the flat plate, is worth our attention. By and large, everything is the same as its counterpart, cooler fluid flowing parallel to a hot plate. The obvious difference is the temperature profile, such as that shown in Fig. 16.9, which is "flipped." In spite of this flip,

---

3. While some standard textbooks show a larger h_{conv} at the start of the turbulent regime compared to that at the leading edge, Lienhard V (2020) does not. That being the case, this point demands further study; understandably the specifics depend on fluid viscosity, flow velocity, and other parameters.

the temperature difference between the plate and the freestream is the same, and everything else appears to be the same. As such, we would expect the same heat transfer characteristics for the same hot and cold temperatures compared to its hot-plate counterpart. There is, however, one subtle distinction in the boundary layer, especially the region very near the plate. This near-plate layer that imposes the main resistance to heat transfer is less perturbed for the cold-plate case. For the laminar boundary layer segment, the main effect is a slightly reduced thermal conductivity that corresponds to less energetic molecular motion at the lower temperature. As a consequence, the convection heat transfer coefficient is slightly lower than the hot-plate counterpart. The difference is more pronounced in the turbulent regime because not only the molecular motion is lessened by the lower temperature, so also are the turbulent motions involving fluid elements, each of which comprises a large number of molecules.

16.4   Primary parameters in forced convection

This is the opportune point to take a close look and figure out the key parameters involved in forced convection. Strictly speaking, natural convection has nothing to do with forced convection, as the (external) forcing is zero. When they do come together, we have *mixed convection*, where both the self-driven mechanism and that caused by an external force are important. This intermarrying between natural and forced convection takes place when the Grashof number is of the order of the square of the Reynolds number. Namely,

$$Gr/Re^2 \sim 1. \tag{16.8}$$

The Grashof number signifies the significance of the buoyancy force with respect to viscous force, that is,

$$Gr = g\beta(T_s - T_\infty)L^3/\nu^2, \tag{16.9}$$

where g is gravity, β is the thermal expansion coefficient, L is the characteristic length, and ν is the kinematic viscosity. As mixed convection is covered as a special case of forced convection, the parameters of importance in natural convection are thus also included in forced convection. Let us examine these parameters one factor at a time.

Thermal conductivity: We have been enlightened that convection naturally involves conduction. In point of fact, conduction is the decisive parameter at a solid surface, where the fluid halts as it sticks to the surface. As such, heat is transported in this infinitely thin stagnant film via conduction. That being the case, the thermal conductivity of the involved fluid, k, is an ineluctable parameter.

Specific heat capacity at constant pressure: Heat or thermal capacity denotes how much thermal energy or heat is required to raise the temperature of the substance, such as a fluid, of interest. This is most conveniently expressed in terms of the specific heat capacity at constant pressure, c_P. At standard atmospheric temperature and pressure (SATP), the value of the specific heat capacity of water, 4.2×10^3 J/kg/K, is more than four times that of air, 1×10^3 J/ kg/K. Because of that, for every degree increase in its temperature, a unit mass of water can take up four times the amount of thermal energy than that of a unit mass of air. Due to its high specific heat capacity, water is a good medium for thermal energy storage.

Density: For the thermal energy pick-up ability comparison between water and air, density, ρ, should also be considered. At SATP, there is approximately 1000 kg of fluid involved when passing 1 m^3 of water over a hot plate. By comparison, 1 m^3 of air only entails 1.2 kg of air. It follows that when the temperature of 1 m^3 of water is raised by 1 K, roughly 4.2×10^6 J of heat is removed from the hot plate, whereas the corresponding amount of heat is removed by 1 m^3 air undergoing a 1 K temperature increase is only 1×10^3 J. This indicates a three-order-of-magnitude difference in heat removal capacity per unit volume, similar to the difference in density.

Thermal diffusivity: Thermal diffusivity, $\alpha = k/(\rho c_P)$, signifies the proportion of thermal energy conducted through the material (fluid, when dealing with convection) with respect to the amount the material retains. A greedy welfare organization will withhold a large proportion of money it receives from donors for its own exploitation, for example, paying the executives lucrative salaries, leaving little to pass on to the needy ones down the street. So also will a fluid with a small heart, that is, thermal diffusivity. To be specific, a small thermal-diffusivity fluid keeps a large portion of heat within itself and conveys little forward. On the other hand, a fluid with a big and generous heart has a high thermal diffusivity, passing most of the heat onward, into the freestream for the hot plate case.

Dynamic and kinematic viscosities: On the other side of the coin, density, along with dynamic viscosity, μ, decides the Reynolds number, the pivotal nondimensional parameter in forced convection that denotes the inertia force of moving fluid elements with respect to the damping viscous force. Following from the previous paragraph, it is easier for us to invoke the kinematic viscosity, v, which is μ/ρ, covering both density and dynamic viscosity effects. For water at SATP, its kinematic viscosity, v, is 1.0×10^{-6} m^2/s, whereas it is 1.5×10^{-5} m^2/s for air. For a free-stream velocity U over a plate of length L, the Reynolds number based on the length of the plate, Re $= \rho UL/\mu = UL/v$. To compare apples to apples, we need to keep the Reynolds number the same for water versus air heat convection from a hot plate; specifically,

$$Re = U_{water}L/\nu_{water} = U_{air}L/\nu_{air}. \qquad (16.10)$$

Over a unit width of the hot plate with a unit height of fluid flowing over it, fixing the Reynolds number implies

$$U_{water}/1.0 \times 10^{-6} = U_{air}/1.5 \times 10^{-5}. \qquad (16.11)$$

This gives $U_{water} = 0.067 \, U_{air}$. Multiplying the specific heat capacity ratio from the previous paragraph with density, 4.2×10^6 J over 1×10^3 J by 0.067 gives 280. This indicates a water-air thermal energy difference of two orders of magnitude, instead of three.

Summary of key parameters in force convection: In summary, the convection heat transfer coefficient,

$$h_{conv} = f(k, c_p, \rho, \mu, U, L, \text{etc.}), \qquad (16.12)$$

where

k is the thermal conductivity,
c_p is the specific heat capacity at constant pressure,
ρ is the density,
μ is the dynamic viscosity,
U is the velocity,
and L is the length.

What about the "etc."? Recall that the local convection heat transfer coefficient, h_x, depends on the temperature gradient; therefore, the temperature difference, $\Delta T = T_H - T_L$, is part of the etcetera. Also, the fluid properties are a function of the temperature and, hence, the high and low temperatures, T_H and T_L, also come into play. The properties are evaluated at the film temperature, which is the average temperature, $(T_H + T_L)/2$. As pointed out at the beginning of this section, mixed convection may be considered as a special case of forced convection. With that in mind, the etcetera includes those parameters associated with natural convection. The gravitational force, g, is the foremost factor, and it is the only one, in addition to the thermal expansion coefficient, $\beta = -(1/\rho)(\partial\rho/\partial T)_P$, that has not been discussed in this section. To put it succinctly, extending forced convection to include mixed convection, then, the convection heat transfer coefficient,

$$h_{conv} = f(k, c_p, \rho, \mu, U, L, T_H, T_L, g, \beta, ...), \qquad (16.12a)$$

where "..." is inserted to leave room for other parameters which we have not considered and may come into play under certain exceptional conditions.

16.5 Nusselt number, Reynolds number, and Prandtl number

We learned from the dimensional analysis that normalizing the key dimensional players into nondimensional groups is invaluable. First of all, nondimensional parameters allow generalization, collapsing the otherwise "infinitely-large" possible conditions or datasets into simple relations of appropriate nondimensional parameters. For example, deducing the relationship between the normalized heat transfer coefficient, Nusselt number, as a function of Reynolds number and Prandtl number for a uniform flow over a heated flat plate can save an engineer the hassle of having to conduct an experiment and compile experimental data for every fluid of concern at every velocity of interest.

16.5.1 Nusselt–Reynolds encounter[4]

Before we explain what the famed nondimensional parameters are, let us be enlightened by a personal encounter between Wilhelm Nusselt and Osborne Reynolds. As introduced earlier, the Reynolds number denotes the relative importance of the inertial force with respect to the viscous force. It is named after British engineer and physicist, Professor Osborne Reynolds (1842–1912) of the University of Manchester. It is said that back then the young and aspiring Ernst Kraft Wilhelm Nusselt (1882–1957) went and visited Professor Reynolds in his laboratory, hoping to study under him. Apparently, Reynolds rejected Nusselt because Reynolds did not think the young man in front of him looked smart enough to take on the challenge. Dag Heward Mills rightly put it, "One of the greatest forms of direction for your life is rejection." Nussclt must have understood this philosophy or, most likely, he was aware of our giving-up weakness. "Our greatest weakness lies in giving up. The most certain way to succeed is always to try just one more time," asserted Thomas Edison. Nusselt subsequently studied under a relatively unknown professor, O. Knoblauch and, ultimately, earned a big name for himself in heat transfer. It is very likely that Reynolds' rejection contributed to Nusselt's success. As the saying goes, "The most powerful motivation is rejection."

16.5.2 Nusselt number

Other than the temperature gradient, the decisive parameter dictating the rate of heat transfer via convection per unit surface area of interest is the convection heat transfer coefficient. Namely, the convection heat flux,

$$Q'_{conv}/A = h_{conv}(T_H - T_L), \qquad (16.13)$$

---

4. This story was conveyed by Professor K. C. Cheng in a graduate heat transfer class, where the author was privileged to be among the small group of tutees. According to Professor Cheng, this incident was recorded by Professor Reynolds' laboratory assistant, who was present when this encounter took place.

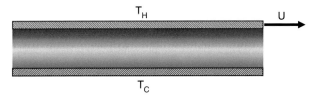

FIGURE 16.10 A moving hot plate over a narrow passage of fluid on top of a fixed cold plate (created by D. Ting). In the limiting condition that the hot top plate is not moving, that is, U = 0, heat is "convected" through the stagnant fluid by conduction, resulting in Nu = 1. Otherwise, the convection is "forced" by the moving top plate that drags the fluid along. As long as there is some fluid motion, the Nusselt number is greater than one, that is, heat convects faster than conduction alone.

where T_H is the higher temperature and T_L is the lower temperature. Let us consider heat convection from a hot plate to a cold plate via a narrow passage, where the fluid in the gap is dragged along by the moving hot plate, as illustrated in Fig. 16.10. A limiting condition occurs when the hot top plate is not moving and the fluid within the gap is stagnant. Under this condition, the heat transfer from the hot plate to the cold plate takes place via conduction through the standing fluid. Specifically, the heat transfer rate per unit area of plate,

$$Q'_{cond}/A = k(T_H - T_L)/L, \qquad (16.14)$$

where k is the thermal conductivity of the fluid in the gap and L is the width of the gap. The Nusselt number is the ratio of convective to conductive heat transfer. This ratio can be obtained by dividing Eq. (16.13) by (16.14), which gives the nondimensional Nusselt number,

$$Nu = h_{conv}L/k. \qquad (16.15)$$

Explicitly, the Nusselt number denotes the heat convection with respect to the heat conduction. Since we have Fig. 16.10 in front of us, we note that the faster the hot top plate moves, the higher the resulting convection heat transfer coefficient, h_{conv}. Instead of moving the hot plate, we can force hot fluid over the cold plate via a fan. The point is that the Nusselt number, Nu, increases with fluid velocity and, thus, Nu is a function of Re. We will expound on this shortly.

16.5.3 Prandtl number

One more nondimensional parameter to be highlighted in this section is the Prandtl number, which signifies momentum with respect to thermal diffusivity. Mathematically, the Prandtl number,

$$Pr = \nu/\alpha = (\mu/\rho)/[k/(c_P\rho)] = c_P\mu/k. \qquad (16.16)$$

The Prandtl number for air at SATP is about 0.7 and, hence, it is sometimes assumed to be one for convenience. An example is estimating the flow turbulence

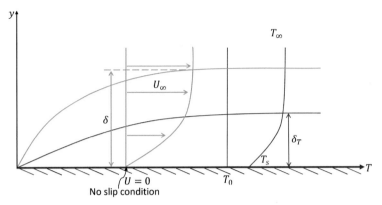

FIGURE 16.11   Prandtl number illustrated for the flow of a hot fluid over a cold plate (created by O. Imafidon). Note that a thinner boundary layer implies faster diffusion. Specifically, the fluid (momentum) over thermal boundary layer ratio varies approximately with the Prandtl number to the power of one-third, that is, $\delta/\delta_T \approx Pr^{1/3}$.

of a flow field with some temperature inhomogeneity via temperature measurements. Namely, a thermocouple with a tiny[5] junction can be used to assess the turbulence in an exhaust stack. The other common fluid is water. For water, the Prandtl number has a value of approximately 7 at SATP and it decreases drastically with increasing temperature.

Since we have been discussing flat plate heat convection, let us illustrate the role of the Prandtl number for this familiar case. Fig. 16.11 depicts the momentum (velocity) boundary layer, controlled by momentum diffusivity, with respect to the thermal boundary layer, dictated by thermal diffusivity, for a fluid with $Pr = 0.1$. Note that thermal energy diffuses seven times slower than momentum in water at SATP. That being the case, heat from the hot freestream diffuses slowly through a thick thermal boundary layer into the cold plate, when the involved fluid is water. It is known that if Pr is not too small,

$$\delta/\delta_T \approx Pr^{1/3}, \tag{16.17}$$

where δ is the momentum boundary layer thickness and δ_T is the thermal boundary layer thickness. On that account, for water with $Pr = 7$ at SATP, δ/δ_T is roughly 2, that is, the thermal boundary layer is about twice the velocity boundary layer thickness. This is because heat diffuses much slower, reaching the cold plate considerably later than the fluid (momentum). For air with Pr of 0.7 at SATP, heat diffuses moderately faster, resulting in a thinner thermal boundary layer than the momentum boundary layer. In short, the smaller the Prandtl number, the faster the heat diffuses with respect to momentum through the fluid. We will look at some typical Nu–Re–Pr relationships next.

---

5. The junction has to be tiny so that the inertia or thermal mass is small, that is, the response is fast enough to capture the fast turbulent motions.

16.6 Nu–Re–Pr relationships

From the material covered above, it is clear that we can express,

$$Nu = hL/k = C_o Re^{C_1} Pr^{C_2}, \qquad (16.18)$$

where the empirical coefficient and exponents, C_o, C_1, and C_2 are real and positive. The average Nusselt number, Nu, can be determined from the Reynolds number based on the entire length of the heating or cooling object, such as the flat plate of concern. In general, the coefficients C_o, C_1, and C_2 are largely independent of the nature of the fluid, that is, Pr. Because of that, the family of straight lines corresponding to different Prandtl numbers can be collapsed into a single straight line of $\log(Nu/Pr^{C_2})$ versus $\log$ Re. The values of these coefficients vary with respect to the particular conditions involved. As we have been considering uniform flow over a hot or cold plate, we will single this out. Even then, there are two ideal situations, that is, (1) constant surface temperature along the plate and, (2) uniform heat flux throughout the entire plate.

16.6.1 Constant temperature flat plate

For a uniform *laminar* flow over a smooth plate with a constant temperature, the local Nusselt number has been found to be

$$Nu_x = h_x x/k = 0.332 \, Re_x^{1/2} Pr^{1/3}, \qquad (16.19)$$

for $Re_x < Re_C \approx 5 \times 10^5$ and Pr > 0.6, where Pr > 0.6 covers common fluids such as air and water and Re_C is the critical Reynolds number[6] below which the flow is laminar. Here, x is the distance from the leading edge of the plate. The one-third power of Pr reflects the velocity-thermal boundary layer thickness. The one-half exponent for the Reynolds number bespeaks that Nu_x, which is $h_x x/k$, is increasing less than linearly with x, or Re_x, which is equal to $\rho U x/\mu$. This is because the local convection heat transfer coefficient, h_x, actually decreases with increasing x due to boundary layer growth; see Fig. 16.8. To put it another way, h_x decreases, x increases, and k remains unchanged, as we move downstream from the leading edge. Collectively, they lead to Nu_x varying with $Re_x^{1/2}$. We will expound further on this as we relate this with the wall shear in a later section. Interested readers are encouraged to look at various Nu_x versus Re_x plots based on different studies; see, for example, those compiled by Lienhard V (2020).

The *average* Nusselt number, Nu, for *laminar* flow can be obtained by integrating the local Nusselt number, Nu_x, over the span of the surface of interest. The average Nusselt number from the leading edge, x = 0, to x distance

---

6. The critical Reynolds number for a uniform flow over a smooth flat plate, $Re_C \approx 5 \times 10^5$; this critical value decreases with increasing flow disturbance and surface roughness.

downstream is

$$Nu = 0.664 \ Re^{1/2}Pr^{1/3}, \tag{16.20}$$

for $Re < 5 \times 10^5$ and $Pr > 0.6$.

Beyond the transitional boundary layer, for the *turbulent* portion of the flat plate, the local Nusselt number,

$$Nu_x = 0.0296 \ Re_x{}^{4/5}Pr^{1/3}, \tag{16.21}$$

for $5 \times 10^5 \leq Re_x < 10^7$ and $0.6 \leq Pr \leq 60$. The exponent of Re_x, C_1, has a value of 4/5, which is larger than the $\frac{1}{2}$ for the laminar portion. This implies that Nu_x increases more rapidly with x when the boundary layer is turbulent. It is clear that turbulence enhances heat convection compared to its laminar counterpart, even though the turbulent boundary layer is thicker than the laminar one upstream.

Similar to the laminar case, the *average* Nusselt number for *turbulent* flow can be obtained by integrating the local Nusselt number, Nu_x, over the turbulent flow region. The average Nusselt number from the transitional boundary layer to x distance downstream is

$$Nu = 0.037 \ Re^{4/5}Pr^{1/3}, \tag{16.22}$$

for $5 \times 10^5 \leq Re < 10^7$ and $0.6 \leq Pr \leq 60$. In practice, however, the *average* Nusselt number of the *entire plate* is usually sought. For this case, the average Nusselt number,

$$Nu = \left(0.037 \ Re^{4/5} - 871\right)Pr^{1/3}, \tag{16.23}$$

for $5 \times 10^5 \leq Re < 10^7$ and $0.6 \leq Pr \leq 60$.

Example 16.2 Solar photovoltaic panel cooling by forced convection.
Given: A typical 1 m by 1.7 m solar photovoltaic panel is subjected to atmospheric wind at 6 m/s. The panel is at 45°C while the wind is at 25°C.
Find: The heat removal rate by forced convection imposed by the wind, if the wind blows across the (a) 1 m width and (b) 1.7 m length of the panel.

Solution: For wind blowing at 6 m/s in the width-wise direction, the corresponding Reynolds number,

$$Re = UL/\nu = 6(1)/1.5 \times 10^{-5} = 4.00 \times 10^5,$$

where we have utilized $\nu = 1.5 \times 10^{-5}$ m²/s for air at SATP. Since this Reynolds number is less than the critical Reynolds number, the flow can be treated as laminar. It is important to note that in real-life obstacles tend to disturb the fluctuating wind such that the flow becomes turbulent below a critical Reynolds number of 5×10^5. If we further assume that the surface temperature of the solar panel is approximately constant, we can invoke the laminar flow equation for a constant-temperature surface, that is, the average Nusselt number,

$$Nu = 0.664 \ Re^{1/2}Pr^{1/3} = 0.664\left(4.00 \times 10^5\right)^{1/2}(0.7)^{1/3} = 373.$$

By definition,

$$Nu = hL/k = h(1)/0.026 = 373,$$

which gives the average convective heat transfer coefficient,

$$h = 9.69 W/(m^2 \cdot K).$$

For the 1 m by 1.7 m solar panel, the total convection heat transfer rate,

$$Q'_{conv} = h\, A(T_s - T_\infty) = 9.69(1)(1.7)(45 - 25) = 330\ W.$$

If the 6-m/s wind blows along the length of the panel instead,

$$Re = UL/\nu = 6(1.7)/1.5 \times 10^{-5} = 6.80 \times 10^5.$$

This signifies that the flow is turbulent at the latter portion of the plate and, thus, we employ Eq. (16.23) to calculate the average Nusselt number over the entire plate,

$$\begin{aligned}
Nu &= (0.037\, Re^{4/5} - 871)Pr^{1/3} \\
&= \left[0.037(6.80 \times 10^5)^{4/5} - 871\right](0.7)^{1/3} = 749.
\end{aligned}$$

Thus, the average convective heat transfer coefficient is

$$h = k\, Nu/L = 11.46\ W/(m^2 \cdot K).$$

For the entire 1 m by 1.7 m panel, the heat loss rate via convection,

$$Q'_{conv} = h\, A(T_s - T_\infty) = 11.46(1)(1.7)(45 - 25) = 390\ W.$$

We see a significant increase in heat convection from laminar to combined laminar and turbulent flow simply by a change in the wind direction with respect to the panel.

It should be mentioned that enhancing forced convection heat transfer from a photovoltaic panel is beneficial in terms of reducing thermal stresses and, more so, for improving the energy conversion efficiency (Yang et al., 2020). For this reason, many innovative turbulence generators have been devised to further forced convection (Yang et al., 2021).

16.6.2 Uniform heat flux flat plate

The constant-temperature condition can be applied in practice to determine forced convection from flat surfaces with nearly homogeneous temperatures. The calculation improves as the surface temperature deviation approaches zero, which is the case when there is a phase change such as condensation on the bottom side of the plate or the surface involved is highly conductive. Another standard condition is a uniform heat flux surface. This can be realized when a uniform heat source is placed underneath the surface. An everyday example is uniform solar radiation hitting and pushing an even amount of heat flux through a surface. This results in an increase in the heat transfer rate, especially through the more thermal-resistive laminar portion of surface. For the *laminar* part of the surface, the local Nusselt number,

$$Nu_x = 0.453 Re_x{}^{1/2} Pr^{1/3}, \tag{16.24}$$

for $Re_x < 5 \times 10^5$ and $Pr > 0.6$. We see that both the coefficient, C_o, and the exponent, C_1, are larger than those of the isothermal (constant temperature) counterpart. Substituting for some typical Re_x values will show that Nu_x for the uniform-heat-flux case is about 36% higher than that of the isothermal plate. The average Nusselt number over the entire plate of length L can be deduced to be

$$Nu = 0.680 \, Re^{1/2} Pr^{1/3}. \tag{16.25}$$

For the *turbulent* portion of the uniform-heat-flux flat plate, the local Nusselt number,

$$Nu_x = 0.0308 \, Re_x{}^{4/5} Pr^{1/3}, \tag{16.26}$$

for $5 \times 10^5 \le Re_x < 10^7$ and $0.6 \le Pr \le 60$. As the turbulent boundary layer is relatively effective in convecting thermal energy away from the hot surface, applying a uniform heat flux through the plane only result in a 4% increase in Nu_x compared to the isothermal counterpart. If we average this over the whole length, we get

$$Nu = 0.0363 \, Re^{4/5} Pr^{1/3}. \tag{16.27}$$

Example 16.3 Solar thermal panel subjected to atmospheric wind forced convection.

Given: A commercial 1.22 m by 3.05 m solar thermal panel is placed on a thick roof for the purpose of heating water for hot-water usage. On a typical summer day, the panel surface is at 85°C while the 25°C atmospheric air is moving at 6 m/s.

Find: The convection heat transfer rate brought about by the wind blowing in the (a) widthwise and (b) lengthwise direction over the panel.

Solution: For wind blowing at 6 m/s in the width-wise direction, the corresponding Reynolds number,

$$Re = UL/\nu = 6(1.22)/1.5 \times 10^{-5} = 4.88 \times 10^5.$$

This suggests that the flow is largely laminar. The solar thermal panel receives heat predominantly from direct solar radiation and, thus, can be regarded as having a uniform heat flux. With that in mind, we call upon Eq. (16.25) to determine the average Nusselt number over the entire panel,

$$Nu = 0.680 \, Re^{1/2} Pr^{1/3} = 0.680(4.88 \times 10^5)^{1/2} (0.7)^{1/3} = 422.$$

The average convection heat transfer coefficient can thus be deduced,

$$h = k\,Nu/L = 0.026(422)/(1.22) = 8.99\,W/(m^2 \cdot K),$$

which is larger than the value for the constant surface temperature condition. For the 1.22 m by 3.05 m solar thermal panel, the total heat convection rate,

$$Q'_{conv} = h\,A\,(T_s - T_\infty) = 8.99\,(1.22)(3.05)\,(85 - 25) = 2007\,W.$$

For wind blowing at 6 m/s along the 3.05 m long panel, the Reynolds number is

$$Re = UL/\nu = 6(3.05)/1.5 \times 10^{-5} = 1.22 \times 10^6.$$

This indicates turbulent flow on the downstream portion of the panel and, thus, we utilize Eq. (16.27) to deduce the average Nusselt number over the entire panel,

$$Nu = 0.0363\,Re^{4/5}Pr^{1/3} = 0.0363\big(1.22 \times 10^6\big)^{4/5}\,(0.7)^{1/3} = 2384.$$

For this notably larger Nu, the average convective heat transfer coefficient,

$$h = k\,Nu/L = 0.026(2384)/3.05 = 20.33\,W/(m^2 \cdot K),$$

which is more than two times larger than its laminar counterpart. For the entire 1.22 m by 3.05 m panel, the total convective heat loss rate,

$$Q'_{conv} = h\,A\,(T_s - T_\infty) = 20.33\,(1.22)(3.05)\,(85 - 25) = 4538\,W.$$

In short, we see a more than doubling of convection heat transfer from laminar to largely turbulent flow, when the wind changes its direction from widthwise to lengthwise.

Needless to say, there is also radiation heat transfer between a typical solar panel and the sky, in addition to the convective heat exchange considered above. Furthermore, when the system is running, the largest portion of the solar energy gained is transported out of the panel via the working fluid for utility, for example, water heating. In any case, this example demonstrates the deduction of the heat loss due to wind-induced forced convection to atmospheric air. Unlike the photovoltaic case, a reduction in this forced convective loss is desirable, granting more thermal energy to be harnessed for the intended application, for example, supplying hot water.

16.7   Relating heat convection with flow shear at the wall

As discussed earlier, the local heat transfer coefficient, h_x, at the wall is proportional to the temperature gradient and the temperature gradient, in turn, is proportional to the velocity gradient, that is,

$$h_x \propto \partial T/\partial y\big|_{y=0} \propto \partial U/\partial y\big|_{y=0}. \tag{16.28}$$

The shear stress imposed by the velocity gradient for a Newtonian fluid is

$$\tau = \mu \, \partial U/\partial y. \tag{16.29}$$

Along the surface, the wall shear stress,

$$\tau_w = \mu(\partial U/\partial y)_{y=0}. \tag{16.30}$$

We may express this shear stress at the surface in the form,

$$\tau_w = C_f(\rho U_\infty^2)/2, \tag{16.31}$$

where the friction or drag coefficient, C_f, is determined experimentally. So, why do we bother with the wall shear stress? Because it provides the velocity gradient at the wall and, therefore, the temperature gradient and, thereby, the forced convective heat transfer rate. Imagine the heat from a hot wall passing through a series of thermal resistors starting with the thin layer of stagnant fluid at the wall, a resistor with the highest thermal resistance per unit length.

In short, the local Nusselt number can be expressed in terms of the dimensionless temperature gradient at the surface, that is,

$$Nu_x = h_x x/k = \partial T * /\partial y * |_{y*=0}, \tag{16.32}$$

where $T^* \equiv (T_s - T)/(T_s - T_\infty)$ and $y^* \equiv y/L$. At any location, x distance from the leading edge of a flat plate, the local Nusselt number, Nu_x, is a function of the local temperature gradient at the plate. This local temperature gradient at the plate can be inferred from the velocity gradient at $y = 0$ and, thus, the wall shear stress, τ_w. As it is more convenient to work with parameters in their normalized form, we call upon the friction coefficient, C_f.

For *laminar* flow, the local friction coefficient is

$$C_{f,x} = 0.664/Re_x^{1/2}. \tag{16.33}$$

From Eq. (16.28), the local convection heat transfer coefficient, h_x, is proportional to the velocity gradient at the wall, $\propto \partial U/\partial y|_{y=0}$, and, from Eq. (16.30), the wall shear, τ_w, is equal to dynamic viscosity times the velocity gradient at the wall, $\mu(\partial U/\partial y)_{y=0}$, therefore,

$$h_x \propto (\partial U/\partial y)_{y=0} = \tau_w/\mu. \tag{16.34}$$

Substituting for $\tau_w = \frac{1}{2}C_f(\rho U_\infty^2)$ from Eq. (16.31), we can write

$$h_x \propto \frac{1}{2}(0.664/Re_x^{1/2})(\rho U_\infty^2)/\mu = (0.332/Re_x^{1/2})(\rho U_\infty^2)/\mu. \tag{16.35}$$

This can be manipulated into

$$h_x \propto (0.332/Re_x^{1/2})[(\rho^2 U_\infty^2)/\mu^2](\mu/\rho) = 0.332\,Re_x^{1/2}(\mu/\rho). \tag{16.35a}$$

Succinctly,

$$h_x \propto 0.332\,Re_x^{1/2}. \tag{16.35b}$$

The manner in which Nu_x depends on Re_x is consistent with what is given in Eq. (16.19). The average friction coefficient over the entire plate is

$$C_f = 1.328/Re^{1/2}. \tag{16.36}$$

The corresponding average Nusselt number is described by Eq. (16.20).

For *turbulent* flow, the local friction coefficient is

$$C_{f,x} = 0.0592/Re_x^{1/5}, \tag{16.37}$$

for $5 \times 10^5 \leq Re_x \leq 10^7$. The local convection heat transfer coefficient,

$$h_x \propto (\partial U/\partial y)_{y=0} = \tau_w/\mu = \tfrac{1}{2}C_f(\rho U_\infty^2)/\mu = (0.0296/Re_x^{1/5})(\rho U_\infty^2)/\mu. \tag{16.38}$$

This can be simplified into

$$h_x \propto (0.0296/Re_x^{1/5})[(\rho^2 U_\infty^2)/\mu^2](\mu/\rho) = 0.0296 \, Re_x^{4/5}(\mu/\rho), \tag{16.38a}$$

that is,

$$h_x \propto 0.0296 \, Re_x^{4/5}. \tag{16.38b}$$

What remains is to account for the difference in the thermal boundary layer thickness, which is defined by the thermal diffusivity, with respect to the velocity boundary layer thickness, which is prescribed by the momentum diffusivity. Doing so we get Eq. (16.21), which characterizes the local Nusselt number for turbulent flow over a constant-surface-temperature flat plate. The corresponding average friction coefficient over the turbulent portion of the plate is

$$C_f = 0.074/Re^{1/5}, \tag{16.39}$$

and Eq. (16.22) describes the average Nusselt number that is related to $Re^{4/5}$. The higher exponent of the Reynolds number associated with Nu than that associated with C_f can be attributed to the added influence of turbulent fluid inertia that is analogously brought about by an energetic poking stick in promoting sand falling through the hourglass.

For *combined laminar and turbulent* flow (that is, laminar to transition to turbulent), the average friction coefficient over a flat plate of length L is

$$C_f = 0.074/Re^{1/5} - 1742/Re, \tag{16.40}$$

for $5 \times 10^5 \leq Re \leq 10^7$. The last term accounts for the lower friction coefficient value associated with the laminar portion of the flow. The corresponding average Nusselt number over the constant-surface-temperature flat plate of length L is described by Eq. (16.23). Table 16.1 summarizes the expressions for Nu and C_f for forced convection over a horizontal flat plate, for both the constant-surface-temperature and the uniform-heat-flux boundary conditions.

The procedure for solving a forced convection problem over a flat plate is summarized below:

1) Check to see if Re is smaller or larger than $Re_c(\approx 5 \times 10^5)$.

TABLE 16.1 Nusselt number and friction coefficient for forced convection over a horizontal flat plate.

Flow regime and C_f	Thermal boundary	Nu expression
Laminar, $Re_x < 5 \times 10^5$ & $Pr > 0.6$, $C_{f,x} = 0.664/Re_x^{1/2}$, $C_f = 1.328/Re^{1/2}$	Constant surface temperature	$Nu_x = 0.332\ Re_x^{1/2}\ Pr^{1/3}$ $Nu = 0.664\ Re^{1/2}\ Pr^{1/3}$
	Uniform heat flux	$Nu_x = 0.453\ Re_x^{1/2}\ Pr^{1/3}$ $Nu = 0.680\ Re^{1/2}\ Pr^{1/3}$
Combined laminar & turbulent, $5 \times 10^5 \le Re < 10^7$ & $0.6 \le Pr \le 60$, $C_f = 0.074/Re^{1/5} - 1742/Re$	Constant surface temperature	From leading edge, $Nu = (0.037\ Re^{4/5} - 871)\ Pr^{1/3}$
Turbulent, $5 \times 10^5 \le Re < 10^7$ & $0.6 \le Pr \le 60$, $C_f = 0.074/Re^{1/5}$	Constant surface temperature	$Nu_x = 0.0296\ Re_x^{4/5}\ Pr^{1/3}$ From critical x onward, $Nu = 0.037\ Re^{4/5}\ Pr^{1/3}$
	Uniform heat flux	$Nu_x = 0.0308\ Re_x^{4/5}\ Pr^{1/3}$ $Nu = 0.0363\ Re^{4/5}\ Pr^{1/3}$

2) If Re is less than Re_c, then solve the problem as a laminar one.

If Re is larger than Re_c, then check if the length L of the plate is much larger than the distance from the leading edge at which flow transition occurs.
If L is much larger than the transitional distance, then solve the problem as a turbulent problem. Otherwise, solve it as a combined laminar and turbulent problem.

1) Solve for the average convection heat transfer coefficient, and the friction coefficient, C_f, if it is also of interest.
2) Solve for the total heat transfer rate, $Q' = h\,A\,(T_s - T_\infty)$, and the total drag force, $F_D = \frac{1}{2}C_f A \rho U_\infty^2$, where A is the area of the plate and U_∞ is the freestream velocity.

16.8   Forced convection around a circular cylinder

Cylindrical tubes are routinely present in heat exchangers and, thus, forced convection around a circular cylinder is worth highlighting. Fig. 16.12 illustrates the local Nusselt number, Nu_θ, around a circular cylinder. We see that the local Nusselt number, Nu_θ, starts up high at the stagnation point ($\theta = 0°$) and decreases with increasing θ as the boundary layer thickens. After the separation point, where a Nu_θ minimum occurs, Nu_θ increases with θ, due to intense mixing. The higher Reynolds number curves ($Re = 1.9 \times 10^5$, 2.2×10^5) have two minima. The sharp increase in Nu_θ at $\theta \approx 90°$ is due to transition from laminar to turbulent flow. Then, Nu_θ decreases again due to the thickening of the turbulent boundary layer. It reaches its second minimum at $\theta \approx 140°$, which is the flow separation point of the turbulent boundary layer, and it increases with θ, due to forceful

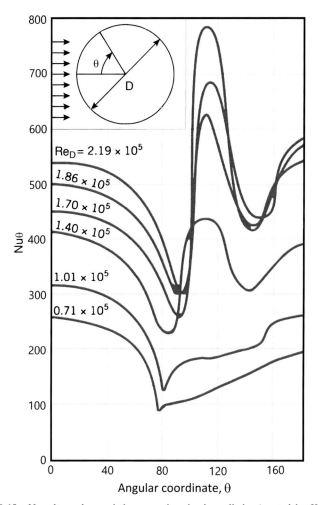

FIGURE 16.12 Nusselt number variation around a circular cylinder (created by K. Esmaeli-foomani).

mixing in the turbulent wake region. The average Nusselt number for crossflow over a circular cylinder can be expressed as

$$Nu = \frac{hD}{k} = 0.3 + \frac{0.62Re^{1/2}Pr^{1/3}}{1 + (0.4/Pr^{2/3})^{1/4}}\left[1 + \left(\frac{Re}{28200}\right)^{5/8}\right]^{4/5}, \qquad (16.41)$$

for Re Pr > 0.2, where D is the diameter of the cylinder.

Since we are discussing common bluff bodies, let us include other typical shapes. The average Nusselt number for cross flow over a sphere can be described by

$$\text{Nu} = \frac{hD}{k} = 2 + \left[0.4\text{Re}^{1/2} + 0.06\text{Re}^{2/3}\right]\text{Pr}^{0.4}\left(\frac{\mu_\infty}{\mu_s}\right)^{1/4}, \qquad (16.42)$$

for $3.5 \leq \text{Re} \leq 80,000$ and $0.7 \leq \text{Pr} \leq 380$. Subscripts "$\infty$" and "$s$" represent freestream fluid and surface, respectively, at which temperature the corresponding fluid viscosity is to be based. The average Nusselt number for flow across a single cylinder of different geometries can be described by the general expression, $\text{Nu} = C_o\text{Re}^{C_1}\text{Pr}^{C_2}$. Typical values of the empirical coefficients C_o, C_1, and C_2 are tabulated in Table 16.2, where all fluid properties are evaluated at the film temperature $T_f = (T_\infty + T_s)/2$.

16.9 Other nondimensional parameters of forced convection

The principal dependent nondimensional parameter for convection heat transfer is the Nusselt number, Nu. For forced convection, the Nusselt number expressed as a function of the Reynolds number and the Prandtl number is the most potent and commonly applied formula; specifically, $\text{Nu} = C_o\text{Re}^{C_1}\text{Pr}^{C_2}$. Accordingly, Re and Pr are the key independent nondimensional variables. Nevertheless, there are alternate ways to express Nu in terms of the influencing factors as functions of other nondimensional parameters. One way or another, these less common nondimensional parameters are a combination of Re, Pr, and even Nu.

Péclet number: The Péclet number denotes the significance of the advective transport rate with respect to the diffusive transport rate. In convection heat transfer, the Péclet number represents heat convection (characterized by an appropriate velocity, U, times a pertinent distance, L) relative to heat diffusion (characterized by thermal diffusivity, α), that is,

$$\text{Pe} = LU/\alpha = LU/(\rho c_P/k). \qquad (16.43)$$

The underlying physics of Pe is somewhat analogous to Nu, which is the ratio of heat convection to heat diffusion. Explicitly, however, Pe restricts heat convection to that brought about by velocity, U, over a distance, L; because of this, it excludes the roles played by the boundary layer, flow turbulence, etc. As such, Pe represents some of the key independent factors involved, making possible convection heat transfer to take place, it does not really portray the outcome. To put it another way, Pe describes the conditions of a forced convection and Nu represents the results. With $\text{Re} = \rho UL/\mu$ and $\text{Pr} = \nu/\alpha = (\mu/\rho)/[k/(c_P\rho)] = c_P\mu/k$, we have

$$\text{Re Pr} = (\rho UL/\mu)(c_P\mu/k) = UL/(\rho c_P/k) = \text{Pe}. \qquad (16.44)$$

If we choose Pe as the independent nondimensional variable, we can write

$$\text{Nu} = C_3\text{Pe}^{C_4} = C_3\text{Re}^{C_4}\,\text{Pr}^{C_4}, \qquad (16.45)$$

TABLE 16.2 Nusselt number expressions for forced convection across cylinders of different shapes (created by X. Wang).

Cross-section of the cylinder	Fluid	Range of Re	Nusselt number
Circle	Gas or liquid	0.4–4 4–40 40–4000 4000–40,000 40000–400,000	$Nu = 0.989\ Re^{0.330}\ Pr^{1/3}$ $Nu = 0.911\ Re^{0.385}\ Pr^{1/3}$ $Nu = 0.683\ Re^{0.466}\ Pr^{1/3}$ $Nu = 0.193\ Re^{0.618}\ Pr^{1/3}$ $Nu = 0.027\ Re^{0.805}\ Pr^{1/3}$
Square	Gas	3900–79,000	$Nu = 0.094\ Re^{0.675}\ Pr^{1/3}$
Square (titled 45°)	Gas	5600–111,000	$Nu = 0.258\ Re^{0.588}\ Pr^{1/3}$
Hexagon	Gas	4500–90,700	$Nu = 0.148\ Re^{0.638}\ Pr^{1/3}$
Hexagon (titled 45°)	Gas	5200–20,400 20400–105,000	$Nu = 0.162\ Re^{0.638}\ Pr^{1/3}$ $Nu = 0.039\ Re^{0.782}\ Pr^{1/3}$
Vertical plate	Gas	6300–23,600	$Nu = 0.257\ Re^{0.731}\ Pr^{1/3}$
Ellipse	Gas	1400–8200	$Nu = 0.197\ Re^{0.612}\ Pr^{1/3}$

$Nu = hL/k = C_o\ Re^{C_1}\ Pr^{C_2}$.

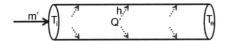

FIGURE 16.13 Forced convection from a hot fluid flowing in a pipe (created by D. Ting). The fluid convects heat to the inner wall of the pipe along the passage and, thus, its temperature at the pipe exit, T_e, is less than that at the inlet, T_i.

which restricts the effect of Re to be the same as Pr. We understand that there is no reason for this to be the case except under some very special conditions. It is probably because of this that the more versatile form, $Nu = C_o Re^{C_1} Pr^{C_2}$, is customarily adopted in practice.

Stanton number: The Stanton number, credited to British engineer and scientist Thomas Stanton [1865-1931], is the ratio of heat transferred into a fluid over the thermal capacity of the fluid. It is similar to the thermal diffusivity that denotes the amount of thermal energy diffusing through the fluid with respect to the portion absorbed by the fluid, with a noted difference that thermal diffusion is realized by convection characterized by the convection heat transfer coefficient, h. Specifically,

$$St = h/(\rho U c_P). \tag{16.46}$$

We see that the Stanton number is actually Nu divided by Re and Pr, that is,

$$St = Nu/(Re\ Pr) = Nu/Pe. \tag{16.47}$$

Note that Nu appears because St is deduced via geometric similarity of the momentum (velocity) boundary layer and the thermal boundary layer; namely, the wall shear decides the heat transfer at the wall and, thus, the heat convection into the moving stream. Eq. (16.47) shows that St has the same drawback as Pe in the sense that the exponents for both Re and Pr are implicitly the identical. It is more restrictive than Pe because St is derived based on the assumption that the wall shear dictates the convection of heat.

16.10 Internal forced convection

For a steady flow of hot fluid in a pipe, conservation of energy states that the thermal energy lost by the fluid is equal to that gained by the pipe. This is illustrated in Fig. 16.13, where the heat transfer rate from the fluid to the pipe,

$$Q' = m'\ c_P(T_e - T_i). \tag{16.48}$$

Here, m' is the mass flow rate, T_e is the temperature of the exiting fluid, and T_i is the temperature of the fluid at the inlet. For a hot fluid moving in a cooler pipe, T_e is less than T_i and, thus, Q' is negative, signifying the fluid is losing thermal energy. A positive heat gain by the fluid occurs when the pipe is hotter than the fluid. For the typical moving fluid condition, the heat transfer between the fluid and the pipe is dominated by forced convection. Therefore, we can write

$$Q' = h \ A \ \Delta T, \qquad (16.49)$$

where h is the average convective heat transfer coefficient for the section of pipe of interest, A is the internal surface area of the pipe over which heat convection takes place, and ΔT is the average temperature difference between the fluid and the inner surface of the pipe.

16.10.1 Pipe flow regimes

Before we move further, let us revisit flow in a pipe, specifically, the flow regime and the entrance length. Flow over a flat plate generally starts out as laminar because the Reynolds number, whose characteristic length is the distance from the leading edge, is initially small. With increasing distance, the Reynolds number increases and, thus, the flow, in due course, undergoes transition into the turbulent domain. For flow in a pipe, the characteristic length is the pipe diameter and, hence, the Reynolds number is a constant for a constant-diameter pipe. Because of this, the flow is either laminar or turbulent over the entire length of the pipe. There is, nevertheless, the entrance length, the beginning section of the pipe where the boundary layer develops.

16.10.2 Hydrodynamic and thermal entrance lengths

Recall that the hydrodynamic entrance length is the region from the tube inlet to the point at which the boundary layer merges at the centerline. It follows that the hydrodynamically developed region is the region beyond the hydrodynamic entry region in which the velocity profile is fully developed and remains unchanged from that point onward. Similarly, the thermal entrance length is the region of flow over which the thermal boundary layer develops and reaches the tube center. The thermal entrance length is equal to the hydrodynamic entrance length, when the Prandtl number is one. In convection heat transfer, a fully developed flow is the region in which the flow is both hydrodynamically and thermally developed.

For laminar flow, the hydrodynamic entrance length,

$$L_{hydro} \approx 0.05 \ Re \ D. \qquad (16.50)$$

The corresponding thermal entrance length is

$$L_{thermal} \approx 0.05 \ Re \ Pr \ D. \qquad (16.51)$$

For turbulent pipe flow,

$$L_{hydro} \approx L_{thermal} \approx 10D. \qquad (16.52)$$

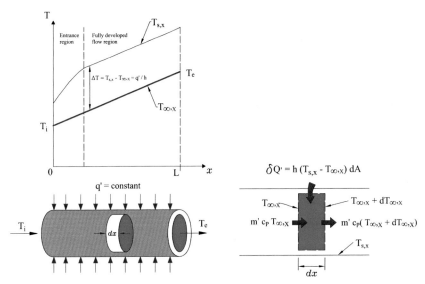

FIGURE 16.14   Forced convection inside a uniform-heat-flux pipe (created by F. Kermanshahi). Note that beyond the entrance length, the fluid temperature increases linearly in the flow direction.

We see that the Prandtl number does not come into play when the flow is turbulent. This is because, in turbulent flow, heat convection is dominated by mixing of lumps of fluid encompassing a great number of molecules, rather than the movement of the molecules themselves, as in diffusion. It is also worth mentioning that the friction and the heat transfer coefficients in the fully developed flow region remain constant.

16.10.3   Uniform-heat-flux pipe

For the situation that the pipe is wrapped and furnished with a good heating blanket, we can have an ideal condition called uniform heat flux. In this case, the rate of heat entering the fluid can be expressed as

$$Q' = q' \, A = m' c_P (T_e - T_i), \tag{16.53}$$

where q' is the heat flux. We can express this equation in terms of the convection heat transfer coefficient, h, and the temperature difference between the fluid and the pipe wall, ΔT, as

$$Q' = q' \, A = \int h_x \left(T_{s,x} - T_{\infty,x} \right) dA = h \, A \, \Delta T. \tag{16.54}$$

As depicted in Fig. 16.14, h_x is the local heat convection coefficient, $T_{s,x}$ is the corresponding pipe surface temperature, and $T_{\infty,x}$ is the local fluid temperature. For most practical purposes, c_P is a constant. With no changes in the fluid mechanics after the entrance length, the local heat transfer coefficient, h_x, decreases to an asymptotic value when the flow becomes fully developed. The initial decrease in h_x in the entrance region is caused by the thickening of the boundary layer. Consider the differential fluid element shown in Fig. 16.14, where

$$\delta Q' = h_x \left(T_{s,x} - T_{\infty,x} \right) dA, \qquad (16.55)$$

we see that a constant h_A, along with a constant heat flux, leads to a constant local temperature difference, $T_{s,x} - T_{\infty,x}$. From Eq. (16.53), it follows that for a given inlet temperature, T_i, the exit fluid temperature, T_e, increases proportionally with area, A. For a constant cross-section pipe, we have the fluid temperature, $T_{\infty,x}$, increasing linearly along the pipe. In short, $T_{s,x}$, $T_{\infty,x}$, and $T_{s,x} - T_{\infty,x}$ all vary linearly with increasing distance beyond the entrance length, as exhibited in Fig. 16.14.

16.10.4 Constant-surface-temperature pipe

For a constant-surface-temperature pipe that is being heated by a hot fluid running through it, the rate of heat loss by a small sectional element of fluid moving in the pipe can be expressed as

$$\delta Q' = m' c_P dT = h \, dA \, (T - T_s), \qquad (16.56)$$

where T is used to denote the (mean) fluid temperature, the only variable in a fully developed constant-diameter pipe flow, and T_s is the constant surface temperature of the pipe. The equation can be recast as

$$[1/(T - T_s)] \, dT = \left[h/(m' c_P) \right] dA. \qquad (16.57)$$

Integrating from inlet to exit gives

$$T_e - T_s = (T_i - T_s) \exp\left\{ -\left[hA/(m' c_P) \right] \right\}. \qquad (16.58)$$

This evinces that the temperature varies in an exponential manner, not linearly as in the uniform heat flux case. For the entire pipe or a section of it, the total heat transfer rate,

$$Q' = h \, A \, \Delta T_{ln}, \qquad (16.59)$$

where the *logarithmic mean temperature difference* $\Delta T_{ln} = (T_e - T_i)/\ln\{(T_s - T_e)/(T_s - T_i)\}$.

16.10.5 Nusselt number and pumping cost for laminar and turbulent forced convection in a pipe

For a given problem in practice, we wish to achieve the maximum heat transfer per unit pumping cost. To put it another way, a large Nu and a low friction are sought. The flow conditions are usually set and, hence, the relevant expressions for developing and fully developed laminar and turbulent pipe flows are required.

The pumping cost is decided by the pressure drop,

$$\Delta P = f \frac{L}{D} \frac{\rho U^2}{2} \qquad (16.60)$$

The required power to overcome the pressure drop is

$$\dot{W}_{pump} = \dot{V}\Delta P = \frac{\dot{m}\Delta P}{\rho} \qquad (16.61)$$

where $\dot{V}$ is the volume flow rate.

For developing laminar flow with Prandtl greater than 0.5, the Nusselt number,

$$Nu = 1.86 \left(\frac{RePrD}{L}\right)^{1/3} \left(\frac{\mu_b}{\mu_s}\right)^{0.14}. \qquad (16.62)$$

Here, subscript "b" designates "at bulk mean fluid temperature," where the bulk mean fluid temperature is the average between the inlet and outlet temperatures, that is, $T_b = \frac{1}{2}(T_i + T_e)$.

For fully developed laminar pipe flow, the friction factor,

$$f = 4\,C_f = 64/Re. \qquad (16.63)$$

For constant surface temperature, the Nusselt number,

$$Nu = 3.66. \qquad (16.64)$$

For fully developed laminar pipe flow with uniform heat flux,

$$Nu = 4.36. \qquad (16.65)$$

The Graetz number is a dimensionless number that describes developing laminar flow in a tube. Namely, the Graetz number,

$$Gz = (D/x)\,Re\,Pr, \qquad (16.66)$$

where x is the distance from the pipe entrance. Fig. 16.15 shows that the laminar flow becomes fully developed when 1/Gz is greater than 0.05, for both constant surface temperature and uniform heat flux cases. Once again, we see that a uniform heat flux leads to a larger Nu value than when there is a constant surface temperature boundary.

When the Reynolds number is larger than approximately 2300, the flow in a pipe is turbulent. For fully developed turbulent flow in a smooth pipe, the friction factor,

$$f = 0.184\,Re^{-0.2}. \qquad (16.67)$$

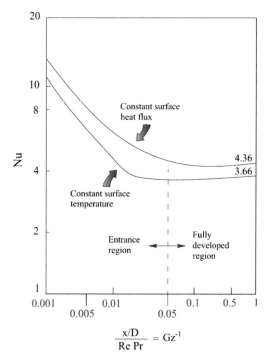

$$\frac{x/D}{Re\ Pr} = Gz^{-1}$$

FIGURE 16.15 Nu versus 1/Gr for developing laminar forced convection in a pipe (created by F. Kermanshahi). Note that beyond the entrance length, the fluid temperature increases linearly in the flow direction.

The corresponding Nusselt number for $0.7 \le Pr \le 160$ and $Re > 10{,}000$,

$$Nu = 0.023\ Re^{0.8}\ Pr^{C_2}, \tag{16.68}$$

where C_2 is 0.4 for heating and 0.3 for cooling of the fluid flowing through the tube; the fluid properties are evaluated at the bulk mean fluid temperature, T_b. This is applicable to both constant-surface-temperature and uniform-heat-flux cases in a smooth pipe. For fully developed turbulent flow in rough tubes,

$$Nu = 0.125\ f\ Re\ Pr^{1/3}, \tag{16.69}$$

where the value of friction factor, f, can be obtained from the Moody chart.

Table 16.3 tabulates the Nusselt number and friction factor for fully developed laminar flow in a tube of common shapes. The Reynolds number for noncircular pipes is based on the equivalent or hydraulic diameter,

$$D_h = 4\ A_c/p_r, \tag{16.70}$$

where A_c is the cross-sectional area and p_r is the perimeter.

TABLE 16.3 Nusselt number and friction factor for fully developed laminar flow in tubes of different cross-sectional shapes (created by X. Wang).

Tube geometry	a/b or $\theta°$	Nusselt number		Friction factor
		T_s = const.	$\dot{q}_s$ = const.	
Circle	–	3.66	4.36	64.00/Re
Rectangle	a/b			
	1	2.98	3.61	56.92/Re
	2	3.39	4.12	62.20/Re
	3	3.96	4.79	68.36/Re
	4	4.44	5.33	72.92/Re
	6	5.14	6.05	73.80/Re
	8	5.60	6.49	32.32/Re
	∞	7.54	8.24	96.00/Re
Ellipse	a/b			
	1	3.66	4.36	64.00/Re
	2	3.74	4.56	67.28/Re
	4	3.79	4.88	72.96/Re
	8	3.72	5.09	76.60/Re
	16	3.65	5.18	78.16/Re
Isosceles Triangle	$\theta°$			
	10°	1.61	2.45	50.80/Re
	30°	2.26	2.91	52.28/Re
	60°	2.47	3.11	53.32/Re
	90°	2.34	2.98	52.60/Re
	120°	2.00	2.68	50.96/Re

Problems

16.1 Convection heat transfer mechanisms.

What are the three heat transfer mechanisms? Convection heat transfer involves which of these three heat transfer mechanisms?

16.2 The role of Prandtl number.

Under what Prandtl number (small, medium, large) is convection heat transfer realized mostly via fluid motion? Why?

354 PART | 4 Ecophysiology-flavored engineering heat transfer

16.3 Solar-power-assisted moving train.

Solar photovoltaic panels placed on top of a moving passenger train are subject to augmented convective cooling and, hence, improved energy conversion efficiency compared to a stationary roof. You are asked to estimate the temperature of solar photovoltaic panels on top of a 7 m long passenger car moving at 120 km/h. What is the panel temperature in the presence of 500 W/m² of radiation?

16.4 Increase flow rate to enhance heat convection in a conduit.

The design group that you belong to is thinking about doubling the velocity in a smooth conduit to boost forced convection. Deduce the change in pumping power and heat transfer rate for

A) an isothermal circular conduit,

B) a uniform-heat-flux circular conduit,

C) an isothermal rectangular conduit with a width-to-height ratio of 2, and

D) a uniform-heat-flux rectangular conduit with a width-to-height ratio of 2.

16.5 Metal forming of a steel plate.

A 2 m wide, 7 m long, smooth steel plate at a constant surface temperature of 357°C is moving at a certain velocity, U, into still air at 27°C. The local convection heat transfer coefficient, h, as a function of the distance from the leading edge, x, is estimated to be

x [m]	0–1	1–2	2–3	3–4	4–5	5–6	6–7
h[W/(m² · K)]	27	24	23	22	21	39	31

A) Calculate the total heat transfer rate from the plate into the still air.

B) Determine the velocity of the moving plate.

16.6 Cold air delivery through a duct.

Cold air at 10°C is conveyed through a 15.24 cm diameter circular duct at a mass flow rate of 0.1 kg/s over a distance of 17 m.

A) Estimate the temperature of the duct if air exits at 14°C.

B) If the surface of the duct is roughened, how would the exit air temperature change?

16.7 Transporting chilled water across a river.

Chilled water at 5°C is transported through a cast-iron pipe at a velocity of 2 m/s across a 13 m wide river. The river water at 11°C moves at 4 km/h across the pipe. For reducing flow-induced vibration caused by vortex shedding, the pipe is pressed into one that has an elliptical cross-section (10 cm by 5 cm), with the longer dimension parallel to the flow stream. The thickness of the elliptical pipe is 0.635 cm. The thermal conductivity of the cast iron pipe, $k = 52$ W/(m²·K). Estimate the chilled water exit temperature.

References

Ancel, A., Gilbert, C., Poulin, N., Beaulieu, M., Thierry, B., 2015. New insights into the huddling dynamics of emperor penguins. Anim. Behav. 110, 91–98.

Anderson, K., 2021. "Dinosaur tissue: a biochemical challenge to the evolutionary timescale," https://answersingenesis.org/fossils/dinosaur-tissue/. (accessed May 8, 2021).

Farlow, J.O., Thompson, C.V., Rosner, D.E., 1976. Plates of the dinosaur stegosaurus: forced convection heat loss fins? Science 192 (4244), 1123–1125.

Gennev, 2020. https://gennev.com/education/menopause-hot-flashes-part-2, (accessed November 1, 2020).

Gerum, R.C., Fabry, B., Metzner, C., Beaulieu, M., Ancel, A., Zitterbart, D.P., 2013. The origin of traveling waves in an emperor penguin huddle. New J. Phys. 15, 125022.

Gilbert, C., Blanc, S., Maho, Y.Le, Ancel, A., 2008. Energy saving processes in huddling emperor penguins: from experiments to theory. J. Exp. Biol. 211, 1–8.

Lienhard V, J.H., 2020. Heat transfer in flat-plate boundary layers: a correlation for laminar, transition, and turbulent flow. J. Heat Transfer 142, 061805 June.

Muth, F., 2013. Cooling down in honeybees is affected by what others are doing. Scientific Am. https://blogs.scientificamerican.com/not-bad-science/cooling-down-in-honeybees-is-affected-by-what-others-are-doing/. (accessed November 1, 2020).

Peters, J.M., Peleg, O., Mahadevan, L., 2019. Collective ventilation in honeybee nests. J. R. Soc., Interface 16, 20180561.

Ting, D.S-K., 2020. Lecture notes on engineering human thermal comfort. Singapore, World Scientific.

Wright, P.G., 1984. Why do elephants flap their ears? South African J. Zool. 19 (4), 266–269.

Yang, Y., Ting, D.S-K., Ray, S., 2020. On flexible rectangular strip height on flat plate heat convection. Int. J. Heat Mass Transfer 150, 1–13 119269.

Yang, Y., Ting, D.S-K., Ray, S., 2021. Chapter 4: Improving heat transfer efficiency with innovative turbulence generators. In: Stagner, J.A., Ting, D.S-K. (Eds.), Climate Change and Pragmatic Engineering Mitigation. Singapore, Jenny Stanford Publishing.

Chapter 17

Thermal radiation

"Mars has global warming, but without a greenhouse and without the participation of Martians. These parallel global warmings – observed simultaneously on Mars and on Earth – can only be a straight-line consequence of the effect of the one same factor: a long-time change in solar irradiance."

Khabibullo Abdusamatov

Chapter Objectives

- Appreciate the radiating sun.
- Comprehend black and gray body radiation.
- Learn to perform radiation heat transfer calculations.
- Be familiar with absorptivity, transmissivity, and reflectivity.
- Recognize view or shape factor.

Nomenclature

A	area; A_s is surface area
E	emissive power; E_b is the total blackbody emissive power
F	view or shape factor; F_{1-2} is the fraction of radiation leaving Surface 1 that strikes Surface 2
h_{rad}	radiation heat transfer coefficient
Q	heat transfer; Q' is heat transfer rate, $Q'_{emit\,max}$ is the maximum heat transfer rate emitting from a black body, Q'_{rad} is the radiation heat transfer rate
R_{rad}	radiation resistance
T	temperature; T_{avg} is the average temperature, T_s is the surface temperature, T_{surr} is the temperature of the surroundings

Greek and other symbols

α_{rad}	absorptivity
ρ_{rad}	reflectivity
$\tau_{rad},$	transmissivity
ε	emissivity; ε_{eff} is the effective emittance

17.1 The radiating Sun

Imagine the good sun gives up on us and stops radiating? No light, no photo-synthesis, and planet Earth briskly freezes into a massive ice ball. Sunlight and

Thermodynamics from Science to Engineering. DOI: https://doi.org/10.1016/B978-0-323-90626-5.00012-4

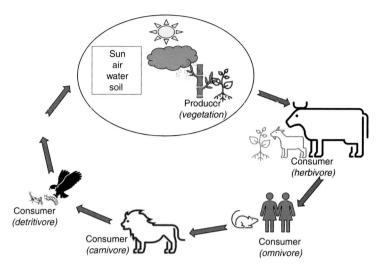

FIGURE 17.1   The food cycle of life sustained by the sun (created by O. Imafidon). Directly or indirectly, all living things feed on solar energy.

solar thermal energy are indispensable. Fig. 17.1 illustrates that the sun provides the needed energy to set in motion and to sustain the cycle of life. It provides the energy for vegetation to bloom via photosynthesis. The vegetation feeds all kinds of herbivores including grasshoppers, lambs, and cattle. The herbivores, along with the greens, nurture the omnivores. Most homo sapiens are omnivores, our diets include both vegetables and meat, including other omnivores such as chicken and catfish. Mice and rats are omnivores that fatten hawks and eagles, some of the most enviable carnivores at the top of the food chain. The death of these predatory creatures, including the ferocious lions, closes the cycle of life or food chain by providing food to detritivores, who decompose the dead meat into nutrients for the plants to thrive under the sun.

The other indispensable aspect of the radiating sun, furnishing thermal energy, is much more recognizable and appreciated. This is because climate change and/or global warming is "the talk around the globe." We should never make any excuses for not being responsible as cohabitants and costewards of Earth. Nevertheless, there are differing views regarding the major cause of global warming or climate change and the number and quantity of uncertainties associated with various extrapolations. For example, astrophysicist Khabibullo Abdusamatov who supervised the Astrometria project of the Russian section of the International Space Station and headed the Space research laboratory at the Saint Petersburg-based Pulkovo Observatory of the Russian Academy of Sciences observed that "Global warming results not from the emission of greenhouse gases into the atmosphere, but from an unusually high level of solar radiation and a lengthy—almost throughout the last century—growth in its intensity. It is no secret that when they go up, temperatures in the world's oceans

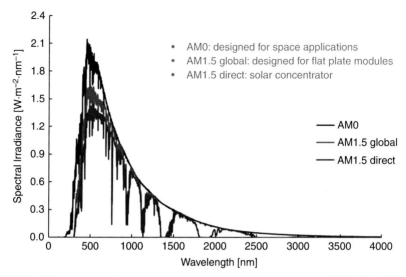

FIGURE 17.2 An approximate portrayal of the solar spectrum based on (PVCDROM, 2021) (created by X. Wang). Note that the visible (to human eyes) range corresponds to the highest intensity of the solar spectrum.

trigger the emission of large amounts of carbon dioxide into the atmosphere. So, the common view that man's industrial activity is a deciding factor in global warming has emerged from a misinterpretation of cause and effect relations." Notwithstanding that, there is much we can do to live in harmony with the environment. Interested readers can refer to (Vasel-be-Hagh and Ting, 2020; Ting and Carriveau, 2021; Ting and Stagner, 2021; Stagner and Ting, 2022).

Succinctly, solar radiation is the transport of solar energy via electromagnetic waves or photons caused by variations in the electronic configurations of the atoms or molecules; this discovery is credited to James Clerk Maxwell (1831–1879). By this very nature, radiation is most effectively transmitted in the absence of an interfering medium. Due to this fact, the largely nothingness of space presents the ideal passage for solar thermal energy to travel far and wide, furnishing just the right amount of thermal energy to keep our planet warm. How fast does light travel in a vacuum? 2.9979×10^8 m/s!

Being transmitted by electromagnetic waves, radiation has a spectrum of wavelengths and/or frequencies. To this end, radiation heat transfer occurs by electromagnetic waves with a strong spectral dependence. Fig. 17.2 portrays the approximate solar spectrum. Roughly 7% of solar radiation is below the visible wavelength of 380 nm; note that 1 nm is 1×10^{-9} m. The remaining solar radiation is more or less divided between the visible range, which goes up to a wavelength of around 780 nm, and the infrared range, in decreasing intensity level up to about 1 mm. It is worth stressing that the visible range corresponds to the most intense solar radiation (power). It is also interesting to note that while

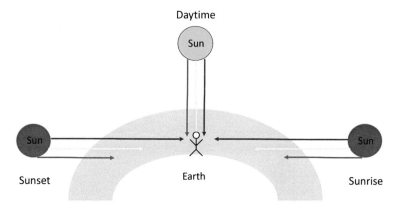

FIGURE 17.3   Blue sky and reddish sunrise and sunset (created by A. Raj). Blue is most visible to human eyes. The least scattered wavelengths when the sun is low correspond to yellow, orange, and red, furnishing beautiful, reddish skyline at sunrise and sunset.

human visibility is limited to the visible range, thermal radiation covers a much wider range, from part of the ultraviolet to infrared, that is, from 1×10^3 to 1×10^6 nm. Isaac Newton put it rightly, "When I look at the solar system, I see the earth at the right distance from the sun to receive the proper amounts of heat and light. This did not happen by chance." At a mean distance of 1.5×10^8 km from earth, the Sun, with a surface temperature near 6000 K, furnishes a total solar irradiance of close to 1400 W/m². The providence of the protective atmosphere aptly attenuates it down to a comfortable level for the habitants of Earth to thrive. As proclaimed by National Geographic (2021), the atmosphere shields us from incoming ultraviolet radiation, keeps our planet warm through insulation, and attenuates the extremes between day and night temperatures.

Blue Sky and Reddish Sunrise and Sunset?

Why is the sky blue except at sunrise or more so at sunset, when it is red? The sky is blue because the shorter wavelengths of visible light, purple (violet and indigo), blue, and some green, are scattered in the atmosphere, with blue most visible to human eyes; see Fig. 17.3. With these shorter wavelengths scattered, the high sun appears yellow, the hottest color of the sun. With decreasing altitude, the sunray passes through progressively thicker atmosphere as shown in Fig. 17.3. As a result, an increasing portion of the purple, blue, and green light is scattered. The mostly unscattered yellow, orange, and red wavelengths pass through the atmosphere. With increasing particles such as water vapor, dust, and soot in the atmosphere, an increasing amount of yellow, followed by orange light, is scattered. This brings forth spectacular orange–red sunrises and sunsets. Michael Josephson put these natural beauties into proper perspective when he testified that "The world has enough beautiful mountains and meadows, spectacular skies and serene lakes. It has enough lush forests, flowered fields, and

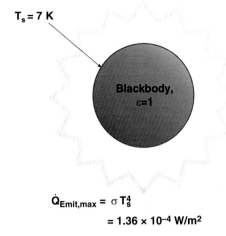

$$\dot{Q}_{\text{Emit,max}} = \sigma T_s^4$$
$$= 1.36 \times 10^{-4} \text{ W/m}^2$$

FIGURE 17.4 A blackbody at 7 K radiating heat (created by X. Wang). Only a body at absolute zero does not radiate heat. It follows that every one of us is meant to radiate heat.

sandy beaches. It has plenty of stars and the promise of a new sunrise and sunset every day. What the world needs more of is people to appreciate and enjoy it."

17.2 All bodies above absolute zero radiate heat

A body does not have to be scorching like the sun, with a surface temperature in the thousands of degrees, to have the qualification to radiate heat. Every surface or body that is above absolute zero (0 K) radiates thermal energy. All solids, liquids, and gases emit, absorb, or transmit radiation in varying degrees. A blackbody is an ideal body that radiates perfectly, that is, it gives out the maximum rate of radiation. This maximum radiation rate of a blackbody is described by the Stefan–Boltzmann Law, specifically,

$$Q'_{\text{emitmax}} = \sigma A_s T_s^{\,4} = E_b, \tag{17.1}$$

where the Stefan–Boltzmann constant, $\sigma = 5.669 \times 10^{-8}$ W/(m²·K⁴), A_s is the surface area of the blackbody, T_s is its surface temperature in degree Kelvin (K) or Rankine (°R), and E_b is the total blackbody emissive power. Fig. 17.4 illustrates such an ideal surface (blackbody) at a mere 7 K. Even at just a few degrees above absolute zero, the atoms and/or molecules making up the body are still shaking (vibrating) a little, enabling the body to give out some thermal energy.

Example 17.1 The color of hotness.
Given: To illustrate thermal radiation, Dr. JAS turns on four burners of an electric range to varying degrees. Burner 1 grows into bright red, Burner

2 becomes dull red, Burner 3 changes into dull yellow, and Burner 4 glows into bright yellow.

Find: A) The order of hotness of the four burners from the highest temperature to the lowest. B) The approximate temperature of the dull yellow burner.

Solution: From Fig. 17.2, we learn that within the visible range of solar radiation, the intensity increases from red to blue with decreasing wavelength. It follows that Burner 4 with the bright yellow is the hottest, followed by the dull yellow Burner 3, bright red Burner 1, and dull red Burner 2.

The dull yellow burner corresponds to a wavelength of approximately 570 nm, the corresponding temperature is in the neighborhood of 1000°C.

Gray surfaces

All real surfaces emit less radiation than an ideal blackbody. The emissive power of a real body (for transparent or, more correctly, translucent substances such as a liquid or a gas) or surface (for an opaque body), E, with respect to that of its perfectly black counterpart, E_b, is called the emissivity, ε. To put it another way, the emissivity of a surface (or body or volume) is defined as

$$\varepsilon = E/E_b. \tag{17.2}$$

Surfaces for which the emissivity is constant are called gray surfaces. With the introduction of emissivity, ε, to account for non-ideal, non-black-body real surfaces or bodies, the radiation emitted from a real surface can be expressed as

$$Q'_{emit} = \varepsilon \, \sigma \, A_s T_s^4. \tag{17.3}$$

It is important to stress that T_s is the absolute surface temperature and ε is between zero and one. The emissivity of most non-shiny surfaces, including human skin, is around 0.9.

17.3 Absorptivity, transmissivity, and reflectivity

The surface or volume of a material also has radiation properties described as absorptivity, α_{rad}, transmissivity, τ_{rad}, and reflectivity, ρ_{rad}. These properties, along with emissivity, ε, characterize the radiation heat transfer behavior of the object involved. According to the conservation of energy, the amount of energy absorbed, plus that transmitted through, and the portion reflected equal to the energy radiated on the object, that is,

$$\alpha_{rad} + \tau_{rad} + \rho_{rad} = 1. \tag{17.4}$$

This is portrayed in Fig. 17.5. Strictly speaking, Eq. (17.4) is derived for a single wavelength. It is, however, also applicable for piecewise gray surfaces, as long as the wavelength range over which the three properties are calculated is the same.

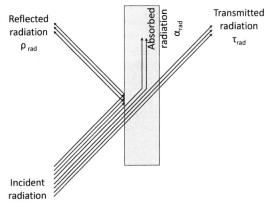

FIGURE 17.5 The sum of absorbed, transmitted, and reflected radiation is equal to the total thermal energy radiated onto the object, that is, $\alpha_{rad} + \tau_{rad} + \rho_{rad} = 1$ (created by A. Raj).

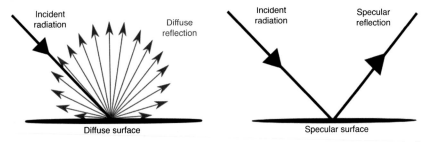

FIGURE 17.6 A diffuse surface versus a specular surface (created by N. Bhoopal). A mirror is possibly the most common specular surface.

The ideal blackbody absorbs and emits every bit of radiation it encounters. Accordingly,

$$\varepsilon = \alpha_{rad} = 1, \tag{17.5}$$

for a blackbody. All real bodies absorb a fraction, never the entirety, of the radiation energy incident on a surface, that is,

$$0 < \alpha_{rad} < 1. \tag{17.6}$$

A surface that scatters an incident ray at a large number of angles, rather than just at one angle as in specular reflection, is a diffuse surface; see Fig. 17.6. For a diffuse surface, or the irradiation striking the surface is diffuse, then, we can invoke Kirchhoff's identity for the gray surface,

$$\varepsilon = \alpha_{rad}. \tag{17.7}$$

Practically, a gray surface emits what it absorbs. This expression is also applicable for non-gray surfaces at a given wavelength; for example when dealing with monochrome irradiation.

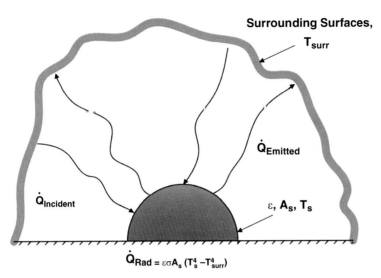

FIGURE 17.7   Radiation exchange between a surface and its surroundings (created by X. Wang). With nothing else present, the surface exchanges heat exclusively with its surroundings.

In real life, a surface is typically exposed to large surroundings. More often, the problem is further complicated by the presence of other objects, and these objects tend to be at different temperatures. For the simplest condition where a surface at temperature, T_s, is exposed only to its surroundings at temperature, T_{surr}, such as that illustrated in Fig. 17.7, the net rate of radiation from the surface to the surroundings is

$$Q'_{rad} = \varepsilon \sigma A_s (T_s^4 - T_{surr}^4). \tag{17.8}$$

It is worth stressing that both T_s and T_{surr} must be expressed in absolute units, that is, either K or °R.

Example 17.2 Old-fashioned versus contemporary fireplaces.
Given: The surface of the long-established metal fireplace, including the long extension up the chimney, for sustaining occupants through the cold winter is plain black because of the wisdom of our ancestors. Regrettably, this sagacious design is being replaced by the modern shiny-finish outfit.

Find: The radiation heat transfer efficiency of a shiny metal surface ($\varepsilon \approx 0.10$) with respect to the plain black one ($\varepsilon \approx 0.95$) for T_s of 700 K and T_{surr} of 290 K.

Solution: According to Eq. (17.8), the radiation heat transfer rate per m^2 surface area is

$$Q'_{rad}/A_s = \varepsilon \sigma (T_s^4 - T_{surr}^4).$$

For the shiny metal surface with emissivity, $\varepsilon = 0.10$, we have

$$Q'_{rad}/A_s = (0.10)\left[5.669 \times 10^{-8}\text{W}/(\text{m}^2 \cdot \text{K}^4)\right]$$
$$\left[(700 \text{ K})^4 - (290 \text{ K})^4\right] = 1.32 \text{ kW/m}^2.$$

For the dull black metal surface with emissivity, $\varepsilon = 0.95$, the heat flux,

$$Q'_{rad}/A_s = (0.95)\left[5.669 \times 10^{-8}\text{W}/(\text{m}^2 \cdot \text{K}^4)\right]$$
$$\left[(700 \text{ K})^4 - (290 \text{ K})^4\right] = 12.5 \text{ kW/m}^2.$$

We see that the plain black surface is 9.5 times better at radiating heat to those who sit around the fireplace than the contemporary shiny one.

The message is that not everything new is better than its older counterpart. Then again, some of the newer designs recirculate the out-going flue gas, extracting over 70% of the energy content into indoor heat, compared to 10% to 20% for a conventional wood fireplace. The traditional wood fireplace also emits significantly more pollutants. With the "return to natural living," intelligently designed wood fireplaces are as efficient and clean as the leading gas fireplaces.

For a moderately small temperature difference, a few degrees, between the surface and the surroundings, we may linearize the expression into

$$Q'_{rad} = h_{rad} A_s (T_s - T_{surr}), \tag{17.9}$$

where the radiation heat transfer coefficient, h_{rad}, is analogous to the convection heat transfer coefficient. While the convection heat transfer coefficient is strongly dependent on the working flow stream, the linearized radiation heat transfer coefficient, on the other hand, is sensitive to the temperatures of both the involved surface and its surroundings. Expressly,

$$h_{rad} = \varepsilon \, \sigma \, (T_s + T_{surr})\left(T_s^2 + T_{surr}^2\right). \tag{17.10}$$

With this radiation heat transfer coefficient, the radiation heat transfer rate can be expressed in terms of the radiation resistance,

$$Q'_{rad} = (T_s - T_{surr})/R_{rad}, \tag{17.11}$$

where the radiation resistance,

$$R_{rad} = 1/(h_{rad} A_s). \tag{17.12}$$

17.4 View or shape factors

One of the unique features of an ice skating/hockey arena is radiant heating. The radiant heaters, such as those shown in Fig. 17.8, are positioned so that they face the spectators. In a sense, when a fan sees the hot coils, she receives the thermal radiation. On the other hand, the skaters in the rink do not see the heaters and, hence, do not benefit from the soothing warmth radiated from the radiant heater.

FIGURE 17.8 Radiant heaters in an ice skating/hockey arena (created by X. Wang). This is a practical illustration of radiation exchange between a surface and its surroundings. The radiant heater is directed at the fan, exchanging heat at a serious loss with the appreciably cooler fan.

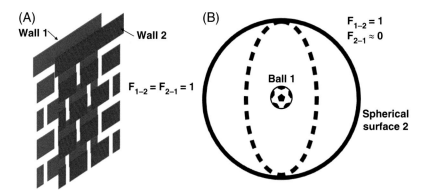

FIGURE 17.9 View factor illustrated (A) two large parallel walls facing one another see only each other and, hence, their view factors, $F_{1-2} = F_{2-1} = 1$, (B) a small ball in the middle of a large spherical enclosure is radiating its heat exclusively to the enclosed surface, that is, $F_{1-2} = 1$, but the much larger inner surface of the enclosure is hardly noticing its radiation to the small ball, that is, $F_{1-2} \approx 0$ (created by O. Imafidon).

The practical reason is not to withhold the heat from the skaters but to not melt the ice.

The ice-skating arena example illustrates something that is called view factor. Radiation shape (or view) factor, F_{1-2}, denotes the fraction of radiation leaving diffuse Surface 1 that strikes Surface 2. The shape (view) factor is strictly geometric and does not depend on surface properties such as emissivity or temperature. Fig. 17.9 portrays two particular cases concerning the view factor. When two large surfaces, Surface 1 and Surface 2, are facing each other, all they see is one another and, hence, both F_{1-2} and F_{2-1} are one. We can deduce the

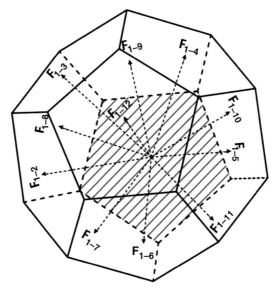

FIGURE 17.10   View factor of a dodecahedron (created by X. Wang). Every inner surface of the 12-equal-sided object has the same view factor, and $F_{1-1} + F_{1-2} + F_{1-3} + F_{1-4} + F_{1-5} + \ldots + F_{1-12} = 1$.

reciprocity relationship to be

$$A_1 F_{1-2} = A_2 F_{2-1}. \tag{17.13}$$

When a small ball with its surface designated as Surface 1 is positioned at the middle of a large spherical enclosure with Surface 2, as depicted in Fig. 17.9B, we see that F_{1-2} is one whereas F_{2-1} is very small.

The first law of thermodynamics states that the total amount of energy must be conserved. Accordingly, the sum of shape factors for a given surface must equal unity, that is,

$$F_{1-1} + F_{1-2} + F_{1-3} + \ldots + F_{1-j} = 1. \tag{17.14}$$

Note that the shape factor F_{1-1} is nonzero only for concave surfaces, that is, surfaces that can "see" themselves. Consider one of the twelve identical inner surfaces of the dodecahedron shown in Fig. 17.10. As the surface is flat, it does not see itself and, thus, it does not radiate to itself. Accordingly, the fraction of radiation leaving Surface 1 that strikes Surfaces 2 to 12 adds up to a total of one.

Example 17.3 A fireball in a spherical enclosure.

Given: To demonstrate the concept of view or shape factor, Dr. JAS suspends a fireball in the middle of a spherical enclosure just like that depicted in Fig. 17.9B. The surface of the hot ball, A_1, is 10% that of the inner surface of the enclosure, A_2, that is, $A_1 = A_2/10$.

Find: The view factors F_{1-2} and F_{2-1}.

(A)   (B)

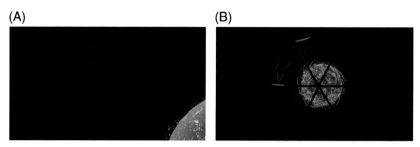

FIGURE 17.11 Solar shield for mitigating climate change (A) a highly porous solar shield at Lagrange L1 equilibrium point of the Earth–Sun system (B) a Dyson sphere that encompasses the Sun and controls its radiation beyond the sphere while capturing a fraction of the radiation to power future civilization (created by X. Wang). Only a small fraction of the solar radiation is to be blocked by the shield, that is, the sunshield view factor, $F_{sun\text{-}shield}$, is around 0.01, so that Earth does not freeze into an lifeless ice ball like the dwarf planet Pluto, with an average temperature of $-233°C$. Note that the debate regarding if Pluto is a dwarf planet has reignited.

Solution: We see that all radiation leaving the surface of the ball lands on the inner surface of the enclosure. Consequently,

$$F_{1-2} = 1$$

According to Eq. (17.13), $A_1 F_{1-2} = A_2 F_{2-1}$. Therefore, we have

$$F_{2-1} = (A_1/A_2) \, F_{1-2} = 0.1 \, F_{1-2} = 0.1$$

It is important to note that shape-factor algebra described here applies only for diffusely emitting and reflecting surfaces. It cannot be used for surfaces such as mirrors that reflect specularly. Providentially, nearly all surfaces in buildings are diffusive.

One proposition to overcome global warming and, thus, the more encompassing climate change, is to deploy a massive solar shield to block solar radiation before it enters the lower atmosphere. We can envision that as a large umbrella protecting the earth from getting sunburned, as sketched in Fig. 17.11. The appropriate position to place the shield is where the gravitational forces of the Sun and Earth are equal to the centripetal force required for the structure (shield) to move with the Earth–Sun system (NASA, 2021). The Lagrange L1 equilibrium point of the Earth-Sun system is most fitting because it offers an uninterrupted view of the sun; see, for example, (Sánchez and McInnes, 2015; Angel, 2006). According to the experts in the field, a small view factor, in the neighborhood of 0.01 from the Sun to the shield, is all that is needed to mitigate climate change. For that reason, the umbrella shown in Fig. 17.11 is about 99% porous. A word of caution, the natural solar–atmosphere system involves many factors in an integrated and complex manner. To that end, human intervention has a higher probability of messing things up, as history has repeatedly proven, rather than resolving the comparatively smaller issue at hand. It is safer, and conceivably more workable, for every one of us to take measures in simplifying our lifestyle, including practicing conservation and nature-friendly engineering.

17.5 Further reading on thermal radiation

Standard heat transfer textbooks such as Bergman et al. (2018), Çengel and Ghajar (2019), and Kreith and Manglik (2017) cover thermal radiation as a standard chapter or two. More specialized monographs including Howell et al. (2021) and Sparrow and Cess (2018) deal with thermal radiation exclusively and thoroughly. Chapter 7 of Ting (2020) details solar radiation fundamentals in heating and cooling calculations.

Problems

17.1 Radiant floor.

A 4 m high room has a 7 m by 10 m radiant floor with a surface temperature of 30°C, while the four walls and the 7 m by 10 m flat ceiling are at 15°C. Assuming the emissivities of all surfaces are 0.9, calculate the rate of radiant heat transfer from the floor.

17.2 Estimating sky temperature.

One way to estimate the sky temperature is to base it on the energy balance of a surface exposed to the sky at night. Gideon placed a wool fleece ($\varepsilon = 0.98$) on the ground during a night when the ambient air was at 12°C. He found that the surface of the wool fleece was at 5°C and the convective heat transfer coefficient was 14 W/(m$^2 \cdot$K). What was the sky temperature? Hint: Assuming the fleece base to be well-insulated, the energy balance requires heat convection from the warmer ambient air to the fleece to be equal to the radiation heat transfer from the fleece to the sky.

17.3 Sun-Roof–Sky radiation exchange.

Solar irradiance of 750 W/m^2 strikes a 10 m by 8 m horizontal roof while the sky is at 10°C and the ambient air is at 30°C. The emissivity of the roof, ε, is 0.95 and its solar absorptivity, α_{rad}, is 0.89. Assume that the underside of the roof is perfectly insulated.

A) What is the surface temperature of the roof if the heat convection by atmospheric air is negligible?

B) What is the surface temperature of the roof if the convection heat transfer coefficient of the atmospheric air is 15 W/(m$^2 \cdot$K)?

17.4 Reduce radiation heat loss via a building envelope cavity.

An air gap of a few centimeters is commonly encountered in a building envelope. Consider a 5-cm air gap in a 100-m^2 wall where the emissivities of both surfaces, ε_1 and ε_2, are 0.95. During a typical winter night, the inner surface is at 15°C while the outer surface is at 5°C.

A) Assuming convection heat transfer through the air gap is negligible, what is the heat loss rate via radiation across the air gap?

B) What is the rate of heat loss, if the convection heat transfer coefficient over both surfaces is 1.8 W/(m$^2 \cdot$°C)?

C) How much heat loss can we reduce, if we coat the two surfaces with shiny silver paint so that their emissivity is lowered to 0.15?

Hint: For heating and cooling load calculations, Eq. (17.10) can be approximated as

$$h_{rad} \approx 4\varepsilon_{eff}\sigma T_{avg}^3, \tag{17.15}$$

where T_{avg} is the average absolute temperature of the two facing surfaces, and the effective emittance,

$$\varepsilon_{eff} \equiv 1/(1/\varepsilon_1 + 1/\varepsilon_2 - 1). \tag{17.16}$$

17.5 Estimate solar radiation from brick wall temperatures.

The sun-facing surface of a 18-cm thick brick wall is at 44°C while the other surface is at 27°C. The thermal conductivity of the brick wall is 0.75 W/(m$\cdot$K), and the emissivity of the surface is 0.9. If convection heat transfer is negligible compared to the direct solar radiation, what is the rate of direct solar radiation on the sun-facing surface per unit area?

17.6 Radiation heat loss in outer space.

T&E Lab member Kaya joined Dr. JAS' team on an outer-space-exploration mission. For some unknown reasons, Kaya was momentarily sucked out of the shower into outer space completely naked. What was the rate of heat loss that he experienced momentarily? Note that Kaya is a medium built 1.7-m tall adult male and outer space is at 2.7 K (Discovery, 2021).

17.7 Solar-powered spacecraft.

A spacecraft orbiting around Earth has a flat solar photovoltaic panel that powers its operation. Side 1 of the panel facing the sun is covered with solar cells, while Side 2 is exposed to outer space at 0 K. The solar radiation is 1353 W/m^2. The absorptivity of Side 1 is 0.8 and that of Side 2 is 0.7. The solar cell solar-to-electricity conversion efficiency is 15%. Answer the following parts.

A) The steady-state temperature of the panel in K.

B) The steady-state temperature of the panel would
   A) decrease,
   B) increase,
   C) remain the same,
if the absorptivity of Side 1 is increased to 0.9.

C) The steady-state temperature of the panel would
   A) decrease,
   B) increase,
   C) remain the same,
if the absorptivity of Side 2 is increased to 0.8.

References

Angel, R., 2006. Feasibility of cooling the earth with a cloud of small spacecraft near the inner Lagrange point (L1). Proc. Natl Acad. Sci. 103 (46), 17184–17189.

Bergman, T.L., Lavine, A.S., Incropera, F.P., DeWitt, D.P., 2018. Fundamentals of Heat and Mass Transfer, 8th ed., Hoboken, John Wiley & Sons.

Çengel, Y.A., Ghajar, A.J., 2019. Heat and Mass Transfer: Fundamentals & Application, 6th ed. McGraw-Hill, New York, NY.

Discovery, 2021. https://www.discovermagazine.com/the-sciences/how-cold-is-it-in-outer-space, (accessed October 15, 2021).

Howell, J.R., Mengüç, M.P., Daun, K., Siegel, R., 2021. Thermal Radiation Heat Transfer, 7th ed. CRC Press, Boca Raton, FL.

Kreith, F., Manglik, R.M., 2017. Principles of Heat Transfer, 8th ed. Cengage Learning, New York, NY.

NASA, 2021. https://map.gsfc.nasa.gov/mission/observatory_l2.html, (accessed October 15, 2021).

National Geographic, 2021. https://www.nationalgeographic.org/topics/resource-library-atmosphere/?q=&page=1&per_page=25, (accessed October 14, 2021).

PVCDROM, 2021. https://www.pveducation.org/pvcdrom/appendices/standard-solar-spectra, (accessed May 4, 2021).

Sánchez, J-P., McInnes, C.R., 2015. Optimal sunshade configurations for space-based geoengineering near the Sun-Earth L1 point. PLoS One 10 (8), e0136648.

Sparrow, E.M., Cess, R.D., 2018. Radiation Heat Transfer, Augmented. CRC Press, Boca Raton, FL.

Stagner, J.A., Ting, D.S-K., 2022. Climate Change and Pragmatic Engineering Mitigation. Jenny Stanford Publishing, Singapore.

Ting, D.S-K., Lecture Notes on engineering human thermal comfort, World Scientifi c, Singapore, 2020.

Ting, D.S-K., Carriveau, R., 2021. Sustaining Tomorrow via Innovative Engineering. World Scientific Publishing, Singapore.

Ting, D.S-K., Stagner, J.A., 2021. Climate Change Science: Causes, Effects and Solutions for Global Warming. Elsevier, Amsterdam.

Vasel-be-Hagh, A., Ting, D.S-K., 2020. Environmental Management of Air, Water, Agriculture, and Energy. CRC Press/Taylor & Francis Group, Boca Raton, FL.

Chapter 18

Heat exchangers

> *"The thermal agency by which mechanical effect may be obtained is the transference of heat from one body to another at a lower temperature."*
>
> Sadi Carnot

Chapter Objectives

- Appreciate intelligently designed heat exchangers for animals to thrive.
- Differentiate indirect (no-contact) heat exchangers from direct heat exchangers.
- Categorize indirect heat exchangers based on the flow configuration, that is, counter-flow, parallel flow, and cross-flow.
- Perform first law thermodynamic analysis on a heat exchanger with a constant-temperature wall.
- Master LMTD (log mean temperature difference) and NTU (number of transfer units) methods.

Nomenclature

A	area; A_c is cross-sectional area
C	heat capacity rate; C_c is the heat capacity rate of the cold stream, C_h is the heat capacity rate of the hot stream, C_{min} is the smaller of C_c and C_h
c_P	specific heat capacity at constant pressure; $c_{P,c}$ is the specific heat capacity at constant pressure of the cold stream, $c_{P,h}$ is the specific heat capacity at constant pressure of the hot stream
F_c	correction factor
h	enthalpy; $h_{c,i}$ is the entering cold stream enthalpy, $h_{c,o}$ is the enthalpy of the cold stream at outlet, $h_{h,i}$ is the inlet hot stream enthalpy, $h_{h,o}$ is the outlet hot stream enthalpy
L	length
LMTD	log mean temperature difference
NTU	number of transfer units
m	mass; m'_c is the mass flow rate of the cold stream, m'_h is the mass flow rate of the hot stream
p_L	perimeter
Q	heat (quantity); Q' is heat transfer rate, Q'_{in} is the rate of heat entering the system, Q'_{max} is the maximum heat transfer rate, Q'_{out} is the rate of heat exiting the system

Essential Engineering Thermodynamics https://doi.org/10.1016/B978-0-323-90626-5.00009-4

R   thermal resistance; R_{cond} is the conduction thermal resistance, $R_{conv,c}$ is the convection thermal resistance on the cold side, $R_{conv,h}$ is the convection thermal resistance on the hot side

T   temperature; T_c is the cold temperature, $T_{c,i}$ is the cold inlet temperature, $T_{c,o}$ is the cold stream outlet temperature, T_h is the hot temperature, $T_{h,i}$ is the hot stream inlet temperature, $T_{h,o}$ is the hot stream outlet temperature, T_i is the inlet temperature, T_o is the outlet temperature, T_w is the wall temperature, dT is differentiation with respect to temperature, ΔT is temperature difference, ΔT_1 is the temperature difference at Position 1, ΔT_2 is the temperature difference at Position 2, ΔT_{lm} is the log mean temperature difference, $\Delta T_{lm,CF}$ is the log mean temperature difference of a counter-flow heat exchanger

t   time or temperature; dt is differentiation with respect to time

U   overall heat transfer coefficient

V   velocity

x   length or distance; dx is a differentiable length or distance

Greek and other symbols

α (non-dimensional) temperature change ratio

β (non-dimensional) temperature change ratio

Δ difference

ρ density

ε heat exchanger effectiveness

$\forall$ volume

18.1 Nature thrives by exploiting effective heat exchangers

The emperor penguin is apparently the only animal that breeds during the merciless Antarctic winter (Ancel et al., 1997). To thrive in the extreme cold, the fathering penguins huddle in a group with specific formation and moving patterns over a four-month incubation fast (Gilbert et al., 2007). Other than conserving a significant amount of metabolic energy via community huddling, individual penguins are also equipped with potent heat exchangers. The most obvious heat exchangers run from their warm body down to their cold feet, which are in direct contact with the frigid ground, as shown in Fig. 18.1. The warm blood moving down to a foot transfers a portion of the thermal energy (heat) to the returning cold blood. This keeps the penguin from getting a chill from the returning cold blood and also reduces the amount of heat loss from the foot to the icy ground. The latter is due to a substantial decrease in the blood temperature reaching the foot which results in a smaller temperature gradient between the foot and the ground underneath and, thus, the heat loss rate.

While animals in cold climates strive to keep their bodies warm via efficient counter-flow heat exchangers, the same counter-flow heat exchangers are used by others to keep their bodies cool. Such is the case for the blood-thirsting kissing bug. Kissing bugs employ a countercurrent heat exchanger along their blood-sucking passage to prevent them from being overheated by the warm blood they

FIGURE 18.1 Countercurrent (counter-flow) blood vessels of a penguin between its body and feet (created by Y. Yang).

are sucking in (Lahondère et al., 2017). A schematic of this counter-flow heat exchanger is depicted in Fig. 18.2. The heat exchanger cools the blood from around 37°C to about 31°C by the time it reaches the head, and to below 27°C before reaching the abdomen of the kissing bug. Interested readers can check out Ask Nature (2021) for other examples.

Example 18.1 Blood cooling.

Given: A female mosquito can suck about 5×10^{-9} m^3 of human blood using its 6-needle proboscis (Quirós, 2016). Assume that a kissing bug can suck the same amount of blood.

Find: The cooling capacity needed by a kissing bug for bringing this amount of blood at 37°C to 27°C.

Solution: The amount of heat that needs to be removed from 5×10^{-9} m^3 of human blood at 37°C to 27°C,

$$Q = m \, c_P \Delta T, \qquad (E18.1.1)$$

where m is the mass of material of concern, c_P is the specific heat capacity, and ΔT is the temperature change. For human blood, the density, ρ, is in the neighborhood of 1060 kg/m^3 and the average c_P is around 3.6 kJ/kg/°C.

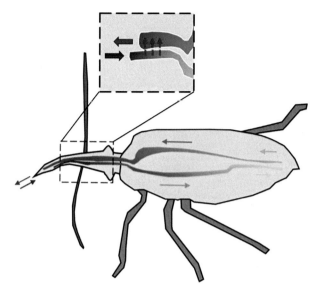

FIGURE 18.2 Countercurrent (counter-flow) heat exchanger of a kissing bug for cooling the warm blood intake (created by J. King).

Substituting for these values, we obtain the amount of heat to be

$$Q = \rho \forall \, c_P \Delta T = 1060 \left(5 \times 10^{-9}\right)(3600)(10) = 0.19 \text{ J}.$$

This amount of cooling is quite significant, even for a 2-cm kissing bug that is ten times larger than a mosquito.

Blood is needed for the mother mosquito to make eggs. One serving of 5×10^{-9} m^3 is larger than her abdomen and thus, she passes out water as she feasts on the appetizing human blood. Thank heavens, kissing bugs, unlike mosquitoes, do not have a preference for human blood.

18.1.1 Indirect (noncontact) heat exchanger

We see that counter-flow heat exchangers are prevalent in the animal kingdom. Both of the foregoing examples involve two counter-flowing fluids that do not mix when heat is transferred from the warm stream to the cold stream. This kind of heat exchanger, where the two streams do not mix or come into direct contact with each other, is called an *Indirect Heat Exchanger*. To put it concisely, a heat exchanger is a device that facilitates the exchange of thermal energy, heat, from a warmer fluid to a cooler one. For the more prevalent non-contact or indirect heat exchangers, the thermal energy is typically transmitted through a solid wall separating the two flow streams.

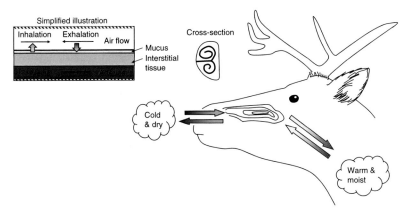

FIGURE 18.3   Reindeer nasal cavity, a direct contact heat exchanger for conserving both heat and moisture (created by Y. Yang).

18.1.2   Direct contact heat exchanger

Animal breathing, especially in a harsh environment, relies on potent *Direct Contact Heat Exchangers*, where the hot and cold streams come into contact with one another. Rudolph the red nose reindeer residing with Santa in the North Pole, along with un-imaginary reindeers in the arctic region, conserve both body heat and water via their uniquely designed nasal cavities (Magnanelli et al., 2016). The intelligently designed nasal cavity, shown in Fig. 18.3, trap both heat and water of the outgoing warm and moist air from the reindeer lungs, minimizing both thermal energy and water loss. When they inhale, the incoming cold and dry air pick up the entrapped heat and water, warming and moisturizing the incoming fresh air before it reaches the lungs.

A more common direct contact heat exchanger is illustrated by a panting dog. With its mouth open and tongue sticking out, the dog dissipates sensible and, more so, latent heat effectively via evaporative cooling of the saliva. In engineering practice, cooling towers are direct contact heat exchangers for dissipating waste heat from a power plant. Fig. 18.4 shows the removal of thermal energy from the hot water primarily via latent heat of vaporization. Specifically, the cooler (and dryer in terms of absolute humidity) atmospheric air is drawn around the lower part of the cooling tower and moves upward, in the opposite direction of the descending hot droplets from the spray at the top. The up-drafting cool air enhances the evaporation. Even though only a small portion of the liquid from the droplets is evaporated, the associated thermal energy removal is substantial because of the weighty latent heat of vaporization.

In engineering, indirect or non-contact heat exchangers are more common. That being the case, moving forward we will limit this introductory textbook to indirect heat exchangers. We have a couple of specialized monographs for the keen minds who wish to delve into heat exchangers; the heat exchanger

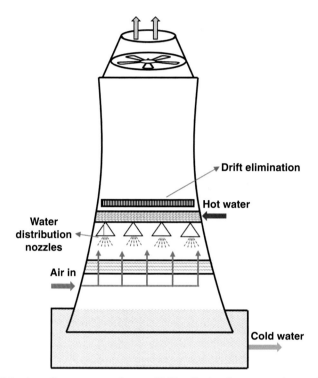

FIGURE 18.4   A cooling tower operating as a highly effective direct contact heat exchanger that capitalizes the large latent heat of vaporization (created by M. Babaei Jamnani).

design handbook by Thulukkanam (2013) is a very comprehensive and relatively up-to-date authoritative reference. Also of interest is the book by Kakaç et al. (2020). Let us proceed and categorize indirect heat exchangers based on the flow arrangement.

18.2   Counter-flow, parallel-flow, and crossflow heat exchangers

Thus far, we have been enlightened that counter-flow (countercurrent) heat exchangers are widespread in nature. Attested by Barry Commoner, "Nature knows best." To that end, we expect counter-flow heat exchangers to outperform their parallel-flow and crossflow counterparts in general. Fig. 18.5 differentiates these three most common types of heat exchangers: the counter-flow, the parallel-flow, and the crossflow heat exchangers. In a counter-flow heat exchanger, the warm stream moves opposite to the cold stream. If the two streams are made up of the same fluid, for example, water, then, the temperature difference between the two streams remains roughly the same throughout the flow passage. That being the case, a substantive temperature difference is maintained for effective

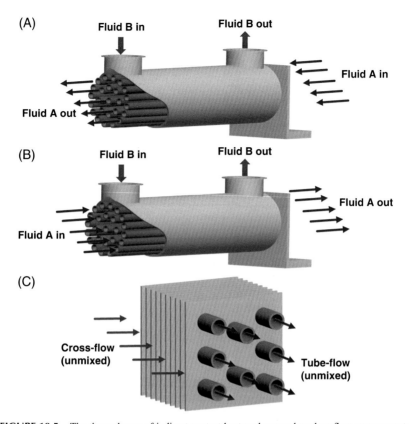

FIGURE 18.5 The three classes of indirect contact heat exchangers based on flow arrangement: (A) counter-flow heat exchanger, (B) parallel-flow heat exchanger, and (C) crossflow heat exchanger (created by X. Wang).

heat transfer along the entire path. Thereupon, counter-flow heat exchangers are appreciably efficient, as nature has already informed us so. Fig. 18.5A shows that the cold stream can be heated above the hot stream exit temperature, and the hot stream can be cooled below the exiting cold stream temperature, for a sound counter-flow heat exchanger. In other words, for the same flow passage, counter-flow heat exchangers give rise to the largest change in temperature from inlet to outlet. This is true for both hot and cold streams.

When both streams move in the same direction, parallel to one another, we have a parallel-flow heat exchanger. From Fig. 18.5B, we can infer that the temperature difference between the hot and cold streams is large at the entrance, but this difference diminishes rapidly. As a result, the initial efficacious heat transfer furnished by the large temperature gradient is short-lived. This is why parallel-flow heat exchangers are not common in practice.

Between counter-flow and parallel-flow heat exchangers, we have the cross-flow heat exchanger, as depicted in Fig. 18.5C. The temperature profiles of

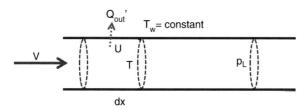

FIGURE 18.6 A fluid moving along a constant-temperature passage (created by D. Ting). For the case shown, the fluid is at a higher temperature than the wall of the passage. As a consequence, the fluid loses thermal energy as it flows through the passage. The rate of heat transfer, however, decreases along the flow passage because of decreasing temperature difference between the fluid and the conduit.

both hot and cold streams after entering the heat exchanger are nonuniform. Irrespective of that, the temperature difference between the hot and cold streams is relatively high and, thus, also the effectiveness of the heat exchanger. This solid performance, along with the flexibility to easily accommodate various physical or space constraints and working fluids, make crossflow exchangers a very popular engineering system. Two everyday examples are the radiator for keeping the internal combustion engine of a car cool, and the radiator for keeping an indoor space warm in the winter. The substantial difference in the thermal properties of the two fluids, typically water and air, calls for heat fins on the airside to even out the heat transmission of the two sides.

18.3 Moving along a constant-temperature passage

Let us look at a simple heat exchange situation, as illustrated in Fig. 18.6. The fluid is at a higher temperature than the wall of the passage, and the wall is maintained at a constant temperature, T_w, throughout the entire passage. Consider the section where the flow characteristics stay roughly the same along the passage, for example, far from the entrance where the steady flow is fully developed. For a moderate temperature difference and variation, the changes in the fluid properties are negligible. That being the case, the overall heat transfer coefficient, U, for the most part, remains largely unaltered. To sum up, the assumptions invoked in our analysis are:

1) the wall temperature of the passage, T_w, is fixed,
2) the flow is steady and the flow characteristics stay the same along the pathway,
3) the properties of the fluid remain constant, and
4) the overall heat transfer coefficient, U, remains unaltered throughout the passage.

The first law of thermodynamics concerns the conservation of energy. For the control volume encompassed by the dashed cylinder in Fig. 18.6, we have

$$m\, c_P dT/dt = Q'_{in} - Q'_{out}, \qquad (18.1)$$

where m is the mass of the enclosed fluid, c_P is the specific heat capacity of the fluid, T is the temperature of the fluid, t is time, Q'_{in} is the rate of heat entering, and Q'_{out} is the rate of heat exiting the control volume. Eq. (18.1) says that the rate of energy stored in the control volume is equal to the rate of energy entering it minus the rate of energy exiting it. The mass of the fluid element can be determined from the fluid density, ρ, multiplied by the volume, that is,

$$m = \rho \, A_c dx, \tag{18.2}$$

where A_c is the cross-sectional area and dx is the length of the control volume section. As the control volume of hot fluid moves along the conduit, it gains no heat but loses heat to the cooler wall at a rate defined by

$$Q'_{out} = U \, p_L dx \, (T - T_w), \tag{18.3}$$

where U is the heat transfer coefficient and p_L is the perimeter of the conduit. Substituting Eqs. (18.2) and (18.3) into Eq. (18.1), we get

$$(\rho \, A_c dx) \, c_P dT/dt = 0 - U \, p_L dx \, (T - T_w). \tag{18.4}$$

Time, t, denotes the elapsed time as the control volume moves along the passage, that is, $t = x/V$, where x is the distance along the conduit, and V is the velocity of the control volume with respect to the conduit or the flow velocity. The mass flow rate can be expressed as

$$m' = \rho \, A_c V. \tag{18.5}$$

Differentiating $(T - T_w)$ gives

$$d(T - T_w) = dT. \tag{18.6}$$

The velocity, $V = dx/dt$, can be rearranged to give $dt = dx/V$. Substituting this and Eq. (18.6) into Eq. (18.4) lead to

$$(\rho \, A_c dx \, c_P) \, d(T - T_w)/(dx/V) = -U \, p_L dx(T - T_w). \tag{18.7}$$

Rearranging, we get

$$d(T - T_w)/(dx/V) = -U \, p_L dx(T - T_w)/ (\rho \, A_c dx \, c_P), \tag{18.8}$$

which can be simplified into

$$d(T - T_w)/dx = -U \, p_L(T - T_w)/ (\rho \, A_c c_P V). \tag{18.9}$$

Multiplying both sides by the conduit length, L, and noting that the mass flow rate, $m' = \rho A_c V$, we have

$$L \, d(T - T_w)/dx = -U \, p_L L(T - T_w)/ (m' \, c_P). \tag{18.10}$$

This can be rewritten as

$$d(T - T_w)/(dx/L) = -UA(T - T_w)/(m' c_P), \tag{18.11}$$

where A, the product of perimeter and conduit length, is the heat transfer area.

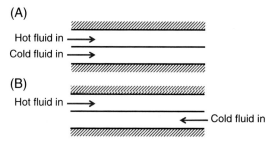

FIGURE 18.7 Heat transfer from a hot stream to a cold stream in (A) parallel flow and (B) counter-flow configurations (created by D. Ting).

Eq. (18.11) can be rearranged for integration from the entrance (inlet) of the conduit to a distant x from the conduit entrance, that is,

$$\int_{T_i-T_w}^{T-T_w} \frac{1}{T - T_w} d(T - T_w) = \int_0^x -\frac{UA}{\dot{m}c_P} d(x/L), \tag{18.12}$$

where T_i is the temperature of the fluid at inlet. Carrying out the integration, we get

$$T - T_w = (T_i - T_w) \exp\{-[UA/(m'c_P)][x/L]\}. \tag{18.13}$$

The temperature of the fluid at the exit or outlet, $x = L$, with respect to the wall, is

$$T_o - T_w = (T_i - T_w) \exp\{-[UA/(m'c_P)]\}. \tag{18.14}$$

The term inside the square brackets is called the number of transfer units (NTU). We will expound on this in a latter section.

18.4 Heat exchange between a hot stream and a cold stream

Fig. 18.7 portrays the heat exchange between a hot stream and a cold stream. Fig. 18.7A shows the case where the two streams are flowing parallel to each other, and Fig. 18.7B illustrates the flow configuration where the hot and cold streams flow opposite to one another. For steady flow with no heat loss to the surroundings, the energy lost by the hot fluid is equal to that gained by the cold fluid, that is, the total heat transfer rate over the entire pipe section,

$$Q' = m'_h (h_{h,i} - h_{h,o}) = m'_c (h_{c,o} - h_{c,i}), \tag{18.15}$$

where m'_h is the mass flow rate of the hot stream, $h_{h,i}$ is the inlet hot stream enthalpy, $h_{h,o}$ is the outlet hot stream enthalpy, m'_c is the mass flow rate of the cold stream, $h_{c,o}$ is the enthalpy of the cold stream at its outlet, and $h_{c,i}$ is the entering cold stream enthalpy. The specific heat capacity, c_P, can be assumed to be constant if the change in the fluid temperature is not severe. In that event, the enthalpy can be replaced by the product of heat capacity and temperature and, hence, we have

$$Q' = m'_h \, c_{P,h}\left(T_{h,i} - T_{h,o}\right) = m'_c \, c_{P,c}\left(T_{c,o} - T_{c,i}\right). \tag{18.16}$$

18.4.1   Heat capacity rate

We can further reduce the number of terms by introducing the *Heat Capacity Rate*, defined as

$$C = m'c_P. \tag{18.17}$$

With this, the heat transfer rate from the hot to cold stream, Eq. (18.16), can be expressed as

$$Q' = C_h\left(T_{h,i} - T_{h,o}\right) = C_c\left(T_{c,o} - T_{c,i}\right). \tag{18.18}$$

We see that the heat capacity rate denotes the required heat transfer rate for increasing or decreasing the fluid temperature by one degree. Note that the decrease in the hot stream temperature is equal to the increase in the cold stream temperature, $\Delta T_h = \Delta T_c$, when the heat capacity rates of the two streams are equal, that is, $C_h = C_c$. In general, the heat capacity rate, C, of a liquid is significantly greater than that of a gas. This is the case with water and air as the two mediums involved. The reason is that around atmospheric conditions (temperature and pressure), the heat capacity of water is more than four times than that of air, and the density of water is about 800 times that of air. Because of this, the factor limiting the rate of heat transfer from the hot to the cold stream, such as that shown in Fig. 18.7, is the lower heat capacity rate stream. This is the bottleneck of the heat transfer traffic. To overcome this slowdown, the side with the lower heat capacity rate is furnished with an extra heat transfer area via heat fins. In nature, desert foxes capitalize the extra heat transfer surface outfitted by their large ears to keep them cool in the scorching environment; see Fig. 18.8. The blood, being primarily a liquid, can effectively convey the heat generated via metabolism to the body surface. For desert foxes, they have many vessels circulating blood to and back from their big ears. The atmospheric air has much lower heat capacity rate. Therefore, the extra surface provided by their big ears augments the heat capacity rate via increasing mass flow rate, explicitly,

$$C = m' \, c_P = (\rho \, V \, A) \, c_P. \tag{18.19}$$

Here, ρ denotes the density of air, V, the velocity of air, and A is the heat transfer surface area furnished by the ears. Readers interested in the multifaceted thermoregulation of the Fennec fox can refer to Maloiy et al. (1982).

FIGURE 18.8 Desert foxes are gifted with large ears interlined with blood vessels to keep them cool (created by X. Wang).

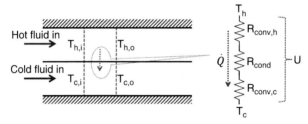

FIGURE 18.9 Hot to cold stream heat transfer for a differential element of a parallel-flow heat exchanger (created by D. Ting). The hot fluid enters at $T_{h,i}$ and leaves at $T_{h,o}$, whereas the cold fluid enters at $T_{c,i}$ and leaves at $T_{c,o}$.

18.5 Log mean temperature difference

The total heat transfer rate of a heat exchanger can be calculated from Eq. (18.16) or (18.18) when the values of all or one less than all of the parameters, m'_h, $c_{P,h}$, $T_{h,i}$, $T_{h,o}$, m'_c, $c_{P,c}$, $T_{c,o}$, and $T_{c,i}$, are known. This is not the case when an engineer is tasked to solve a practical problem and needs to choose an appropriate heat exchanger for the intended application. To put it another way, typically there are two or more unknowns. The log mean temperature difference (LMTD) method is one of the two standard approaches to call upon to resolve this practical challenge. The LMTD method is favored when all the temperatures are known. Let us go through the derivation of this method.

Consider a differentiable element (control volume) of the heat exchanger, as shown in Fig. 18.9. The heat transfer rate from the hot to the cold element is

$$dQ' = m'_h \, c_{P,h}\left(T_{h,o} - T_{h,i}\right) = -C_h dT_h = m'_c \, c_{P,c}\left(T_{c,o} - T_{c,i}\right) = C_c dT_c.$$
$$(18.20)$$

The negative sign associated with C_h dT_h signifies a loss, that is, $T_{h,o}$ minus $T_{h,i}$ is negative. Fig. 18.9 shows that the heat transfer from the hot fluid to the cold fluid takes place via convection from the hot stream to the wall, conduction through the wall, followed by convection from the wall to the cold fluid. The three thermal resistors in series can be combined into an *overall heat transfer coefficient*, that is,

$$U = 1/R_{conv,h} + 1/R_{cond} + 1/R_{conv,c}, \tag{18.21}$$

where $R_{conv,h}$ is the convection thermal resistance on the hot side, R_{cond} is the conduction thermal resistance of the wall, and $R_{conv,c}$ is the convection thermal resistance on the cold side. For the steady-state heat transfer shown, the heat that passes through $R_{conv,h}$ also passes through R_{cond} and $R_{conv,c}$. Mathematically, the differential heat transfer rate can be expressed as,

$$dQ' = -C_h dT_h = C_c dT_c = U\ \Delta T\ dA, \tag{18.22}$$

where dA is the differential wall area for heat transfer and the local temperature difference,

$$\Delta T \equiv T_h - T_c. \tag{18.23}$$

Differentiating this local temperature difference between the hot and cold streams gives

$$d\Delta T = dT_h - dT_c = -dQ'/C_h - dQ'/C_c. \tag{18.24}$$

This differential local temperature difference can be rewritten as

$$d\Delta T = -dQ'\ (1/C_h + 1/C_c) = -U\ \Delta T\ dA\ (1/C_h + 1/C_c). \tag{18.25}$$

Eq. 18.25 can be rearranged into

$$1/\Delta T\ d\Delta T = -U(1/C_h + 1/C_c)\ dA. \tag{18.26}$$

We can integrate the temperature difference from the entrance to the exit, that is, Location 1 to Location 2,

$$\int_1^2 \frac{1}{\Delta T}\ d\Delta T = \int_1^2 -U\ (1/C_h + 1/C_c)\ dA. \tag{18.27}$$

For moderate changes in temperatures, the heat capacity rates, C_h and C_c, remain unaltered along the channel. In practice, the flow characteristics along the passage tend to stay roughly constant and, thus, also the local heat transfer coefficient. The other inherited assumptions include the heat exchanger as an adiabatic system, that is, it is well insulated and, thus, it does not exchange heat with the surroundings. Additionally, the axial conduction along the conduit is assumed to be negligible, and so are changes in potential and kinetic energy. Under these circumstances, we have

$$\ln \Delta T_2/\Delta T_1 = -UA\ (1/C_h + 1/C_c), \tag{18.28}$$

where A is the total heat transfer area. We note that ΔT_1 signifies the temperature difference between the hot and cold fluids at Location 1, that is, at the entrance, and ΔT_2 represents the temperature difference at Location 2, that is, at the exit. Replacing the heat capacity rates with Q' and the inlet and out temperatures according to Eq. (18.18), we get

$$\ln \Delta T_2/\Delta T_1 = -UA\left[(T_{h,i} - T_{h,o})/Q'\right] + UA\left[(T_{c,o} - T_{c,i})/Q'\right]. \quad (18.29)$$

This can be simplified into

$$\ln \Delta T_2/\Delta T_1 = -(UA/Q')\left[(T_{h,i} - T_{c,i}) - (T_{h,o} - T_{c,o})\right], \quad (18.30)$$

which is

$$\ln \Delta T_2/\Delta T_1 = -(UA/Q')(\Delta T_1 - \Delta T_2). \quad (18.31)$$

This can be reorganized into

$$Q' = UA[(\Delta T_1 - \Delta T_2)/\ln(\Delta T_1/\Delta T_2)]. \quad (18.32)$$

The term in square brackets is the LMTD, that is,

$$\Delta T_{lm} = (\Delta T_1 - \Delta T_2)/\ln(\Delta T_1/\Delta T_2). \quad (18.33)$$

Using this short form in Eq. (18.32) gives

$$Q' = UA\ \Delta T_{lm}. \quad (18.34)$$

The "log mean" takes care of the not linear, approximately exponential, variation of the temperature difference along the heat exchanger. The arithmetic mean temperature difference overestimates the average temperature difference. The LMTD accounts for the nonlinearly varying temperature difference along the flow path by deducing a logarithmic average of the temperature difference between the hot and cold feeds at each end of the heat exchanger, giving the actual average temperature difference.

18.5.1 Parallel-flow heat exchanger

The equations derived above apply to both parallel-flow and counter-flow heat exchangers. The difference is in the temperatures at the two ends of the heat exchanger. For a parallel-flow heat exchanger, Location 1 is the entrance for both hot and cold streams, where the temperature difference,

$$\Delta T_1 \equiv T_{h,1} - T_{c,1} = T_{h,i} - T_{c,i}. \quad (18.35)$$

Since the hot stream meets the cold stream at this location, the temperature difference, ΔT_1, is large; see Fig. 18.5A. Because of this, the heat transfer rate from the hot to the cold stream is appreciable. The temperature difference decreases rapidly along the flow path and, thus, also the heat transfer rate. At the exit, the temperature difference between the two streams is described mathematically as

$$\Delta T_2 \equiv T_{h,2} - T_{c,2} = T_{h,o} - T_{c,o}. \qquad (18.36)$$

18.5.2   Counter-flow heat exchanger

For the counter-flow heat exchanger, it is convenient to designate the entrance of the hot stream as Location 1. To that end, Location 1 also corresponds to the exit or outlet of the cold stream. Specifically, the temperature difference at Location 1,

$$\Delta T_1 \equiv T_{h,1} - T_{c,1} = T_{h,i} - T_{c,o}. \qquad (18.37)$$

This temperature difference is relatively moderate. More importantly, this reasonable temperature difference is maintained throughout the entire heat exchanging path. At the outlet of the hot stream, the temperature difference between the exiting hot stream and the incoming cold stream is described by

$$\Delta T_2 \equiv T_{h,2} - T_{c,2} = T_{h,o} - T_{c,i}. \qquad (18.38)$$

Due to the fact that a reasonably significant temperature difference is sustained from one end, Location 1, to the other, Location 2, a reasonable amount of heat transfer rate prevails over the full length of the flow passage. Consequently, a counter-flow heat exchanger furnishes a larger overall heat transfer rate, compared to its parallel-flow counterpart. For given inlet and outlet temperatures, the LMTD for a counter-flow heat exchanger is invariably larger than its parallel-flow counterpart. To put it another way, a smaller counter-flow heat exchanger can accomplish the same amount of heat transfer as a larger parallel-flow heat exchanger. That is why a parallel-flow heat exchanger is only resorted to when the application involved does not permit the use of the more-efficient, counter-flow heat exchanger.

Example 18.2 Heat exchange involves phase change.
Given: Dr. JAS makes use of the Turbulence and Energy Laboratory heat exchanger to illustrate heat exchange involving phase change to her class. Steam at 150°C condenses on the shell side of a heat exchanger. Water enters the 5 cm diameter, 7 m long inner tubes at 10°C and leaves at 90°C. The mass flow rate of the cold water through the inner tubes is 0.5 kg/s.
Find: The rate of condensation of the steam. Assume the overall heat transfer coefficient, $U = 1500$ W/(m$^2 \cdot$K).

Solution: According to Eq. (18.34), the rate of heat transfer,

$$Q' = UA\ \Delta T_{lm},$$

where the LMTD,

$$\Delta T_{lm} = (\Delta T_1 - \Delta T_2)/\ln(\Delta T_1/\Delta T_2).$$

If we regard the heat exchanger as a parallel-flow one, then, according to Eqs. (18.35) and (18.36), the temperature differences at Location 1 and Location 2 are, respectively,

$$\Delta T_1 \equiv T_{h,1} - T_{c,1} = T_{h,i} - T_{c,i} = 150 - 10 = 140°C$$

and

$$\Delta T_2 \equiv T_{h,2} - T_{c,2} = T_{h,o} - T_{c,o} = 150 - 90 = 60°C.$$

Substituting these for the heat transfer rate, we get

$$Q' = UA\,\Delta T_{lm} = 1500[\pi(0.05)(7)](140 - 60)/\ln(140/60) = 155.7 \text{ kW.}$$

If we consider it as a counter-flow heat exchanger instead, then, we can invoke Eqs. (18.37) and (18.38) for the temperature differences at Location 1 and Location 2, respectively,

$$\Delta T_1 \equiv T_{h,1} - T_{c,1} = T_{h,i} - T_{c,o} = 150 - 90 = 60°C$$

and

$$\Delta T_2 \equiv T_{h,2} - T_{c,2} = T_{h,o} - T_{c,i} = 150 - 10 = 140°C.$$

The corresponding heat transfer rate,

$$Q' = UA\,\Delta T_{lm} = 1500[\pi(0.05)(7)](60 - 140)/\ln(60/140) = 155.7 \text{kW.}$$

We get the same answer whether we look at the problem as a parallel-flow or counter-flow heat exchanger. This is because one stream, the hot one that is undergoing a phase change, is at a constant temperature. It follows that reversing its flow direction does not change the condition.

18.5.3 Correction factor

We have derived the LMTD for the idealized counter-flow and parallel-flow heat exchangers. To develop the LMTD from first principles for more complex heat exchangers, such as crossflow and multi-pass heat exchangers, is a daunting task. Even if we undertake and overcome the taxing exercise to fortify our resilience, the resulting expressions will be too intricate for practical usage. A superior solution is to make use of the LMTD equations derived for counter-flow heat exchangers via an appropriate correction factor. This correction factor is a function of the heat transfer geometry and temperatures, that is,

$$F_C = f\left(\text{geometry}, \ T_{h,i}, \ T_{h,o}, \ T_{c,i}, \ T_{c,o}\right). \tag{18.39}$$

In a nutshell, the LMTD for crossflow and multipass heat exchangers can be expressed as

$$\Delta T_{lm} = F_C \Delta T_{lm,CF}, \tag{18.40}$$

where $\Delta T_{lm,CF}$ is the LMTD of the counter-flow heat exchanger. The correction factor communicates the deviation of the LMTD from the corresponding values for the counter-flow case. Having been illuminated by the intelligent designs revealed from penguin feet to kissing bug proboscis, we recognize the efficacy of counter-flow, or counter-current, heat exchangers. To put it another way, crossflow heat exchangers are expected to be less efficient and, hence, the value of the correction factor is naturally less than one. Fig. 18.10 shows the change in correction factor with respect to temperature and geometry of some of the most common heat exchangers.

18.6 Heat exchanger effectiveness and number of transfer units

The other common method for heat exchanger analysis is called the NTU (number of transfer units) method. It is the preferred method when one or more temperatures are not known a priori, that is, there is insufficient information to calculate the LMTD. For example, if we wish to deduce the outlet temperatures of the two streams, we will call upon the NTU method. The effectiveness of a heat exchanger can be defined as the actual heat transfer rate with respect to the maximum possible heat transfer rate. Namely, the *Heat Exchanger Effectiveness*,

$$\varepsilon \equiv Q'/Q'_{max}. \qquad (18.41)$$

The maximum possible heat transfer rate, Q'_{max}, is achieved when the heat transfer area is extended to infinity. For a parallel-flow heat exchanger, this would imply that the outlet temperature of the hot stream is equal to that of the cold stream, that is, $T_{h,o} = T_{c,o}$. For a counter-flow heat exchanger, this corresponds to $T_{h,i} = T_{c,o}$, or, $T_{h,o} = T_{c,i}$, depending on the heat capacity rate. The stream with the lower heat capacity rate would undergo a larger temperature variation and, thus, reaches the maximum possible temperature change. Succinctly,

$$Q'_{max} = C_{min}\left(T_{h,i} - T_{c,i}\right), \qquad (18.42)$$

where C_{min} is the smaller of C_h and C_c.

The heat exchanger effectiveness can be expressed as

$$\varepsilon \equiv Q'/Q'_{max} = 1 - \exp(-NTU), \qquad (18.43)$$

where the NTU, is the ratio of the overall thermal conductance with respect to the smaller heat capacity rate,

$$NTU = UA/C_{min}. \qquad (18.44)$$

To put it another way, the NTU, is a non-dimensional parameter designating the rate of heat transfer. The larger the NTU, the higher the heat transfer rate.

Sample plots of heat exchanger effectiveness versus NTU for some common heat exchangers are shown in Fig. 18.11. With NTU defined as UA/C_{min}, it

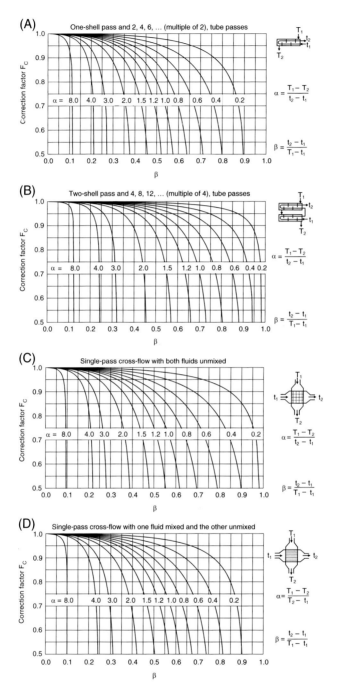

FIGURE 18.10 Correction factor versus temperature ratio for common tube-and-shell and cross-flow heat exchangers (created by Y. Yang). The respective nondimensional temperature change ratios are defined by α and β.

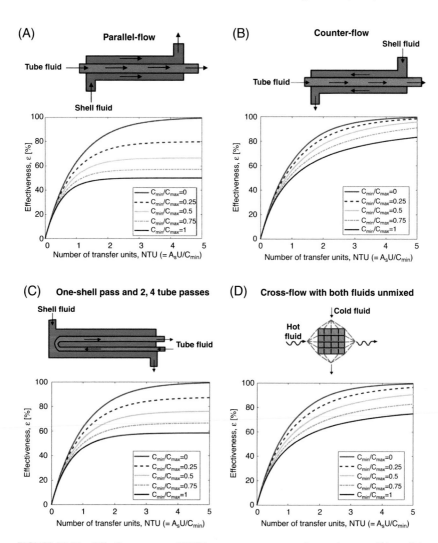

FIGURE 18.11 Effectiveness versus NTU for common non-contact heat exchangers: (A) parallel-flow, (B) counter-flow, (C) one-shell pass and 2, 4 tube passes, and (D) crossflow (created by X. Wang).

is a yardstick for the area available for transferring heat. Increasing the heat transfer area, A, increases NTU. Practically, however, there is an upper limit beyond which any further increase in the heat exchanger size (area) does not justify the extra cost. Fig. 18.11 illustrates that an initial increase in NTU results in a substantial improvement in the heat exchanger effectiveness, ε. The increase slows down rapidly with further increase in NTU. Beyond an NTU value of approximately 1.5, it is not cost-effective to further increase NTU, for the diminishing gain in ε cannot be justified by the additional cost.

Problems

18.1 Parallel-flow heat exchanger.

A parallel-flow heat exchanger is used for heating water (c_p = 4.18 kJ/kg·K) from 15°C to 70°C at a rate of 0.2 kg/s. The heat is provided by 140°C geothermal water (c_p=4.31 kJ/kg·K) that flows parallel at 0.3 kg/s in a thin-walled inner tube with a diameter of 2.54 cm. The overall heat transfer coefficient is 485 W/m²·C. What is the required length?

18.2 Maximum heat transfer rate of a counter-flow heat exchanger.

A counter-flow heat exchanger uses 2.1 kg/s of hot water at 85°C to warm up cold water at 5°C at a rate of 1.7 kg/s. Assume the specific heat of water for both streams to be 4.18 kJ/kg·K.

A) What is the maximum heat transfer rate?

B) What are the outlet temperatures of the cold- and the hot-water streams for the ideal case where the maximum heat transfer rate occurs?

18.3 The better counter-flow heat exchanger.

Two counter-flow heat exchangers are available for you to choose from. One heat exchanger has an overall heat transfer coefficient of 570 W/m²·°C and a heat transfer area of 0.47 m². The other heat exchange has an overall heat transfer coefficient of 370 W/m²·°C and a surface area of 0.94 m². Assume the hot stream is comprised of oil with a specific heat capacity of 2.1 kJ/kg·°C flowing at 198 kg/h, entering the heat exchanger at 180°C. Which one of these would you choose for heating 222 kg/h of water from 15°C to 89°C?

18.4 The heat transfer rate and the exiting air temperature of a radiator.

Atmospheric air (c_p=1.01 kJ/kg·K) at 18°C is drawn across a radiator at 7 kg/s. The radiator is a crossflow heat exchanger with UA = 10 kW/K. The coolant (c_p=4.00 kJ/kg·K) at 90°C flows through the radiator at 3.5 kg/s. The effectiveness of the radiator, ε = 0.4.

A) What is the outgoing air temperature?

B) What is the heat transfer rate?

18.5 LMTD of a car radiator.

The coolant (c_p = 1.0 Btu/lbm·F) of a car radiator at 195°F is cooled down to 135°F via a single-passage, crossflow HEX with air (c_p = 0.245 Btu/lbm·F) at 75°F. The coolant is flowing at 92,000 lbm/h, while the air is flowing at 400,000 lbm/h. What is the LMTD?

18.6 When the heat capacity rates are equal.

Consider the special case that the heat capacity rates of the hot and cold streams of a heat exchanger are equal.

A) What can you say about the temperature drop of the hot fluid from entrance to exit of a heat exchanger with respect to the temperature rise of the cold stream?

B) What can you say about the temperature difference along the flow passage for a counter-flow heat exchanger with equal heat capacity rates?

18.7 One large or two small heat exchangers?

Glycerin flowing at 1.4 kg/s is to be heated from 12°C to 48°C. This can be realized by a heat exchanger with an overall heat transfer coefficient of 895 W/m²·°C, where hot water enters at 95°C and leaves at 50°C. Alternatively, we can utilize two smaller equal-surface-area heat exchangers in series, where 60% of the hot water goes through the first heat exchanger and the remaining 40% goes through the second heat exchanger. The overall heat transfer coefficient of the two smaller heat-exchanger systems is the same as that of the large, single heat exchanger. Compare the options by deducing their heat transfer effectiveness, NTU, and surface area. Which option is preferable? Why?

18.8 Shell-and-tube heat exchanger.

A shell-and-tube heat exchanger is used to heat water from 20°C to 85°C at a rate of 125,000 kg/h by means of steam condensing at 100°C on the outside of the tubes. The heat exchanger has 500 stainless steel tubes (thermal conductivity, $k = 14$ W/m·°C; inner diameter, $D_i = 2$ cm; outer diameter, $D_o = 2.4$ cm) in a tube bundle that is 10 m long. Assume that changes in water properties with moderate changes in temperature are negligible.

A) What is the heat transfer coefficient on the water side (h_i)?

B) What is the exit temperature of the water if its mass flow rate is doubled (assume that h_i remains unchanged with the doubling of mass flow rate)?

C) The temperature of the water at the tube exit can best be increased by

I) increasing the mass flow rate of steam on the shell side.

II) increasing the mass flow rate of water in the tubes.

III) increasing the heat transfer coefficient on the water side (h_i).

IV) increasing the thermal conductivity of the tube.

Explain your choice.

References

Ancel, A., Visser, H., Handrich, Y., Masman, D., Maho, Y.L., 1997. Energy saving in huddling penguins. Nature 385, 304–305.

Ask Nature, 2021. https://asknature.org/collection/cooling-down-in-the-heat/, (accessed April 27, 2021).

Gilbert, C., Maho, Y.L., Perret, M., Ancel, A., 2007. Body temperature changes induced by huddling in breeding male emperor penguins. Am. J. Physiol. Regul. Integr. Comp. Physiol. 292 (1), R176–R185.

Kakaç, S., Liu, H., Pramuanjaroenkij, A., 2020. Heat Exchangers: Selection, Rating, and Thermal Design, Fourth Edition CRC Press, Boca Raton, FL.

Lahondère, C., Insausti, T.C., Paim, R.M.M., Luan, X., Belev, G., Pereira, M.H., Ianowski, J.P., Lazzari, C.R., 2017. Countercurrent heat exchange and thermoregulation during blood-feeding in kissing bugs. eLife 6, e26107 1–19.

Magnanelli, E., Wilhelmsen, Ø., Acquarone, M., Folkow, L.P., Kjelstrup, S., 2016. The nasal geometry of the reindeer gives energy-efficient respiration. J. NonEquilib. Thermodyn. 42 (1), 59–78.

Maloiy, G.M.O., Kamau, J.M.Z., Shkolnik, A., Meir, M., Arieli, R., 1982. Thermoregulation and metabolism in small desert carnivore: the Fennex fox (Fennecus zerda) (Mammalia). J. Zool. 198, 279–291.

Quirós, G., 2021. Mosquitoes use 6 needles to suck your blood. KQED. https://www.kqed.org/science/728086/how-mosquitoes-use-six-needles-to-suck-your-blood. (accessed October 15, 2021).

Thulukkanam, K., 2013. Heat Exchanger Design Handbook, 2nd ed. CRC Press, Boca Raton, FL.

Index

Page numbers followed by "*f*" and "*t*" indicate, figures and tables respectively.

Printed in the United States
by Baker & Taylor Publisher Services